Disaster Risk Reduction

Methods, Approaches and Practices

Series Editor

Rajib Shaw, Keio University, Shonan Fujisawa Campus, Fujisawa, Japan

Disaster risk reduction is a process that leads to the safety of communities and nations. After the 2005 World Conference on Disaster Reduction, held in Kobe, Japan, the Hyogo Framework for Action (HFA) was adopted as a framework for risk reduction. The academic research and higher education in disaster risk reduction has made, and continues to make, a gradual shift from pure basic research to applied, implementation-oriented research. More emphasis is being given to multi-stakeholder collaboration and multi-disciplinary research. Emerging university networks in Asia, Europe, Africa, and the Americas have urged process-oriented research in the disaster risk reduction field. With this in mind, this new series will promote the output of action research on disaster risk reduction, which will be useful for a wide range of stakeholders including academicians, professionals, practitioners, and students and researchers in related fields. The series will focus on emerging needs in the risk reduction field, starting from climate change adaptation, urban ecosystem, coastal risk reduction, education for sustainable development, community-based practices, risk communication, and human security, among other areas. Through academic review, this series will encourage young researchers and practitioners to analyze field practices and link them to theory and policies with logic, data, and evidence. In this way, the series will emphasize evidence-based risk reduction methods, approaches, and practices.

All submitted proposals are carefully reviewed by the series editor.

Raju Sarkar · Sunil Saha · Prateek Sharma ·
Rajib Shaw
Editors

Climate Change, Environmental Hazards and Community-Based Resilience

Springer

Editors
Raju Sarkar
Department of Civil Engineering
Centre of Excellence in Disaster Risk Reduction
Delhi Technological University
New Delhi, Delhi, India

Prateek Sharma
Delhi Technological University
New Delhi, Delhi, India

Sunil Saha
Department of Geography
University of Gour Banga
Malda, West Bengal, India

Rajib Shaw
Graduate School of Media and Governance
Keio University
Fujisawa, Kanagawa, Japan

ISSN 2196-4106 ISSN 2196-4114 (electronic)
Disaster Risk Reduction
ISBN 978-981-95-9484-9 ISBN 978-981-95-9485-6 (eBook)
https://doi.org/10.1007/978-981-95-9485-6

This Springer imprint is published by the registered company Springer Nature Singapore Pte Ltd.
The registered company address is: 152 Beach Road, #21-01/04 Gateway East, Singapore 189721, Singapore

Preface

As climate change continues, the location of weather-related hazards may shift, creating new risks. Problems such as floods, landslides, soil erosion, heat waves, droughts, land degradation, and pollution are becoming more common due to human activities and climate change, causing major damage to property and loss of life. For decades, many countries have been looking for ways to deal with the effects of climate change. These efforts provide valuable lessons, especially for developing countries that are now facing the growing threat of climate-related disasters. Today, building community resilience to handle these environmental dangers has become a global priority. The people who are least able to handle these problems are the most vulnerable.

The book is organized into three sections comprising total 16 chapters, each of which has emphasized to the explore the patterns of climate change, the spread of environmental disasters, and highlight how local communities' knowledge can be used to build resilience against these challenges.

Each chapter written by scholars and/or practitioners with acknowledged expertise in the field and with adequate experience of working in Asian region. The book has cover all dimensions of climate change, environmental dangers, and how local communities are responding to these problems.

The book is sub-divided in three parts and they are:

Part I: Climate Change

Part I has explored the spatio-temporal approach to estimate patterns of climate change, global climate modeling and assessment of climate change impacts, vulnerability to climate and climate change, urbanization and climate change, land use dynamics and climate change, spatio-temporal prediction of temperatures, and spatio-temporal patterns of precipitation.

Part II: Environmental Hazards

This part has explored the different natural disasters, viz., landslide, flood, drought response to climate change, spatial and temporal evaluation of soil erosion, land degradation and regional air pollution for climate change, water quality status during climate change, and tropical cyclones amid climate change.

Part III: Community-Based Resilience

This part has explored the techniques and method that will help for management of micro-climate change in urban area, climate change and society adaptation and development, localized disaster-resilience to coastal communities, community resilience to flood hazards, community-based landslide risk reduction, community resilience to heatwaves, and community resilience to drought hazard.

New Delhi, India Raju Sarkar
Malda, India Sunil Saha
New Delhi, India Prateek Sharma
Fujisawa, Japan Rajib Shaw

About This Book

One of the main reasons for the rise in environmental hazards is climate change. The fourth assessment report from the Intergovernmental Panel on Climate Change (IPCC) predicted a rise in the frequency, severity, and intensity of hazards caused by climate change. Many weather-related hazards are already changing in frequency and intensity due to climate change, and exposed communities that relied on arable land, water availability, and consistent mean temperatures and rainfall are becoming more vulnerable and less resilient. The geographic distribution of weather-related hazards may shift as a result of climate change, potentially creating new risk patterns. Different environmental hazards like flood, landslide, soil erosion, heatwave, drought, land degradation, and air and water pollution are gradually increasing due increasing human interference with nature and associated climate change and causing huge loss of human properties and lives. This makes it challenging for environmental planners and policymakers to create effective plans for handling environmental dangers. Advanced AI techniques and geospatial technology have brought about significant change and made managing environmental difficulties easier.

Nowadays, community-based resilience to withstand environmental hazards brought on by climate change is becoming an important issue worldwide. Those who are least able to cope are the most vulnerable. For these vulnerable communities, adaptation, which is directly tied to fostering community resilience, is a need, not an option. Resilience thinking has the potential to lower risks associated with environmental hazards and climate change, but its operationalization has not received enough attention. To deal with and recover from the effects of climate change and disasters, communities that are resilient and institutions that are reinforced are essential. Thus, there will be more community involvement in the use of common property resources, better sustainable use of ecosystems and biodiversity resources, and higher resistance to climate change and natural disasters.

The book is organized into three parts: Part I has incorporated the modeling of climate change. In Part II, modeling of environmental hazards and its relation with climate change have been discussed, and in Part III, community-based resilience to climate change and different environmental hazards has been discussed.

Contents

Editors and Contributors

About the Editors

Raju Sarkar is a Professor in Department of Civil Engineering and Coordinator of Centre of Excellence in Disaster Risk Reduction (CoEDRR), Delhi Technological University (DTU). Prior to joining back to his parent organization DTU, Dr. Raju was working as Professor in Department of Civil Engineering and Architecture, College of Science and Technology, Royal University of Bhutan, Bhutan, under the Ministry of External Affairs, Government of India Deputation to Bhutan. During his tenure in Bhutan, he has established the research Center for Disaster Risk Reduction and Community Development Studies and also worked as Team Leader to start the new undergraduate program—Engineering Geology in Royal University of Bhutan. Presently, he is also Chair, Commission on Education and Outreach and Co-Chair, Commission on Earthquake Hazard, Risk and Strong Ground Motion, International Association of Seismology and Physics of the Earth's Interior (IASPEI)—IUGG. He has vast experience to work in Hindu-Kush Himalayas region both at government and community levels. He has published quite a good number of original research articles in peer-reviewed journals, books, book chapters, and proceedings of international societies. Dr. Raju is collaborating in a number of research projects funded by ICSU, GCRF, World Bank, RAS-UK, DHI-RGoB, and MoEF. He has keen interest on Geo-hazards Risk Managements, Landslide, Seismology, Construction Engineering, Community Resilience, etc.

Sunil Saha is an Assistant Professor of Geography at the University of Gour Banga and is specialized in the field of Environmental Geography. His research interests are toward understanding fluvial landforms, natural hazards (gully erosion, landslide, and flood), surface and sub-surface hydrological conditions and their impacts on land uses regarding both time and space. He is also interested in applications of Remote Sensing and Geographical Information System in Geo-hazards modeling and management. He has more than 10 years of teaching and research experience. His name has been enlisted in the top 2% world scientist list in 2021, 2022, 2023,

2024 and 2025 published by the Stanford University, USA. Under his supervision, four Ph.D. research scholars and six M.Phil. scholars completed their work in the field of Geo-environmental hazard Management. More than 42 students have completed their dissertations under his guidance at postgraduation level. He has published more than 80 peer-reviewed research papers in international journals of Elsevier, Springer, Taylor and Francis, and MDPI. He has also published 17 book chapters and 1 book. Apart from this, he has presented his research outcomes as many as 12 national and international conferences. He has reviewed so many research articles assigned by the leading international journals like *Journal of Cleaner Production*, *Science of the Total Environment*, *Ecological Indicator*, *Catena*, *Environmental Development and Sustainability*, *Geocarto International*, *Geomatics*, and *Natural Hazards and Risk*.

Prateek Sharma is the Vice Chancellor of Delhi Technological University and specialist in Environmental Engineering. Prof. Sharma pursued his Ph.D. from the Department of Civil Engineering, Indian Institute of Technology, New Delhi; M.E. (Hydraulics) and B.E. (Civil) from Civil Engineering Department, Delhi College of Engineering, Delhi. He has more than 30 years of research and teaching experience. His general research interest focusses on environmental systems modeling, statistical applications in environmental and water resources engineering and, environmental risk assessment. He is a member of several professional societies. He has been admitted as a Fellow of Wessex Institute of Great Britain in 2004, in recognition of outstanding scholarly work in the area of air quality modeling. He is a member of the National Knowledge Network (NKN) constituted by the Central Pollution Control Board (CPCB) to support the activities of the National Clean Air Programme (NCAP) launched by the Ministry of Environment Forests and Climate Change (MoEF&CC).

Rajib Shaw is a Japanese national of Indian origin, is the Professor in Graduate School of Media and Governance in Keio University's Shonan Fujisawa Campus (SFC). Earlier, he was the Executive Director of the Integrated Research on Disaster Risk (IRDR), a decade-long research program co-sponsored by the International Council for Science (ICSU), the International Social Science Council (ISSC), and the United Nations International Strategy for Disaster Reduction (UNISDR). He is also the Senior Fellow of Institute of Global Environmental Strategies (IGES) Japan, and the Chairperson of SEEDS Asia and CWS Japan, two Japanese NGOs. Previously, he served as a Professor in the Graduate School of Global Environmental Studies of Kyoto University. His expertise includes community-based disaster risk management, climate change adaptation, urban risk management, and disaster and environmental education. Professor Shaw is the Chair of the United Nations Science Technology Advisory Group (STAG) for disaster risk reduction; and also the Co-Chair of the Asia Science Technology Academic Advisory Group (ASTAAG). He is the editor of a book series on disaster risk reduction, published by Springer. Prof. Shaw has published more than 45 books and over 300 academic papers and book chapters.

Contributors

Saroj Acharya National Center for Hydrology and Meteorology, Royal Government of Bhutan, Thimphu, Bhutan

Vaishnavi Bansal Department of Civil Engineering, JSS Academy of Technical Education, Noida, India

B. Bharath Department of Civil Engineering, JSS Academy of Technical Education, Noida, India

Gaurav Bharti Department of Civil Engineering, JSS Academy of Technical Education, Noida, India

Nirmal Biswas Department of Geography, University of Gour Banga, Malda, West Bengal, India

Chetna Department of Civil Engineering, JSS Academy of Technical Education, Noida, India

Sipra Chowdhury Department of Geography, University of Gour Banga, Malda, West Bengal, India

Morami Dutta Department of African Studies, University of Delhi, Delhi, India

Priyanka Gogoi Department of Geography, University of Gour Banga, Malda, West Bengal, India

Animekh Hazarika Department of Ecology and Environmental Science, Assam University, Silchar, India

Subhendu Jana Department of Geography, University of Gour Banga, Malda, West Bengal, India

Nishant Kumar Department of Civil Engineering, JSS Academy of Technical Education, Noida, India

Barnali Kundu Department of Geography, University of Gour Banga, Malda, India

Udit Mittal Department of Electrical Engineering, JSS Academy of Technical Education, Noida, India

Aditi Nath Department of Social Work, Assam University, Silchar, India

Arun Jyoti Nath Department of Ecology and Environmental Science, Assam University, Silchar, India

Priti Paul Department of Geography, University of Gour Banga, Malda, West Bengal, India

Amit Roy Department of Electrical Engineering, JSS Academy of Technical Education, Noida, India

Anik Saha Department of Geography, Harishchandrapur College, Malda, West Bengal, India

Sunil Saha Department of Geography, University of Gour Banga, Malda, West Bengal, India

Raghunath Prasad Saket Department of Economics, Satyawati College, University of Delhi, Delhi, India

Ruchi Saraswat Department of Civil Engineering, JSS Academy of Technical Education, Noida, India

Suvandan Saraswat Department of Mechanical Engineering, JSS University, Noida, India

Raju Sarkar Centre of Excellence in Disaster Risk Reduction, Delhi Technological University, Delhi, India;
Department of Civil Engineering, Delhi Technological University, New Delhi, India

Sanjit Kumar Shil Sharma Department of Geography, Mathabhanga College, Mathabhanga, West Bengal, India

Amit Tripathi Department of Civil Engineering, JSS University, Noida, India

Part I
Climate Change

Chapter 1
Landslides in a Changing Climate: A Combined Global and High-Mountain Perspective

Vaishnavi Bansal, Amit Tripathi, Gaurav Bharti, and Raju Sarkar

Abstract Landslides are among the most damaging natural hazards, strongly influenced by climatic and hydrological conditions. With global warming altering precipitation patterns, temperature regimes, and cryospheric processes, the frequency, magnitude, and distribution of slope failures are changing worldwide. This chapter synthesizes insights from the global systematic review and the high-mountain case study analysis. It integrates empirical evidence, modeling approaches, and detailed event data to assess how climate change drives landslide activity. The findings reveal increasing rainfall-induced shallow landslides, enhanced cryosphere-related instabilities, and cascading hazards in mountainous terrain. Regional research imbalances and data uncertainties remain key challenges. The chapter concludes by outlining future risks, adaptation strategies, and research priorities for building climate-resilient landscapes.

Keywords Landslides · Climate change · Rainfall extremes · Permafrost degradation · Glacier retreat · High mountains · Risk reduction

V. Bansal · G. Bharti (✉)
Department of Civil Engineering, JSS Academy of Technical Education, Noida, India
e-mail: bharti000@gmail.com

V. Bansal
e-mail: vaishnavibansal@jssaten.ac.in

A. Tripathi
Department of Civil Engineering, JSS University, Noida, India
e-mail: amittripathi@jssaten.ac.in

R. Sarkar
Centre of Excellence in Disaster Risk Reduction, Delhi Technological University, Delhi, India
e-mail: rajusarkar@dce.ac.in

R. Sarkar et al. (eds.), *Climate Change, Environmental Hazards and Community-Based Resilience*, Disaster Risk Reduction, https://doi.org/10.1007/978-981-95-9485-6_1

Introduction

Landslides are a major natural hazard worldwide, responsible for significant loss of life, economic damage, and long-term disruption of landscapes. Unlike earthquakes or volcanic eruptions, which are primarily geological, landslides are strongly influenced by climatic and hydrological conditions. Rainfall intensity, prolonged wet seasons, snowmelt, and changes in ice and permafrost conditions all contribute to slope failures. As global warming accelerates, the frequency and magnitude of these triggers are expected to change, making landslides increasingly recognized as climate-sensitive hazards (Gupta & Nair, 2012; Umar, 2024).

Climate change affects landslides both directly and indirectly. Direct influences include changes in rainfall regimes that increase soil saturation or raise groundwater tables, reducing slope stability. Indirect influences include shifts in vegetation cover, soil properties, and cryospheric dynamics, such as glacial retreat and permafrost degradation (Bharti et al., 2022; Loranty et al., 2018; Yang et al., 2010). These processes operate at different temporal and spatial scales, complicating the attribution of specific landslide events to climate variability or long-term climate change.

Research on the relationship between climate and landslides has expanded substantially since the 1990s (Saraswat et al. 2020). Systematic reviews show that most studies have been concentrated in Europe, especially in the Alps, while large parts of South America, Africa, and Asia remain underrepresented (North et al., 2020; Payne et al., 2020). Methodologically, two approaches dominate: prospective modeling based on climate projections coupled with slope stability models, and retrospective analyses using historical climate and landslide records. A smaller but important body of work uses paleo-environmental evidence to reconstruct past landslide–climate interactions (Picarelli et al., 2017). Despite progress, uncertainties remain high, particularly due to differences in emission scenarios, global circulation models, and downscaling methods (Fumagalli, 2025).

High-mountain environments present some of the clearest examples of how climate change can destabilize slopes. In these settings, permafrost degradation, glacier thinning, and increased meltwater act as powerful destabilizing factors. Huggel et al. (2012) document a series of catastrophic slope failures in the Alps, Andes, Caucasus, and North America since the 1990s, including rock–ice avalanches and glacier-related debris flows. These events highlight both the sensitivity of mountain slopes to climatic warming and the complex feedbacks and threshold behaviors that can make slope failures abrupt and devastating. Importantly, such failures often show delayed responses, where destabilization processes unfold over years or decades before a catastrophic release occurs.

Uncertainty in attribution remains a central challenge. While shallow rainfall-induced landslides can often be linked statistically to precipitation extremes, larger and deeper failures in permafrost or glaciated terrain are harder to attribute with confidence (Fumagalli, 2025; Coe, 2017). Non-climatic triggers such as earthquakes, rock structure, or human activities must also be excluded before climatic attribution can be made. Nevertheless, both global reviews and high-mountain case studies converge on the conclusion that climate change is likely to increase landslide activity in many regions, especially where extreme precipitation and cryospheric change intensify in the coming decades (Ramya et al., 2023; Yu et al., 2021). By combining broad patterns with localized evidence, this chapter provides a comprehensive overview of landslides in a changing climate. The discussion covers climatic drivers, global findings, high-mountain lessons, data comparisons, and future risks and solutions.

Climate Factors Behind Landslides

Climate stands as a paramount external catalyst for landslides, with geological substrates, lithological attributes, and anthropogenic interventions furnishing preparatory instabilities, while meteorological and hydrological perturbations furnish the precipitating impulses (Asghar et al., 2024). Climate change recalibrates these triggers, modulating landslide incidence, seasonality, and scale across heterogeneous terrains (Crozier, 2010). Predominant climatic influencers encompass precipitation regimes, snowmelt dynamics, thermal oscillations, permafrost dissolution, and glacial recession (Loranty et al., 2018; Lafrenière and Lamoureux, 2019), each exerting context-specific effects contingent on regional hydroclimatology, slope morphology, and temporal horizons.

Precipitation and Extreme Rainfall

Precipitation emerges as the preeminent and universally acknowledged landslide instigator, wherein brief, high-magnitude deluges swiftly inundate regolith, inflate pore pressures, and erode shear resistance (Kaul, 2019), whereas protracted precipitation incrementally elevates phreatic surfaces, undermining superficial and profound-seated masses alike. Empirical inquiries affirm potent linkages between rainfall thresholds and shallow detachments, with intensity-duration envelopes serving as prognostic tools, modulated by edaphic traits and topographic configurations.

In British Columbia, projections posit a 6% escalation in ephemeral precipitation could amplify debris flows by ~30% by century's end. Amid climatic evolution, the crux resides not in volumetric means but in the proliferation of extremes: models anticipate intensified heavy episodes, even amid static or declining annual aggregates (El Kenawy, 2024; Conradie & Koster, 2019), disproportionately fueling shallow slides and debris torrents over ambient norms. Gariano and Guzzetti (2016) underscore this as a resilient projection, poised to directly escalate landslide propensity. In alpine precincts, Huggel et al. (2012) accentuate precipitation's synergy with deglaciation, mobilizing morainal detritus into far-reaching debris flows; the 1998 Peruvian Salcantay cataclysm, displacing ~25 million m^3, exemplifies rainfall's catalysis of paraglacial vulnerabilities.

Snowmelt and Seasonal Effects

Snowpack accretion and ablation cycles profoundly sway slope poise, with accelerated melt—oft conjoined with precipitation—deluging substrates and inciting disequilibrium. Warming precipitates phenological shifts: precocious vernal melt, attenuated snow reservoirs, and augmented rain-on-snow incidences (Grogan et al., 2020; Wrona et al., 2005). This confluence overwhelms permeability, a hallmark of mid-latitude cordilleras where such hybrids have spawned recurrent slides. Though underscrutinized relative to pluvial triggers, forecasts of winter rain dominance herald heightened perils. In lofty elevations, snowmelt interfaces with glacial drawdown and permafrost flux, synergistically destabilizing flanks. Insulative snow shields permafrost; its winter attenuation fosters profundal autumnal gelifraction and brisk springtime liquefaction, enfeebling lithic integrity.

Temperature Variability and Freeze–Thaw Processes

Thermal regimes impinge on stability multifacetedly: elevated means perturb edaphic hydrology via augmented evapotranspiration and vegetal reconfiguration. Pivotal are perizeral oscillations (~0 °C), engendering iterative freeze–thaw that rends bedrock and disintegrates pedons. In alpines, amplified winter cyclicity erodes mass competency, priming failures under ensuing hydrometeors. Gariano and Guzzetti (2016) note ancillary evapotranspiration shifts molding protracted hydrology, salient in xeric-temperate milieus.

Permafrost Degradation

Permafrost—enduringly cryotic substrate—underpins alpine and polar stabilities, its interstitial ice cementing fissures and voids (Subcommittee, 1988). Thermal ascent dissolves this matrix, vitiating cohesion and courting collapse. Cataclysmic rock-ice avalanches in the Alps, Caucasus, and Alaska evince this, as the 2007 Italian Monte Rosa detachment (~0.3 million m^3) from antecedent thaw. Threshold breaches yield precipitous failures, compounded by debuttressing and meltwater ingress; confidently, such losses will redefine regimes in the Alps, Andes, and Himalayas.

Glacier Retreat and Sediment Exposure

Glacial attrition manifests palpably, unmasking oversteepened scarps and friable sediments erstwhile buttressed. Debuttressing reorients stresses, fomenting instabilities, while paraglacial legacies furnish debris reservoirs for remobilization. The 2000 Caucasian Tyrnyauz flow (~10 million m^3) originated thus. Moraine impoundments spawn precarious lakes, vulnerable to landslide-calibrated GLOFs, intertwining geohazards.

What Past Studies Show

Understanding the nexus between climate change and landslides demands synthesizing evidence across varied regions, temporal scales, and investigative methodologies. Over the last three decades, scholarly output on this topic has surged, yielding valuable insights while exposing enduring knowledge deficits. Gariano and Guzzetti (2016) delivered a systematic global review of climate-landslide linkages, complemented by Huggel et al. (2012)'s focused scrutiny of high-mountain case studies. Collectively, these contributions delineate advancements in hazard comprehension alongside persistent hurdles in forging a cohesive narrative for landslides amid climatic flux (Kaul, 2019; Rahman et al., 2023).

Growth of Scientific Literature

Since the early 1990s, scholarly interest in climate-driven landslides has escalated markedly, with Gariano and Guzzetti (2016) documenting a steady rise in peer-reviewed publications and pronounced acceleration post-mid-2000s. This expansion parallels enhancements in climate modeling fidelity, proliferation of satellite and remote sensing capabilities, and heightened global awareness of anthropogenic warming (Yang et al., 2013; Marsh et al., 2023). Their review cataloged over 140 studies explicitly probing landslide-climate interrelations from 1983 to 2016—a respectable tally, though dwarfed by investigations into parallel impacts like floods or droughts. Notably, geographic coverage skews unevenly, favoring Europe, North America, and select Asian locales, while Africa and South America receive scant attention despite prolific landslide hazards in the Andes and East African highlands. Such disparities stem from asymmetries in research infrastructure and data accessibility. Domains boasting longstanding landslide catalogs, dependable meteorological archives, and climate modeling prowess—exemplified by Italy, Switzerland, the UK, and Canada—predominate. Conversely, data-scarce, high-risk terrains languish understudied. Observing that richly documented high-mountain cases cluster in the

European Alps, whereas Himalayan or Andean cataclysms often evade thorough documentation.

Methodological Approaches

Gariano and Guzzetti (2016)'s systematic appraisal delineates three principal methodologies for dissecting climate impacts on landslides: retrospective empirical inquiries, prospective simulations, and paleoenvironmental reconstructions, each bearing distinct merits and constraints.

Retrospective Empirical Analyses

These investigations juxtapose historical landslide datasets against climatic chronicles, discerning statistical ties via rainfall intensity-duration metrics, snowmelt patterns, or thermal anomalies (Fumagalli, 2025). A Norwegian case, for instance, unveiled robust correlations between extreme downpours and debris flows, implying amplified future incidences under projected precipitation intensification. Their forte lies in empirical veracity, yet pitfalls include inventory biases toward impactful events, sparse mountain meteorology, and the fallacy of equating correlation with causation amid multifaceted drivers (Groves, 2025).

Prospective Modeling Studies

Forward-oriented modeling channels downscaled climate projections—encompassing temperature and precipitation—into hydrological or geomechanical frameworks to forecast evolving landslide susceptibilities. Alpine Italian simulations, for example, prognosticate surges in shallow failures from intensified short-burst rains, offset by diminished snowmelt triggers amid shrinking snowpacks. Uncertainties loom large, tethered to emissions pathways, model variances, and downscaling artifacts, often surpassing those in stability computations (Tavoni & Valente, 2022).

Paleo-Environmental Reconstructions

A modest cohort leverages geomorphic proxies—like lacustrine sediments or stratigraphy—to trace antecedent climate-landslide couplings. Holocene records evince regime shifts mirroring paleoclimatic oscillations, furnishing analogs for contemporary projections. Huggel et al. (2012) spotlight glacial-periglacial archives as treasure troves, chronicling post-deglaciation failures spanning millennia.

Regional Patterns in Research

Geographic imbalances pervade the literature. Europe leads, with intensive scrutiny of the Alps and Apennines via retrospective and projective lenses, augmented by UK and Scandinavian emphases on thresholds and debris dynamics (Tolusso, 2019); North American efforts cluster in British Columbia, the Rockies, and Alaska, prioritizing precipitation-cryosphere interfaces; Asia features Japanese, Chinese, and Himalayan vignettes on pluvial and glacial perils, though holistic syntheses falter; South America and Africa, rife with Andean and rift-valley disasters, suffer profound neglect, with Peruvian, Colombian, and Ethiopian events under-analyzed per Huggel et al. (2012).

Insights from High-Mountain Case Studies

Huggel et al. (2012) augment global reviews by dissecting cryosphere-dominated alpine dynamics. Their dissection of the 2002 Kolka-Karmadon rock-ice avalanche in the Caucasus—displacing ~120 million m^3—exemplifies how subtle warming can unleash cataclysmic slope surrenders. Salient takeaways encompass permafrost thaw eroding rock cohesion via ice-fracture liquefaction; debuttressing from glacial thinning exposing sediments; cascading perils like GLOFs; and lagged, nonlinear manifestations confounding direct attribution (Vogel et al., 2024). These microscale validations align with macroscale syntheses, affirming climate's transformative yet labyrinthine imprint on landslide paradigms.

Key Knowledge Gaps

Notwithstanding strides, lacunae endure: geographic voids in high-vulnerability swaths; deficient longitudinal inventories and climatic series; modeling heterogeneities impeding syntheses; attribution quandaries parsing climatic from tectonic-anthropogenic signals; and underexplored cascades interweaving landslides, inundations, and cryospheric fluxes.

Lessons from High Mountains

High-mountain environments rank among the landscapes most acutely vulnerable to climate change impacts, where cryospheric elements—glaciers, perennial snowfields, and permafrost—profoundly govern slope stability. Even modest climatic perturbations can unravel centuries of equilibrium, precipitating mass movements that were previously dormant. Huggel et al. (2012) elucidate how warming not only escalates the incidence of landslides but also transforms their typology, scale, and interlinkages, fostering novel hazard configurations (Capobianco et al., 2025; Munasinghe, 2024).These alpine-derived insights furnish indispensable perspectives on the prospective global ramifications of climate-altered landslide regimes, bridging localized cryospheric dynamics with broader geohazard patterns.

Rock–Ice Avalanches

One of the most striking impacts of climate change in mountains is the occurrence of rock–ice avalanches. These involve massive collapses of rock masses containing ice or underlain by permafrost. Warming temperatures degrade ice-filled fractures, reducing cohesion and promoting detachment.

The 2002 Kolka–Karmadon event in the Russian Caucasus illustrates this phenomenon vividly. A huge rock–ice avalanche, mobilizing about 120 million m^3 of material, rushed down-valley at high velocity, destroying settlements and infrastructure. It argued that permafrost degradation, linked to rising temperatures in the region, was a critical factor. Similar cases have been reported in Alaska, the Alps, and the Himalayas.

These events demonstrate the nonlinear responses of mountain slopes. Permafrost degradation occurs gradually over decades, but once a threshold is crossed, collapse can be sudden and catastrophic. Such dynamics make prediction extremely difficult and complicate the attribution of single events to climate change.

Glacier Retreat and Slope Destabilization

Glaciers exert a dual influence on mountain slope dynamics: during periods of advance, they actively erode and steepen valley walls through glacial abrasion and plucking, yet in their stable phase, they furnish critical buttressing support that reinforces adjacent rock masses and suppresses gravitational instabilities. Glacier retreat disrupts this equilibrium by progressively removing the stabilizing ice load—a phenomenon termed debuttressing—which transfers shear stresses onto newly exposed rock faces and oversteepened slopes, rendering them susceptible to fracturing and failure.

Huggel et al. (2012) catalog numerous instances where such debuttressing has precipitated widespread slope instabilities across glaciated ranges. In the Swiss Alps, decades of accelerated retreat have correlated with surges in rockfalls and debris flows from deglaciated cirques and valley heads; in the Peruvian and Bolivian Andes, rapid ice loss has unveiled expansive fields of unconsolidated moraine deposits, primed for remobilization by convective rainfall; similarly, in the Caucasus Mountains, glacial drawdown has undermined entire valley flanks, culminating in massive debris flows that threaten infrastructure and settlements. These examples illustrate the lagged but profound geomorphic repercussions of warming, where debuttressing not only heightens primary mass movements but also sets the stage for compound hazards in paraglacial environments.

Debris Flows from Deglaciated Terrain

A significant consequence of accelerating glacier retreat in high-mountain regions is the proliferation of debris flows sourced from newly exposed, deglaciated landscapes. As ice masses recede, they unmask vast expanses of poorly consolidated moraines, till, and paraglacial sediments that lack the stabilizing influence of overlying ice. These unconsolidated deposits prove highly prone to mobilization during intense rainfall events or rapid snowmelt episodes, transforming into hypermobile debris flows capable of traversing long distances and inflicting severe downstream damage.

Illustrative examples abound across glaciated ranges. In Peru's Cordillera Blanca, the 1998 Salcantay debris flow entrained approximately 25 million m^3 of loose material from freshly deglaciated slopes, surging through valleys and obliterating communities. Analogous incidents recur in the Himalayas, where monsoon deluges interact with glacial sediments to spawn voluminous debris flows, exacerbating flood risks in densely populated foothills.

Prospective modeling, as synthesized by Gariano and Guzzetti (2016), forecasts a marked upsurge in such events, driven by the convergence of widespread deglaciation and intensifying extreme precipitation in tropical and subtropical highlands. This trend emphasizes the enduring legacy of cryospheric loss, where hazard peaks may lag initial warming by decades, demanding vigilant monitoring of paraglacial zones in hazard zoning and infrastructure planning.

Thresholds, Feedbacks, and Delayed Responses

A pivotal insight from high-mountain landslide investigations centers on the critical roles of thresholds, feedback loops, and temporal lags in mediating climate-driven instabilities. These nonlinear dynamics reveal that alpine slope systems can endure prolonged gradual warming without overt failure, only to succumb abruptly once tipping points—such as intensified permafrost degradation or hydrological saturation—are surpassed, precipitating large-scale collapses.

Feedback mechanisms further exacerbate vulnerabilities by creating self-reinforcing cycles. For instance, the failure of a rock wall may newly expose underlying permafrost to solar insolation, hastening thaw rates and priming adjacent sectors for successive detachments; likewise, debris flows can denude protective vegetation cover, thereby elevating erosion potential and susceptibility to recurrent landsliding.

Compounding these processes are lag effects, which obscure direct causal linkages and demand long-term perspectives. Contemporary landslides often represent deferred consequences of prior cryospheric shifts, such as glacier retreat from decades past that has progressively debuttressed slopes. As Huggel et al. (2012) contend, while such delays hinder attribution to specific climate events, they unequivocally reshape the prevailing hazard landscape, necessitating anticipatory modeling that incorporates historical precedents and future projections.

Lessons for Global Understanding

High-mountain case studies yield several generalizable insights that enhance broader comprehension of climate-driven landslide dynamics, bridging local observations with global patterns (Vashistha et al., 2025; Barwa and Singh, 2024). These lessons emphasize the non-linear nature of slope destabilization, the pivotal role of cryospheric changes, and the imperative for integrated hazard frameworks (Khali et al., 2025).

1. Nonlinear Slope Stability Responses: Climate change disrupts slope equilibrium in unpredictable, non-linear fashions, where incremental warming accumulates until critical thresholds trigger sudden, catastrophic failures.
2. Dominance of Cryospheric Processes: Permafrost thaw and glacier retreat emerge as foundational drivers of landslide initiation and propagation in alpine settings, fundamentally altering rock and soil mechanics.
3. Amplification via Cascading Hazards: Initial landslides frequently spawn secondary perils, such as debris flows or outburst floods, exponentially magnifying overall risk profiles.
4. Challenges in Attribution Amid Clear Trends: Pinpointing climate causation for discrete events proves intricate, yet aggregate evidence reveals a discernible uptick in landslide frequency and intensity under progressive warming.
5. Global Applicability of Patterns: While many paradigmatic cases derive from the Alps, analogous mechanisms pervade diverse ranges including the Andes, Himalayas, and Alaska, affirming the universal pertinence of these findings.

These principles advocate for adaptive strategies that transcend regional boundaries, prioritizing threshold monitoring and multi-hazard modeling to anticipate escalating threats worldwide.

Data and Comparisons

Data-driven synthesis is essential to understand how landslides respond to climate change. While Gariano and Guzzetti (2016) provide a systematic global review, offering statistical insights into methods, regions, and projections, Huggel et al.

(2012) supply case-based data on high-mountain slope failures. By combining these perspectives, a more integrated view of climate–landslide interactions can be achieved.

Systematic Review Evidence

Gariano and Guzzetti (2016) conducted a comprehensive systematic review of over 140 peer-reviewed studies on climate-landslide interactions, spanning publications from 1983 to 2016. By categorizing research according to methodological approaches, geographic emphases, and dominant climatic drivers, their analysis reveals a field heavily oriented toward empirical assessments of historical data, with growing but uncertain reliance on forward projections. This body of work establishes robust evidence for rainfall extremes as primary triggers but highlights gaps in modeling future scenarios and integrating long-term paleoenvironmental insights. The Table 1.1, based on their findings and restructured for clarity, delineates the distribution of methodological approaches, illustrating the predominance of retrospective analyses alongside emerging prospective tools essential for adaptation.

This distribution underscores that nearly half of studies adopt backward-looking empirical methods, while prospective modeling gains traction for policy relevance. However, the limited exploration of paleo records represents a critical shortfall in capturing multi-centennial climate-landslide dynamics, impeding deeper analogs for projected warming.

Table 1.1 Distribution of methodological approaches in global climate–landslide studies (1983–2016)

Approach	Share of studies (%)	Characteristics	Strengths	Limitations
Retrospective empirical studies	~45	Compare historical landslide inventories with climate records (rainfall, snow)	Grounded in real events; highlight rainfall thresholds	Limited by data quality; correlations not always causal
Prospective modeling studies	~40	Use climate projections in slope stability or hydrological models	Forward-looking; useful for adaptation planning	High uncertainty from climate models and downscaling
Paleo-environmental studies	~15	Use stratigraphic/ lake sediment records to link past climate to landslides	Long timescales; provide analogues for future changes	Sparse; indirect evidence; difficult to generalize regionally

Table 1.2 Geographic distribution of climate–landslide studies (1983–2016)

Region	Share of studies (%)	Key findings	Data gaps
Europe	~50	Strong focus on Alps, Apennines, UK; rainfall thresholds and modeling dominate	Overrepresentation; less focus on Eastern Europe
North America	~20	Studies in BC, Alaska, Rockies; rainfall–landslide links and cryospheric focus	Sparse data in Mexico and Central America
Asia	~ 20%	Japan, Himalayas, China; mix of rainfall and cryosphere studies	Limited consistent datasets in Himalayas
South America	<5	Some work in Andes (Peru, Colombia, Chile)	Major underrepresentation
Africa	<3	Very few studies, despite known landslide activity (e.g., East African rift)	Severe data scarcity
Oceania	<2	Scattered work in New Zealand	Limited broader coverage

Geographic Distribution of Studies

Gariano and Guzzetti (2016) underscore the stark unevenness in the global distribution of climate-landslide research, with a disproportionate emphasis on well-resourced regions and significant neglect of vulnerable areas. This bias arises from disparities in data availability, funding, and research infrastructure, leading to over-reliance on European datasets for broader generalizations. Consequently, projections for underrepresented regions—like parts of Africa and South America—carry higher uncertainties, potentially underestimating risks in rapidly warming tropical and subtropical mountains. The Table 1.2, adapted from Gariano and Guzzetti (2016), quantifies this distribution across studies from 1983 to 2016, revealing dominant foci on rainfall thresholds and modeling in Europe, contrasted with sparse cryospheric and rainfall analyses elsewhere.

Europe's dominance in climate-landslide research skews global insights, while Africa and South America languish with minimal coverage, exacerbating biases in worldwide risk assessments.

High-Mountain Case Evidence

This study provided an in-depth analyses of catastrophic landslide events in high-mountain environments, highlighting the role of climate-driven processes such as permafrost degradation and glacier retreat. These case studies illustrate how warming temperatures destabilize steep alpine slopes, often leading to cascading hazards like rock-ice avalanches and debris flows (Deline et al., 2015; Haeberli & Whiteman, 2021). By examining specific incidents across diverse ranges—including

Table 1.3 Selected high-mountain landslide events linked to climate drivers (adapted from Bansal and Sarkar 2023)

Event & location	Year	Estimated volume (M m^3)	Main trigger	Climate connection	Consequences
Kolka–Karmadon, Caucasus	2002	~120	Rock–ice avalanche	Permafrost degradation due to warming	>100 fatalities; valley destruction
Monte Rosa, Alps (Italy)	2007	~0.3	Rock avalanche	Summer heatwave → permafrost thaw	No casualties; major geomorphic change
Salcantay, Peru (Andes)	1998	~25	Debris flow from deglaciated slope	Rainfall mobilizing loose moraine	Villages destroyed; fatalities reported
Tyrnyauz, Caucasus	2000	~10	Debris flow from moraines	Glacier retreat exposing unconsolidated sediment	Infrastructure damaged
Mount Cook, New Zealand	1991	~12	Rock avalanche onto glacier	Warming weakening permafrost	Glacier surge triggered

the Caucasus, Alps, Andes, and Southern Alps—these examples reveal patterns of cryospheric instability interacting with extreme weather, underscoring the urgent need for region-specific risk assessments. The Table 1.3 summarizes selected events demonstrating the variability in triggers, scales, and impacts.

These cases exemplify the diverse climate-related triggers in high-mountain settings, including permafrost thaw, glacier retreat, and rainfall interacting with unconsolidated sediments. Despite varying scales and outcomes, each event underscores the heightened sensitivity of alpine slopes to warming-induced processes.

Integrated Comparison

To link systematic review evidence with case studies, it is useful to compare the types of climatic drivers emphasized globally versus those dominant in mountain environments mentioned in Table 1.4.

This table highlights the differing emphases between broad-scale global analyses and targeted high-mountain investigations, revealing gaps in global coverage for cryospheric processes while underscoring their prominence in alpine contexts.

Table 1.4 Comparative emphasis: global studies versus high-mountain cases

Climate driver	Evidence from global review	Evidence from high-mountain cases
Rainfall extremes	Most widely studied; strong empirical support	Important as a trigger of debris flows in deglaciated terrain
Snowmelt	Moderate focus; linked to seasonal landslides	Critical when combined with glacier retreat and rainfall
Temperature shifts	Indirectly studied (vegetation, evapotranspiration)	Directly linked to freeze–thaw and permafrost degradation
Permafrost thaw	Minor focus in global studies; few regions studied	Central driver of catastrophic rock–ice avalanches
Glacier retreat	Rarely modeled globally; data sparse	Key factor in slope debuttressing and moraine instability

Technical Insights from Data

Synthesizing the datasets from systematic global reviews and high-mountain case studies unveils critical technical patterns that illuminate the strengths and limitations of each approach, while highlighting convergent evidence on climate-landslide dynamics. These insights underscore the complementary nature of broad-scale statistical analyses and localized mechanistic probes, revealing how disparate scales and foci contribute to a more nuanced understanding of evolving hazards (Golmohammadi, 2025; Tatli, 2025).

1. **Scale of Analysis**: Global reviews excel at distilling statistical trends across diverse climates and topographies but frequently underrepresent hyper-local processes, such as nuanced permafrost thaw dynamics or site-specific hydrological feedbacks. In contrast, high-mountain case studies deliver granular mechanistic details—detailing failure initiation, propagation, and cryospheric interactions—but their findings often resist broad extrapolation due to unique topographic and climatic contexts. Bridging these scales through multi-resolution modeling could enhance predictive fidelity.
2. **Types of Hazards**: Empirical global datasets predominantly feature shallow landslides and rainfall-induced debris flows, reflecting their prevalence in populated low- to mid-elevation terrains and the relative ease of inventorying such events (Bansal & Sarkar, 2023). High-mountain investigations, however, spotlight large-volume catastrophic avalanches, rock-ice detachments, and moraine instabilities, which involve deeper-seated, cryosphere-mediated failures with outsized downstream impacts. This divergence signals a need for integrated typologies that encompass both frequent, localized perils and rare, high-consequence events.
3. **Data Quality and Temporal Coverage**: Robust correlations between extreme rainfall and landslide triggers are well-substantiated through extensive historical records and threshold analyses. Yet, cryospheric-linked phenomena—permafrost

degradation and glacier retreat—grapple with sparse long-term monitoring, fragmented inventories, and challenges in reconstructing pre-instrumental baselines, impeding quantitative attribution. Advancements in remote sensing, geophysical surveys, and proxy reconstructions are imperative to fortify these datasets.

4. **Regional Bias and Global Relevance**: European-centric evidence—particularly from the Alps—propels both systematic reviews and case narratives, benefiting from superior data infrastructure and research capacity. Nonetheless, analogous high-mountain processes manifest globally across the Andes, Himalayas, and Alaska, with comparable debuttressing and thaw mechanisms amplifying risks in understudied ranges. Mitigating this Eurocentrism demands expanded international collaborations and data-sharing initiatives to validate universal patterns.

These technical revelations advocate for hybrid methodologies that fuse global breadth with local depth, thereby refining hazard models and informing equitable risk governance.

Future Risks and Solutions

Converging evidence from systematic global syntheses and high-mountain case dissections unequivocally demonstrates that climate change will profoundly reconfigure landslide hazards on a worldwide scale, transcending mere intensification to encompass shifts in geographic hotspots, seasonal windows, and failure modalities. Prospective trajectories portend not only elevated frequencies in precipitation-vulnerable realms but also emergent perils from cryospheric legacies, such as lagged debuttressing and permafrost-driven instabilities, which could spawn novel compound events with cascading societal repercussions. Grappling with these transformations is paramount for crafting resilient adaptation and mitigation paradigms that safeguard human settlements, critical infrastructure, and ecological integrity amid escalating uncertainties (Elsamadony et al., 2025; Prabhakar, 2025). This necessitates proactive integration of probabilistic forecasting, dynamic zoning, and transdisciplinary interventions to preemptively curb exposure and vulnerability in an era of irreversible climatic flux.

Projected Trends in Landslide Activity

Increased Frequency of Rainfall-Induced Landslides

One of the most evident projections is the escalation in rainfall-triggered landslides, driven by more intense and frequent precipitation events. Climate models forecast a rise in extreme rainfall—characterized by short-duration, high-intensity storms—particularly in mid-latitude and tropical regions worldwide (Liu & Zhang, 2025). Areas like southern Europe, parts of Asia, and western North America are anticipated to face heavier storms that surpass current rainfall thresholds for slope instability. Notably, even without an overall increase in annual rainfall totals, shifts toward concentrated, intense bursts will suffice to destabilize slopes.

Changing Seasonality and Snowmelt Effects

Effects In temperate and alpine zones, snowmelt timing is projected to advance earlier in the season due to warming temperatures. Milder winters will foster more frequent rain-on-snow incidents, elevating the likelihood of landslides triggered by the synergy of meltwater and precipitation (Shijin et al., 2022). This shift also alters soil saturation levels and groundwater recharge patterns, potentially prolonging the active landslide period throughout the year.

Expansion of Cryosphere-Related Hazards

High-mountain environments will witness ongoing slope destabilization from permafrost thaw and glacier recession. These cryospheric changes are effectively irreversible on human timescales, fostering novel hazard zones in presently stable terrains. For instance, newly exposed deglaciated slopes laden with loose sediments may persist in an unstable state for decades.

Shifting Hazard Zones

Climate shifts will reshape the spatial distribution of landslide risks. Previously low-risk areas could see heightened activity, whereas others might experience declines (Rosser et al., 2021). Permafrost-dominated regions, for example, may confront surging instability, while arid zones could witness slight reductions in landsliding amid drying trends. Such transformations underscore the imperative for regular revisions to hazard maps and planning instruments.

Increasing Exposure and Vulnerability

Although climatic drivers initiate landslides, societal risk hinges on exposure—encompassing people and assets within hazard-prone areas—and vulnerability to resultant damage. Swift urbanization, infrastructure proliferation, and population surges in mountainous terrains are outpacing hazard mitigation initiatives, thereby amplifying exposure. In numerous developing nations, informal settlements cling to precarious, steep slopes lacking adequate drainage or engineering safeguards (Kumar, 2024). Additionally, tourism ventures and hydropower developments are encroaching on perilous high-mountain domains. The Himalayas and Andes, in particular, are expanding road networks, tunnels, and reservoirs in debris flow- and rockfall-vulnerable locales. Absent the integration of forward-looking climate scenarios in planning, these endeavors risk compounding vulnerabilities.

Uncertainty and Challenges in Prediction

Projections reveal unambiguous trends, yet forecasting precise future landslides grapples with inherent uncertainties. Key challenges include discrepancies across global and regional climate models, yielding varied precipitation and temperature outcomes; limitations in downscaling global data to localized mountain contexts, which engender inaccuracies; sparse long-term landslide records and fine-resolution meteorological datasets in many areas; and intricate nonlinear interactions, feedbacks, and delayed responses that undermine straightforward projections. Nevertheless, advancements in ensemble modeling and probabilistic techniques are refining forecasts, while merging climate scenarios with physics-based slope stability analyses paves the way for resilient risk evaluations (Bansal & Sarkar, 2024a, 2024b, Ye et al., 2025; Singh et al., 2023).

Adaptation and Risk Reduction Strategies

Mitigating escalating landslide threats necessitates a blend of structural, non-structural, and institutional interventions. Insights from global analyses and mountain-focused research outline pivotal approaches.

Monitoring and Early Warning

Robust monitoring networks—employing rainfall gauges, remote sensing, and in-situ sensors—are crucial for identifying failure precursors. In alpine settings, thermal and geophysical tools can track permafrost thaw and glacial dynamics. Proven early warning systems, reliant on rainfall thresholds, operate effectively in nations such as Italy, Japan, and Switzerland, with future iterations requiring climate-calibrated thresholds to reflect evolving precipitation regimes.

Hazard Mapping and Land Use Planning

Hazard and susceptibility mapping must evolve dynamically, embedding climate projections. As risk zones migrate, land-use regulations should curtail development in nascent high-threat areas. In vulnerable mountainous developing contexts, refined planning can shield at-risk populations, bolstered by fusing indigenous knowledge with empirical data to boost uptake and efficacy.

Engineering and Nature-Based Solutions

Conventional engineering—retaining walls, drainage networks, and slope reinforcement—remains indispensable in critical zones. Complementing these, nature-inspired methods like reforestation and bioengineering bolster slope durability and hydrological control. High-mountain engineering must accommodate cryospheric flux, incorporating adaptable designs for glacial lakes and moraine barriers.

Research and Knowledge Gaps

Enhancing adaptation demands tackling gaps pinpointed in this review involves sustained monitoring of slope and climatic variables in key sites; unified models linking climate, hydrology, and geomorphology to simulate chained hazards; refined linkages between cryospheric shifts and landsliding; expanded investigations in data-poor areas like Africa, the Andes, and Himalayas; attribution methods to parse climate signals from tectonic or human influences; and socioeconomic evaluations of vulnerability and adaptability across scenarios. Bridging these voids will yield sharper predictions and potent risk mitigation.

Outlook and Integrated Framework

Synthesizing global reviews with high-mountain cases advocates a multi-tiered framework for climate-altered landslide risk appraisal. At the global level, systematic reviews and models delineate overarching climatic patterns and hotspots; regionally, downscaled projections gauge hazard specifics and incidences; locally, field studies and surveillance reveal bespoke processes and triggers. This stratified methodology balances macro trends with micro nuances, while enabling ongoing risk refinements amid emerging evidence.

Conclusion

Future landslide risk will be profoundly influenced by the interplay of climate change, human development, and geomorphic processes. Systematic reviews and detailed mountain case studies consistently highlight key trends: intensifying rainfall extremes are expected to heighten the frequency and severity of shallow landslides and debris flows; cryospheric transformations, such as permafrost thawing and glacier retreat, will undermine slope stability in high-mountain regions; shifting climate patterns will give rise to entirely new hazard zones; and rising exposure to these risks will escalate without adaptive measures in land-use planning and infrastructure development. Addressing these challenges demands integrated risk management strategies that incorporate advanced monitoring, robust policy frameworks, and active community engagement.

Although uncertainties persist in the specifics of these dynamics, the overarching trajectory is unmistakable: climate change is already amplifying landslide hazards in the present, not merely as a distant threat. Effective preparation hinges on linking global-scale analyses with targeted local interventions, fostering resilient landscapes, and iteratively refining knowledge and tools to keep pace with an evolving climate.

References

Asghar, M., Ayaz, M., Ullah, S., & Ali, S. (2024). Climate disasters and sustainability challenges: A way forward in industry 4.0. In *The Palgrave handbook of green finance for sustainable development* (pp. 155–190). Springer International Publishing.

Bansal, V., & Sarkar, R. (2023). Numerical analysis and implications of fluctuating water levels on slope stability. In *International Conference on Sustainable Infrastructure: Innovation, Opportunities and Challenges* (pp. 407–419). Springer.

Bansal, V., & Sarkar, R. (2024a). Comparative analysis of slope stability for Kalimpong Region under dynamic loading using limit equilibrium method and machine benchmark learning classifiers. *Iranian Journal of Science Technology, Transactions of Civil Engineering, 48*, 2785–2807.

Bansal, V., & Sarkar, R. (2024b). Prophetical modeling using limit equilibrium method and novel machine learning ensemble for slope stability gauging in Kalimpong. *Iranian Journal of Science Technology, Transactions of Civil Engineering, 48*, 411–430.

Barwa, S., & Singh, H. (2024). Climate-induced changes in forest ecosystems of Indian Himalayan region: A meta-analysis. In *Urban forests, climate change and environmental pollution: physio-biochemical and molecular perspectives to enhance urban resilience* (pp. 839–866). Springer Nature Switzerland.

Bharti, G., Shukla, B. K., Srinivasan, V., & Bansal, V. (2022). Experimental study on eccentrically loaded rectangular footing on sand for embedded condition to enhance serviceability of structures. *Materials Today: Proceedings, 61*, 517–522.

Capobianco, V., Choi, C. E., Crosta, G., Hutchinson, D. J., Jaboyedoff, M., Lacasse, S., & Reeves, H. (2025). Effective landslide risk management in era of climate change, demographic change, and evolving societal priorities. *Landslides*, 1–19.

Coe, J. A. (2017). Landslide hazards and climate change: A perspective from the United States. In *Slope safety preparedness for impact of climate change* (pp. 479–523). CRC Press.

Conradie, E. M., & Koster, H. P. (2019). *T&T Clark handbook of Christian theology and climate change*.

Crozier, M. J. (2010). Deciphering the effect of climate change on landslide activity: A review. *Geomorphology, 124*(3–4), 260–267.

Cutter, S. L. (2018). Compound, cascading, or complex disasters: What's in a name? *Environment: Science and Policy for Sustainable Development, 60*(6), 16–25.

David Raj, A., & David Raj, A. (2025). Climate change induced hydro-meteorological extremes in the Himalayan region. In *Climate change: Conflict and resilience in the age of anthropocene* (pp. 33–56). Springer.

Deline, P., Gruber, S., Delaloye, R., Fischer, L., Geertsema, M., Giardino, M.,... & Schoeneich, P. (2015). Ice loss and slope stability in high-mountain regions. In *Snow and ice-related hazards, risks, and disasters* (pp. 521–561). Academic Press.

El Kenawy, A. M. (2024). Hydroclimatic extremes in arid and semi-arid regions: Status, challenges, and future outlook. In *Hydroclimatic extremes in the Middle East and North Africa* (pp. 1–22). Elsevier.

Elsamadony, M., Pileggi, S. F., Liu, J., & Fujii, M. (2025). Crafting a resilient and sustainable future amid major crises. *Journal of Cleaner Production*, 145583.

Fumagalli, M. (2025). Rainfall-dependent susceptibility mapping for shallow landslides: Data-driven and physically-based modelling under climate change.

Gariano, S. L., & Guzzetti, F. (2016). Landslides in a changing climate. In *Earth-Science Reviews, 162*, 227–252.

Golmohammadi, A. H. (2025). Conceptual and practical boundaries of ecological landscape design and landscape ecology.

Grogan, D. S., Burakowski, E. A., & Contosta, A. R. (2020). Snowmelt control on spring hydrology declines as the vernal window lengthens. *Environmental Research Letters, 15*(11), Article 114040.

Groves, M. (2025). *Extreme weather and climate change: A reference handbook*. Bloomsbury Publishing.

Gupta, A. K., & Nair, S. S. (2012). *Environmental extremes disaster risk management-addressing climate change* (p. 40). National Institute of Disaster Management, New Delhi.

Haeberli, W., & Whiteman, C. (2021). Snow and ice-related hazards, risks, and disasters: Facing challenges of rapid change and long-term commitments. In *Snow and ice-related hazards, risks, and disasters* (pp. 1–33). Elsevier.

Huggel, C., Clague, J. J., & Korup, O. (2012). Is climate change responsible for changing landslide activity in high mountains? *Earth Surface Processes and Landforms, 37*(1), 77–91.

Kaul, V. (2019). *Holistically understanding and enhancing the adaptation of remote high-mountain communities to hydrometeorological extremes and associated geohazards in a changing climate*. Doctoral dissertation, University of Sheffield.

Khali, H., Chatterjee, R. S., & Rawat, K. S. (2025). Geohazard linkages with surface geodynamics: A short-term temporal topographic analysis of mass movement and gradation patterns in the upper Bhagirathi basin. *Discover Applied Sciences, 7*(10), 1056.

Kumar, P. (2024). Social and economic impact in the landslide prone zones and related policies. In *Landslides in the Himalayan region: Risk assessment and mitigation strategy for sustainable management* (pp. 499–529). Springer Nature Singapore.

Lafrenière, M. J., & Lamoureux, S. F. (2019). Effects of changing permafrost conditions on hydrological processes and fluvial fluxes. *Earth-Science Reviews, 191*, 212–223.

Liu, S., & Zhang, Y. (2025). Unravelling the relationship between extreme precipitation duration and wet-day frequency at global scale. *Scientific Reports, 15*(1), 22943.

Loranty, M. M., Abbott, B. W., Blok, D., Douglas, T. A., Epstein, H. E., Forbes, B. C., & Walker, D. A. (2018). Reviews and syntheses: Changing ecosystem influences on soil thermal regimes in northern high-latitude permafrost regions. *Biogeosciences, 15*(17), 5287–5313.

Marsh, R., Skliris, N., Tompkins, E. L., Dash, J., Dominguez Almela, V., Tonon, T., … & Webber, M. (2023). Climate-sargassum interactions across scales in the tropical Atlantic. *PLoS Climate, 2*(7), e0000253.

Munasinghe, D. S. (2024). *Risk Propagation simulator for assessing urban risk.* Doctoral dissertation, University of Salford, UK.

North, M. A., Hastie, W. W., & Hoyer, L. (2020). Out of Africa: The underrepresentation of African authors in high-impact geoscience literature. *Earth-Science Reviews, 208*, Article 103262.

Payne, D., Snethlage, M., Geschke, J., Spehn, E. M., & Fischer, M. (2020). Nature and people in the Andes, East African Mountains, European Alps, and Hindu Kush Himalaya: Current research and future directions. *Mountain Research and Development, 40*(2), A1.

Pescaroli, G., & Alexander, D. (2016). Critical infrastructure, panarchies and the vulnerability paths of cascading disasters. *Natural Hazards, 82*(1), 175–192.

Picarelli, L., Comegna, L., Gariano, S. L., Guzzetti, F., Mercogliano, P., Rianna, G., … & Tommasi, P. (2017). Potential climate changes in Italy and consequences for land stability. In *Slope safety preparedness for impact of climate change* (pp. 151–197). CRC Press.

Prabhakar, A. (2025). A sustainable and inclusive economic development: A global imperative. *Journal of Recycling Economy & Sustainability Policy, 4*(1), 1–16.

Rahman, B., Van den Broeck, P., & Parra Novoa, C. (2023). Disaster Risk Reduction (DRR) policy in the making: The socio-ecological dialectics of institutionalised and community responses following the 2010 Merapi Mega-Eruption.

Ramya, A., Poornima, R., Karthikeyan, G., Priyatharshini, S., Thanuja, K. G., & Dhevagi, P. (2023). Climate-induced and geophysical disasters and risk reduction management in mountains regions. In *Climate change adaptation, risk management and sustainable practices in the Himalaya* (pp. 361–405). Springer International Publishing.

Rosser, N., Kincey, M., Oven, K., Densmore, A., Robinson, T., Pujara, D. S., … & Dhital, M. R. (2021). Changing significance of landslide hazard and risk after the 2015 Mw 7.8 Gorkha, Nepal earthquake. *Progress in Disaster Science, 10*, 100159.

Saraswat, R., Parkash, S., & Pal, S. (2020). Landslide susceptibility assessment due to construction of buildings in Garhwal area using GIS. In: S. Ahmed, S. Abbas, H. Zia, H. (Eds.), *Smart cities—Opportunities and challenges.* Lecture Notes in Civil Engineering (Vol. 58). Springer. https://doi.org/10.1007/978-981-15-2545-2_70

Shijin, W., Yuande, Y., & Yanjun, C. (2022). Global snow-and ice-related disaster risk: A review. *Natural Hazards Review, 23*(4), 03122002.

Singh, S., Mishra, K., Chavan, R., & Tiwari, H. L. (2023, December). Advancements and challenges in hydrological modeling: A comprehensive review. In *International Conference on Hydraulics, Water Resources and Coastal Engineering* (pp. 423–442). Springer Nature Singapore.

Subcommittee, P. (1988). Glossary of permafrost and related ground-ice terms. *Associate Committee on Geotechnical Research, National Research Council of Canada, Ottawa, 156*, 63–64.

Tatli, H. (2025). Statistical perspectives on Mediterranean precipitation: Power-law insights in hydro-climatology.

Tavoni, M., & Valente, G. (2022). Uncertainty in integrated assessment modeling of climate change. *Perspectives on Science, 30*(2), 321–351.

Tolusso, E. (2019). Alpine protected areas in a changing climate. A geography of science perspective.

Umar, N. (2024). A geospatial review analysis of increased frequency of large glacier-related landslides in Mountainous regions. In *Landslide: Susceptibility, risk assessment and sustainability: Application of geostatistical and geospatial modeling* (pp. 459–477). Springer Nature Switzerland.

Vashistha, A., Shah, A. A., Batmanathan, N., & Dashora, A. (2025). A data-driven multi-criteria model for GLOF risk in tectonically active Himalayan regions. *Modeling Earth Systems and Environment, 11*(5), 333.

Vogel, K., Sieg, T., Veh, G., Fiedler, B., Moran, T., Peter, M., … & Bronstert, A. (2024). Natural hazards in a changing world: Methods for analyzing trends and non-linear changes. *Earth's Future, 12*(5), e2023EF003553.

Wrona, F. J., Prowse, T. D., Reist, J. D., Beamish, R., Gibson, J. J., Hobbie, J., … & Zhulidov, A. (2005). Freshwater ecosystems and fisheries. In *ACIA. Arctic Climate Impact Assessment* (pp. 353–452).

Yang, J., Gong, P., Fu, R., Zhang, M., Chen, J., Liang, S., … & Dickinson, R. (2013). The role of satellite remote sensing in climate change studies. *Nature Climate Change, 3*(10), 875–883.

Yang, Z. P., Ou, Y. H., Xu, X. L., Zhao, L., Song, M. H., & Zhou, C. P. (2010). Effects of permafrost degradation on ecosystems. *Acta Ecologica Sinica, 30*(1), 33–39.

Ye, C., Wu, H., Oguchi, T., Tang, Y., Pei, X., & Wu, Y. (2025). Physically based and data-driven models for landslide susceptibility assessment: Principles, applications, and challenges. *Remote Sensing, 17*(13), 2280.

Yu, G. A., Yao, W., Huang, H. Q., & Liu, Z. (2021). Debris flows originating in the mountain cryosphere under a changing climate: A review. *Progress in Physical Geography: Earth and Environment, 45*(3), 339–374.

Chapter 2
Impact Assessment of LULC and Climate Change on Landslide Susceptibility Through DLNN and Soil Resistance Meter: Past and Future Perspectives

Sunil Saha, Anik Saha, Raju Sarkar, Priyanka Gogoi, Barnali Kundu, and Saroj Acharya

Abstract The expansion of human settlements modifies the land use pattern. Changing LULC and rainfall patterns have increased the frequency of landslides in Eastern Bhutan. The Markov Chain Model and deep learning algorithms were used for evaluating the past, present, and future landslide susceptibility in parity with changing LULC and rainfall. To achieve the goal, a total 878 landslide locations from different periods were considered. Landslide susceptibility maps (LSMs) for 1990, 2020, 2050, and 2080 were produced using RF, ANN, and DLNN. For the production of future LSMs in 2050 and 2080, LULC and rainfall had been predicted using the Markov Chain Model and ANN. LSMs were evaluated using the ROC, accuracy, F1 measure, Jaccard, MCC, MRSE, and MAE. All the models worked well, but the DLNN model was the most accurate. In some areas having very high and very low landslide risk, an electric resistivity meter was used to check the resistivity and confirm the susceptibility levels. According to the analysis, if the LULC and rainfall pattern change in the present manner, the area of the very high landslide susceptibility zone will increase by 17.40% in 2080. This work can be helpful to planners and used in other similar areas.

Keywords Future landslide susceptibility map · Dynamicity of land use/land cover · Electric resistivity · Markov Chain Model · Deep learning algorithms

---

S. Saha · P. Gogoi · B. Kundu
Department of Geography, University of Gour Banga, Malda, India

A. Saha
Department of Geography, Harishchandrapur College, Malda, India

R. Sarkar (✉)
Department of Civil Engineering, Delhi Technological University, Delhi, India
e-mail: rajusarkar@dce.ac.in

S. Acharya
National Center for Hydrology and Meteorology, Royal Government of Bhutan, Thimphu, Bhutan
e-mail: sacharya@nchm.gov.bt

R. Sarkar et al. (eds.), *Climate Change, Environmental Hazards and Community-Based Resilience*, Disaster Risk Reduction, https://doi.org/10.1007/978-981-95-9485-6_2

Introduction

Slope failures are one of the most common sources of landscape change, and they have a considerable effect on human society and economic activity. Landslides are the consequence of a complicated spatio-temporal association of a number of aggravating and activating variables, and one of those is land cover. Whereas, most geomorphological and geological elements change over longer timeframes, a highly dynamic feature is land cover, which may transform quickly owing to natural and human forces (Fuller & Rutherfurd, 2021). The dynamics of land cover may influence landslide behavior, causing considerable variations in the distribution and frequency of landslides, often over short time frames (Depicker et al., 2021). Numerous investigations have reported how alterations in plant cover alter hydrodynamic and structural mechanisms in the soil. Changes in land cover and the resulting effects on slope strength are mostly human-induced. For example, increased deforestation has been associated with an increase in modest landslides (Chiarelli et al., 2021; Mkuzi, 2020). According to Makwinja et al. (2021) the most vulnerable land cover is left in cultivated areas that progressively transform into natural grassland, bushes, and woodlands. Wildfires are also a key element that can lead to increased surface water runoff (Evelpidou et al., 2022). Furthermore, burned regions can enhance the sediment content of surface runoff, encouraging the formation of debris flows (Jakob, 2022). Alterations in urban regions have occurred to encourage higher efficiency in the utilization of cultivation and livestock lands in some of the most advantageous locations in order to maximize food output while lowering operating expenses. This resulted in a decrease in crops and cattle grazing in elevated rural areas and on higher inclines (Erasu Tufa & Lika Megento, 2022; Marzen et al., 2022). We may infer that land cover changes will occur in the future in any dynamic ecosystem, especially given global climatic and socioeconomic shifts. Climate forecasts in the Mediterranean region reveal that rainfall patterns and severity will probably alter in the coming years, affecting a range of ecosystem functions (Khelifa et al., 2022; Leal Filho & Manolas, 2022). Judgment makers and politicians are concerned about changes globally, particularly their repercussions, which include geo-hydrological dangers.

Greater intensity, or long-term precipitation, is among the key triggering factors for land collapses (Chen et al., 2021a, 2021b). Atmospheric circulations influenced by terrain produce different patterns of rainfall in hilly areas, where slides are common (Zhang et al., 2021). However, it is unclear how much such a geographical change in rainfall influences slope resistivity. If the impact is significant and rainfall changes can be predicted, these predictions could be useful for assessing and forecasting landslide hazards.

Bhutan has been severely impacted by climate anomalies over the last decade, resulting in concentrated rainy seasons and high precipitation concentrations (Dorji et al., 2021). Severe rainfall is occurring more often, raising the danger of landslides (Jones et al., 2021). The prediction of new variations in rainfall due to climate change will be useful as the foundation for landslide hazard assessments. Torrential rain occurrences have been identified as the primary initiating element in the research region.

In recent times, published research has given greater attention to future possibilities, including changes in LULC (Majeed et al., 2021; Chen et al., 2021a, 2021b). However, few studies have examined how changes in land cover affect the vulnerability of slopes to landslides (Depicker et al., 2021; Mondini et al., 2021). Jakob and Owen (2021) asserted that the past and present are the keys to the future and that the landslides of the future will be characterized by the same starting circumstances as previous landslides. As a result, the vulnerability assessment could be used to forecast future geographically positioned landslides. Landslide susceptibility maps may be created in a variety of ways. These are fundamentally classified into two parts: quantitative and qualitative methods (Khaidem & Singh, 2021). The qualitative methods rely on expert opinion, and the input of data to the algorithm is often collected from field studies, with geomorphic and photo-interpretation evaluations when needed (Ahmadi & Pekkan, 2021). Statistical or deterministic analysis is used in quantitative techniques.

Statistical methods have been widely used to assess landslide susceptibility in recent years. However, many methods are available, such as discrimination assessment (Alqadhi et al., 2021; Tanyu et al., 2021); weight of evidence (Ali et al., 2021; Farooq & Akram, 2021; Ali et al., 2021); logistic regression (Tanyu et al., 2021; Alqadhi et al., 2021; Chen & Zhang, 2021); and frequency ratio (Nanda et al., 2021; Shu et al., 2021). One of the disadvantages of these approaches is that they demand a decent dataset of input in terms of numbers and spatial extent, with its reliability and validity heavily reliant on it. Slope resistivity assessment and the safety factor equations are used in deterministic approaches. They necessitate a thorough understanding of the material's mechanical and physical qualities (Azmoon et al., 2021; Ogila, 2021). Different statistical and other models have been used for assessing landslide events. In the present study, two machine learning techniques, i.e., random forest (RF) and artificial neural network (ANN), and one deep learning technique, i.e., deep learning neural network, have been used. Future prediction of LULC and rainfall has been carried out using the Markov Chain Model and ANN.

Different machine learning algorithms have proven excellent accuracy in performing the landslide susceptibility models in mountainous areas. However, a diverse case study is still needed for further advancement. Moreover, a few questions regarding the impact of changing climates and LULC are still unclear. So, the main research questions are: (1) Has the level of landslide susceptibility increased, decreased, or remained the same over time? (2) Is there any influence of the changing pattern of rainfall and LULC on the increasing frequency of landslides? And (3), what will be the scenario of landslide susceptibility in the future in parity with the changing rainfall and LULC? For finding the answers to the following research questions, deep learning algorithms and the Markov chain model have been used, along with the total of 878 landslide locations. For training and testing, the model 70:30 ratio has been used based on the previous literature (Ali et al., 2021; Farooq & Akram, 2021; Ali et al., 2021; Nanda et al., 2021; Shu et al., 2021). Afterward, the prepared models were validated by the accuracy (%), F1 measure, Jaccard, MCC, MAE, and RMSE.

For justifying the causes of the susceptibility level, an electric resistivity meter has been used for measuring the status of soil resistivity up to 30 m below the surface. Considering the aforementioned research questions, the aim of the research work is to assess the past and present landslide susceptibility as well as predict the future landslide susceptible level in parity with the changing LULC and rainfall pattern in Eastern Bhutan.

Description of the Study Area

Our study area is located in eastern Bhutan, between the coordinates 26° 47′ 28.74″ and 27° 19′ 28.46″N and 91° 22′ 27.55″ and 91° 39′ 59.56″E (Fig. 2.1). This block covers approximately 1071 km^2 and has 6327 households with a population of 29,347 people. This region is mostly a hilly Bhutanese unit. Soils such as alisol, acrisol, chernozem, cambisol, and regosol may be found there. This is situated between 137 and 3927 m above sea level. Summer, monsoon, autumn, winter, and spring are the five different seasons in Bhutan. Hot summers and cold winters characterize the climate of eastern Bhutan, which is more temperate and drier than the western part. Temperatures range from 20 to 35 °C in the summer and 5–15 °C in the winter. The slope angle varies from 7.5 to 72.16°. The research area is characterized by an average annual rainfall of 1500–3500 mm, which has a significant influence on landslide incidents. The total number of households in the study is about 6327, as per the NSB. This area is landslide-prone, according to Dikshit et al. (2020). A large number of landslides occur every year and cause large damage to human lives and properties (Tables 2.1 and 2.2). The frequency of landslides in Eastern Bhutan increases year over year. Proper management of landslides is essential for the sake of socio-economic development. In this regard, the preparation of landslide susceptibility maps for the future based on past and present landslide susceptibility conditions is a good way to formulate landslide management strategies.

Materials and Methods

First, we first took into account 12 contributing elements, involving elevation, slope, curvature, aspect, drainage density, sediment transportation index (STI), topographical wetness index (TWI), topographical positioning index (TPI), geology, distance from lineament, soil classification, and road density, in order to create a susceptibility map for landslides for the current research region.

We have examined the influence of landslide confounders using the multi-co-linearity test after including all condition variables. On the other hand, we performed a multi-co-linearity test for each conditioning element to determine its involvement in the modeling of landslide susceptibility. We identified LULC and rainfall as the primary contributing factors to the landslide incidents in this region. We combined

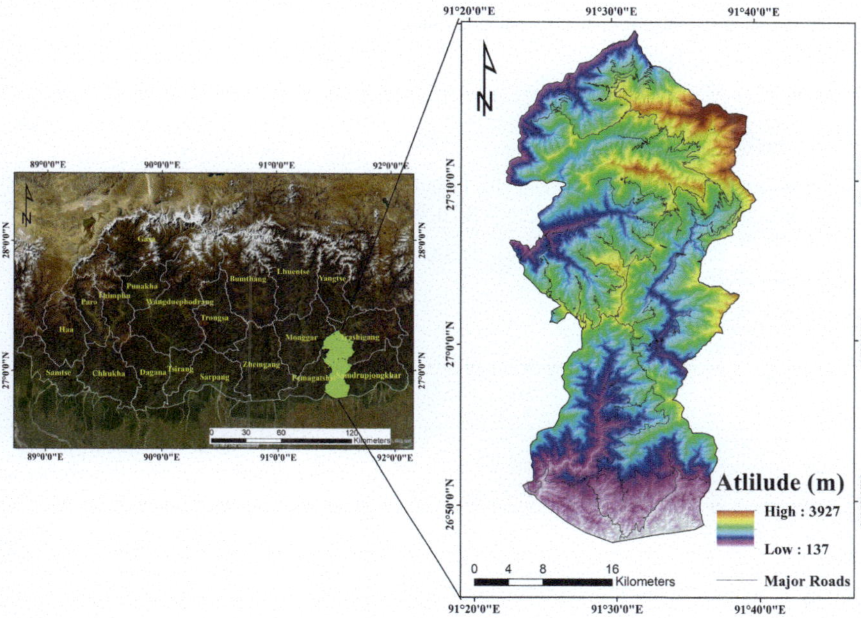

Fig. 2.1 Location of the study area

these two factors to create a susceptibility map for landslides for the years 2050 and 2080. We separated the landslide inventory data into 70 and 30% for training and testing the landslide susceptibility models. We created the landslide susceptibility map (Fig. 2.2) using the RF and DLNN models. We validated the model (Fig. 2.2) using accuracy, specificity, sensitivity, specificity, the Jaccard coefficient, the F1 measure, the Matthews correlation coefficient (MCC), the root mean square error (RMSE), and the area under the receiving curve (AUC).

Data Set Used

Landslide Inventory Data

In the analysis of landslide hazards, inventory mapping plays a critical role as an insight into various approaches (Nath et al., 2022). The landslide inventory map depicts all previous landslide incidents that have had a significant impact on a specific region. These maps serve as the foundation for the subsequent maps. We created a map of landslide inventory based on the goal and feasibility of our current research, the available data, and the size of the survey area. We have collected information about landslide locations from the DoR, regional office, and Google Earth image from 1990 to 2020. We conducted several time-field surveys in 2021 to verify the location

Table 2.1 Locational details of few recent landslides

SI no.	Date of landslide	Location
1	01-12-2021	SJ-Trashigang PNH. (Chainage 79 km from SJ) near Moshi
2	26/09/2020, 25/10/2020, 14/09/2020, 11/09/2020, 7/09/2020, 31/07/2020, 22/07/2020, 28/06/2020	Gyelposhing/Nganglam to Pemagatshel. (Chainage 36 km from Tsebar) at Fadhari
3	11/09/2020, 7/09/2020,	Nganglam–Gyelposhing at CH 9.9
4	11/09/2020, 7/09/2020,	Nganglam–Gyelposhing at CH 10.2
5	09-11-2020	CH 13.10kms from Nganglam
6	7/09/2020,	Dewathang–Samdrupcholing at CH 0.50kms
7	19/08/2020	SamdrupJongkhar– Trashigang at CH 64kms. (4kms from Narphung town)
8	31/07/2020	Pemagatshel–Khothakpa Khar TsebarYurung at CH 30 kms
9	31/07/2020	Nganglam–Gyelposhing at 21 Kms
10	31/07/2020	Nganglam–Gyelposhing at 33.9 Kms
11	31/07/2020, 30/07/2020, 18/07/2020, 10/07/2020,	Nganglam–Gyelposhing at 35.9 Kms
12	31/07/2020, 28/06/2020	Nganglam–Gyelposhing at 38 Kms
13	31/07/2020	Dewathang-Samdrupcholing at CH 18.4
14	31/07/2020	Dewathang-Samdrupcholing at CH 26.2
15	31/07/2020	Dewathang-Samdrupcholing at CH 28.7
16	30/07/2020	Tsebar-Dungmin at 1.7kms
17	30/07/2020,	NganglamPanbang–Nganglam block at 50kms
18	30/07/2020, 19/04/2020	Nganglam–Gyelposhing at 41.5 Kms
19	07-10-2020	Dewathang-Samdrupcholing at 23.5 Kms
20	07-10-2020	Dewathang-Samdrupcholing at 30.0 Kms
21	07-10-2020	Mikuri-Durungri at 36 km
22	1/7/2020, 10/07/2020, 28/06/2020, 25/06/2020, 19/04/2020	Nganglam–Gyelposhing at 41.6 Kms

(continued)

Table 2.1 (continued)

SI no.	Date of landslide	Location
23	28/06/2020	Nganglam-Gyelposhing blocked at 9.2kms
24	28/06/2020	Dewathang-Samdrupcholing blocked at Rekhey, 4.5Kms from Dewathang
25	26/06/2020	Pemagatshel–Nganglam blocked at 25 km from Tshobaley
26	25/06/2020	Nganglam–Gyelposhing at 43.5 Kms
27	06-12-2020	5.5 km from Tshengkhari on NganglamGyelposhing
28	06-02-2020	at chainage 7.06 km from Tshelingkhore
29	12/5/2020, 22/04/2020	Nganglam–Gyalposhing road block at 38.4Kms

Source DoR, Regional office, SamdrupJonkhar

Table 2.2 Major damages caused by landslides in eastern region

SI no.	Date of landslide	Location	Damaged caused
1	19th Sep, 2017	Chhoekhorling Gewog in Pema Gatshel	A man died when the landslide washed away an excavator
2	12th July, 2018	Philooma Chiwog of Orong Gewog in SamdrupJongkhar	Three men died after they were buried by landslide
3	28th May, 2019	Pirmani in Monggar	Damaged a temporary hut
4	14th July, 2019	Chaskhar gewog of Mongar	Destroys three-storey house
5	24th July, 2019	Gomdar Gewog, SamdrupJongkhar	Major damage to the multi-purpose hall. A hut near the hall was also damaged
6	23rd May 2020	Serthig, Lauri and Wangphu Gewogs of SamdrupJongkhar	Gewog centre roads were blocked. Minor damage to one house
7	11th July, 2020	Gomdar Gewog, SamdrupJongkhar	Buried the two-storied traditional gewog office
8	26th August, 2021	PemaGatshel	Landslide killed three people. Landslide washed away the hut they were living in
9	27th August, 2021	Tsalabi village under Kengkhar gewog, Mongar	A shop was damaged by falling boulder

Source Bhutan Broadcasting Services

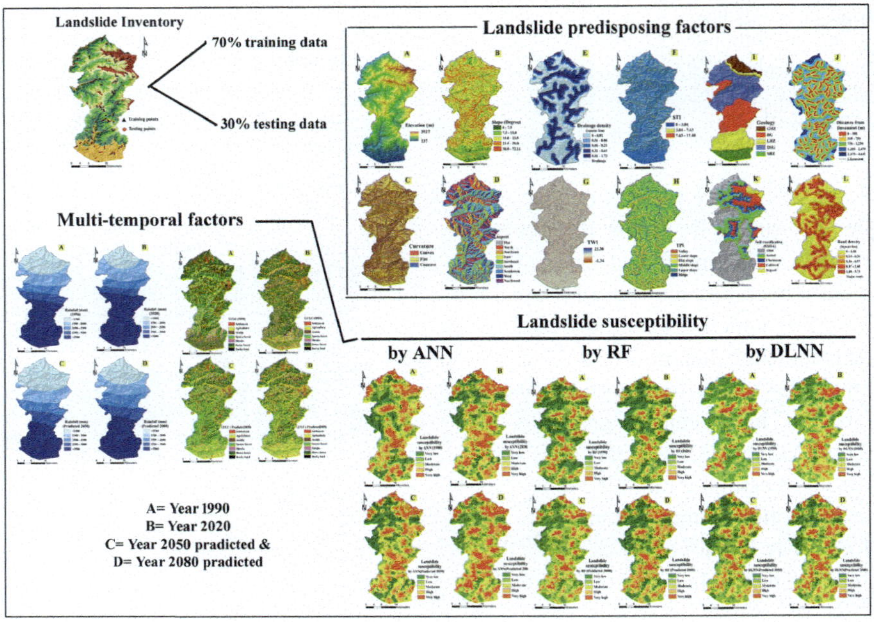

Fig. 2.2 Methodological flow chart

of landslides using handheld GPS (Figs. 2.3 and 2.4). We mapped the landslide susceptibility for 1990 and 2020 using a total of 261 and 540 landslide locations, respectively. For future prediction of landslide susceptibility (2050 and 2080), a total of 878 landslide points were considered. We classified landslide inventory data into training and testing datasets, following the 70:30 ratio from previous literature, to train and test the landslide susceptibility models (Arabameri et al. 2021; Khaidem & Singh, 2021).

Landsat and Climatic Data

In the present study to find out the land use classes since 1990 to 2020, landsat images, which are freely available, have been collected from the USGS Earth Explorer (https://earthexplorer.usgs.gov). Landsat 5 (TM) images of 25/12/1990, 17/12/2000, and 16/12/2010 were used for preparing the LULC maps of 1990, 2000, and 2010. Landslide 8 (OLI) images from 12/18/2020 were used for the LULC map of 2020. Annual rainfall data for Eastern Bhutan since 1990 to 2020 was collected from the meteorological department of Bhutan.

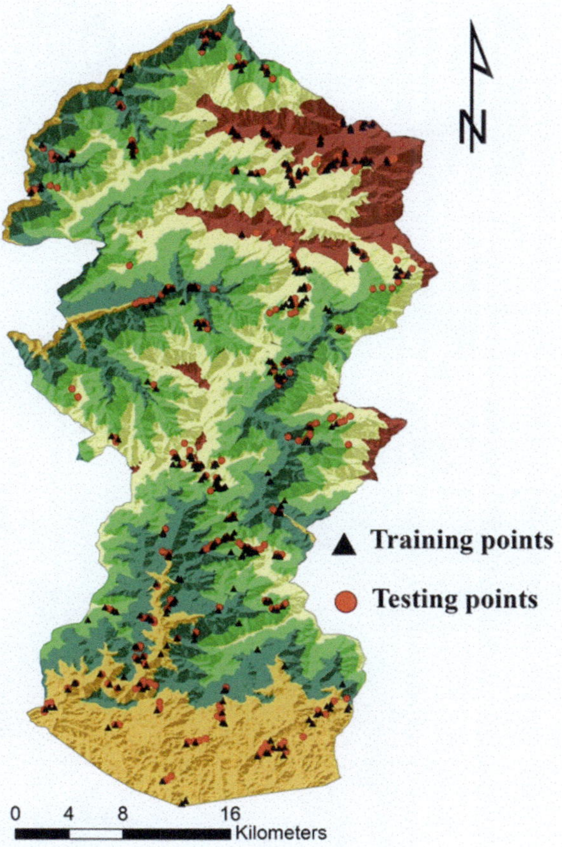

Fig. 2.3 Showing the landslide locations used in the present study for producing the past, present, and future landslide susceptibility maps

Data of Electric Resistivity Meter

In this research, we used an electrical resistivity meter to assess soil resistivity up to 30 m below the surface in areas with high and very low susceptibility, as shown in Fig. 2.5. We used this data to validate our landslide susceptibility results. Furthermore, these findings were helpful when evaluating the past and future landslide susceptibility in relation to changing rainfall and land use.

Multi-collinearity Analysis

A multi-collinearity study in a machine learning model is often the best way to find out how different geo-environmental parameters are related in a straight line (Arabameri et al., 2021). In a study of multiple regression research, this is a statistical method that can uncover two parameters with a significant relationship. As a result, analyzing a

Fig. 2.4 Landslide photos

multi-collinearity model is critical in order to improve outcomes by eliminating large multi-collinearity variables and lowering the model's bias. Academics worldwide have utilized the multi-collinearity test in various domains such as GES mapping (Arabameri et al., 2021), flooding (Saleh et al., 2022), and landslide hazard mapping (Achour et al., 2021). You can assess multi-collinearity (TOL) using the variance inflation factor (VIF) plus tolerability. If the TOL value is <0.10 or 0.20 and the VIF

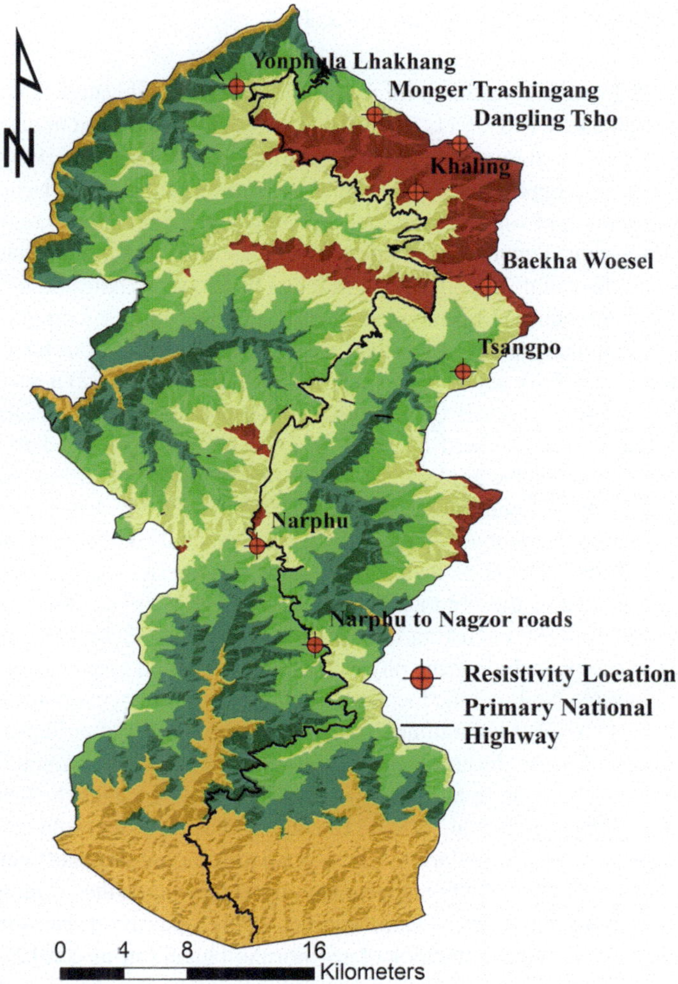

Fig. 2.5 Location of electrical resistivity test

value is >5 or 10, the results reveal strong multi-collinearity between the parameters (Arabameri et al., 2021). Formula for computations of TOL and VIF in a sample are:

$$TOL = 1 - R_j^2 \tag{2.1}$$

$$VIF = \frac{1}{TOL} \tag{2.2}$$

wherein R_j^2 represents the value of regression of j on some other items in the data.

Landslide Conditioning Factors (LCFs)

A total of 14 LCFs were considered for the present study, i.e., elevation, slope, curvature, aspect, drainage density, STI, TWI, TPI, geology, distance from lineament, soil, road density, LULC, and rainfall (Fig. 2.6). Altitude is useful for classifying regions and determining the maximum and minimum elevation within a landscape. In the context of the erosion of soil and gradient mass flow, the height of each region is one of the most important elements (Arabameri et al., 2021). This component has a significant impact on the runoff direction and drainage distribution. Altitude influences a variety of environmental and biological phenomena, including precipitation, heat, forest cover, and human activity. As a result, these physical conditions have compromised the slope's stability and caused a hill collapse. McColl (2022) asserts that the gradient of the terrain primarily determines the risk of a slope failure. According to Davies and Moretti (2022), land failure is more likely when the gradient is steep, but it is least likely when the gradient is mild. Topographic factors frequently impact the likelihood of a land collapse (Lin et al., 2022). The slope aspect governs the abundance of sunlight on any hillside, which is connected to heat and other environmental conditions. The Himalayan slopes with a southern face have seen a large number of slides due to the obvious heavy rainfall (Koley et al., 2022). The likelihood of a land collapse occurring decreases as the value of curvature decreases. Geological processes, including the physical arrangement and materials of the landscape, rock evolution, and specific location processes, primarily govern the resilience of a hill slope (Peiris & Gunarathne 2022). The degree of land collapse vulnerability varies depending on the area's geology composition. The difference between a unit's altitude and the average altitude of its neighbours determines the topographic position index (TPI) (Saleh et al., 2022). To classify current topographic landforms, however, precise levels of boundaries such as gradients, mountains, and troughs must be defined. We employ the TPI for landform analysis, aiming to derive geographical feature classifications through TPI value regions. TPI is a useful variable for material movement and also influences the look of geomorphological settings (valleys, ridges, and slopes, etc.). We also regard this topographic position index as a contributing element to landslides, as we typically observe land collapses on ridges. In hydrology processes, inclination heavily influences the topographic wetness index (TWI). It is a commonly used index for predicting the amount of moisture in the soil in a given region (Arabameri et al., 2021). Because topographical structure has an impact on avalanche danger, indices related to topographical structure are also important for slope instability. Landslides are more likely to occur in areas where there is a lot of rain. The Sediment Transport Index (STI) closely correlates with the flow of particles from upland through a stream during landslide occurrences (Abuzied & Pradhan 2021). The amount of material in a water shadow reflects the amount of material input that is likely to reach the material at the mouth of the catchment. The likelihood of sediment load varies depending on the topography, optimum angle of slope, origin of rock, extent of catchment area, incident duration and magnitude, and length of material movement. The likelihood of landslides increases as the possibility of material movement increases. According to Qi et al. (2021), geological structures such as faults, folds, share zones, fractures, and other geological structures all have a significant impact on the occurrence of landslides. A higher density of lineament increases the likelihood of a landslide, according to Mallick et al. (2021). When a fault line has a greater density, it affects the integrity of rock materials. A lower concentration of fault lines, on the other hand, lessens the likelihood of a land collapse. As a result, the area closer to the lineament will be more susceptible to landslides, and vice versa. Gaidzik and Ramrez-Herrera (2021) found that soil types have a crucial impact on influencing landslide vulnerability. The structure, texture, and other characteristics of the soil substantially control soil erosion. Because of their weak cohesion, granular soil particles are more prone to migration. Finely ground soil particles, on the other hand, are far less prone to erosion since their adhesion is good. According to Chen and Zhang (2021), a high proportion of landslide events occur on untreated cut-slopes caused by road construction. Cut-slopes typically decline at acute angles in the road's face. Slope failures are common on the cut-slopes throughout the wet season. A slope failure occurred on the cut-slopes of the road, progressing to the greater levels of the principal slope section. Verma et al. (2022) calculate the density of drainage in a river basin by dividing the length of all streams by the total basin extent. Simultaneously, an increase in drainage density reduces penetration and accelerates surface water flow (Kumar et al., 2021). The formula is expressed as follows:

$$DrainDen = \frac{\sum_{1}^{n} L}{A} \tag{2.3}$$

where "L" denotes the length of a given stream and "A" denotes the research area,

Rainfall Prediction Using ANN

In our study, an ANN model was used to predict the rainfall (Kumar et al. 2021). The goal of this analysis is to anticipate rainfall one stage ahead of time. The rainfall procedure was considered to become an ANN procedure for the creation of the suggested rainfall prediction model (Gholami & Sahour 2022). That implies that the value of rainfall at a particular point on a spatial and temporal scale is a consequence of a limited collection of past realizations (Ghamariadyan & Imteaz, 2021). This hypothesis allows for the formulation of a model pattern. (t) is a matrix of rainfall values at N different locations at period (t)g is a non-linear method that will be estimated by an ANN; e(t) is a mapping inaccuracy (to be reduced); and k is the (unidentified) number of earlier rainfall realizations that contribute to rainfall at the very next period; Commonly, k causes network lag; only when k = 1, does the next period's rainfall correlate with the current one.

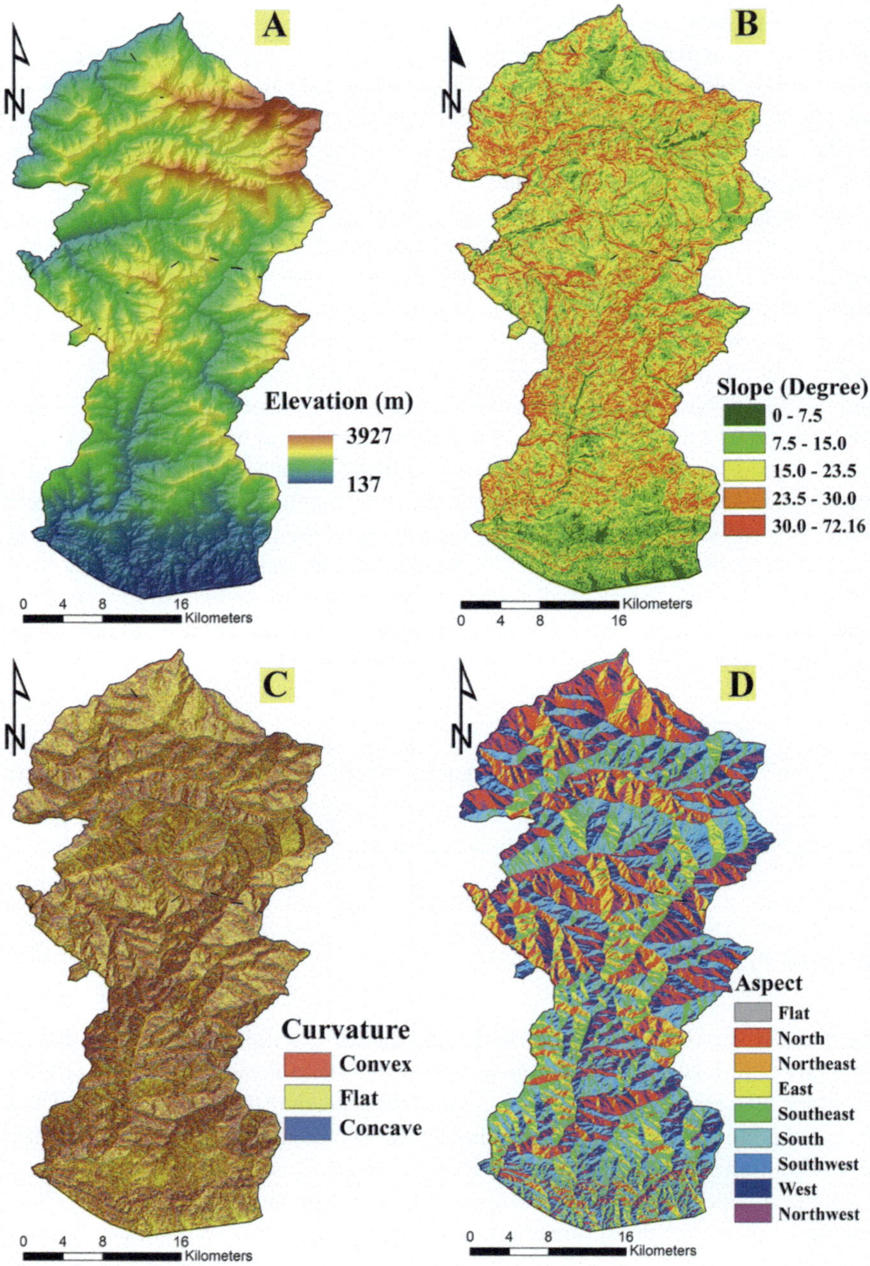

Fig. 2.6 Representing the landslide conditioning factors: A. elevation, B. slope, C. curvature, D. aspect, E. drainage density, F. stream transportation index, G. topographical wetness index, H. topographical positioning index, I. geology, J. distance from lineament, K. soil type, and L. road density

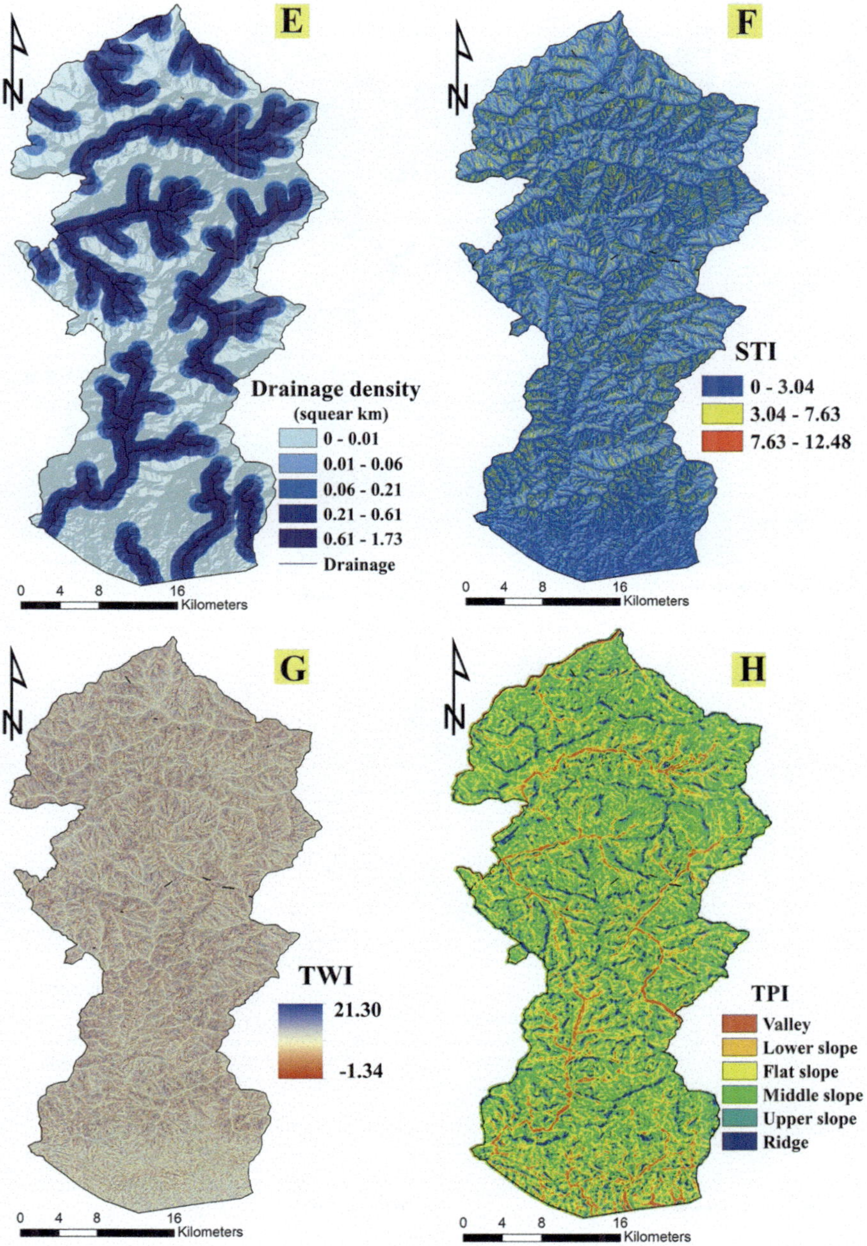

Fig. 2.6 (continued)

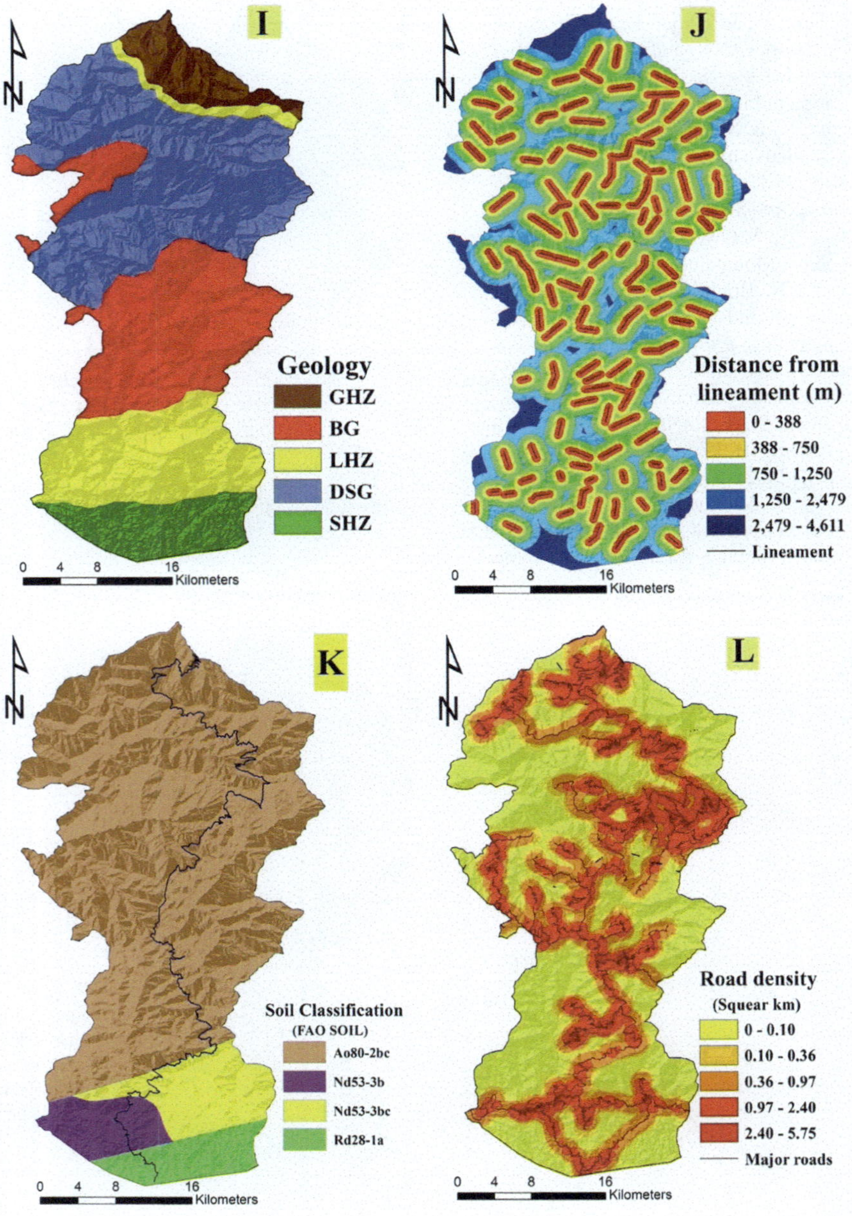

Fig. 2.6 (continued)

The following critical factors must be considered while developing an ANN for forecasting rainfall (Dada et al. 2021):

- To select the best ANN to represent the Markovian function.
- To compute the amount of historical rainfall readings needed to incorporate it as input, calculate the latency for the model of ANN.
- Determine the ANN's ideal complexity for the task, i.e., the number of concealed layers and vertices in each concealed layer.

LULC Change Prediction Using Markov Chain Model

Markov chain (MC) assessment is a stochastic model that considers the prior condition to forecast the future modification of elements through a period and has been frequently used for ecological modeling (Das et al., 2021). It has also been utilized in LULC change detection modeling due to its tremendous capacity to measure the speeds and conditions of transformation across and within classes (Dejene, 2021). It necessitates the use of two sets of LULC data with distinct timescales, where the possibilities of changeover among phases may be calculated. In the modeling of land use and land cover transformation, the MC model often employs the equation.

$$A_{(t+1)} = B_{ij} * A_t; B_{ij} = \begin{pmatrix} B_{11}B_{12} & \dots & B_n \\ B_{21}B_{22} & \dots & B_{2n} \\ \dots\dots & \dots & \dots \\ B_{n1}B_{n2} & \dots & B_{nn} \end{pmatrix}$$

$$= \left(0 =< B_{ij} < 1; \sum_{j=1}^{n} B_{ij} = 1; i = 1, 2, 3, \dots, n \right) \quad (2.4)$$

where $A_{(t+1)}$ and A_t represent LULC at t + 1 and t times, respectively, and B_{ij} represents the matrix of changeover probability.

To initially anticipate the map of LULC, the MC supplied two matrices of transition by cross-tabulating two maps of LULC and a set of compatibility maps (Sankarrao et al. 2021).

Methods for Susceptibility Modelling

Artificial Neural Network (ANN)

Parameters may be used to develop an ANN, which is a mathematical concept having simulation capabilities and structure recognition akin to the brain of a human

(Mouloodi et al., 2021). It uses a nonlinear algorithm to discover the intricate interaction among variables and experimental samples in a network pattern. The widely known ANN consists of three layers: an input (with numerous neurons), a concealed (intrinsic) layer, and an output layer, in which the concealed layer of output neurons multiply every input with a particular weight mechanism, and weight discrepancies are estimated consistently among the parameters and commensurate findings (Pulvermüller et al., 2021). The layer of input (feed-forward of inputs) with 12 nodes of landslide conditioning parameters was supplied into ANN, and the weight values of every component in the concealed layer were estimated by the spine training method to recursively reduce and modify the inaccuracy between the simulation and training sets of data. The concealed layer and the input's output (out) may be estimated as follows:

$$Out = f\left(\sum_{j=1}^{n} \omega_j x_i + \theta_j\right) \tag{2.5}$$

where's x_i the network inflow (causal parameters) as from inputs, θ_j denotes a skew or a value of threshold. ω_j specifies the scaling factor, and $f(x)$ is the transfer method, Usually depicted by the $\left(1 + e^{-2x}\right)^{-1}$ and ω_j is computed by minimising an error function of the match among the output and the feature of target.

Random Forest (RF)

Breiman provides a random forest model. It is a powerful ensemble-learning approach. The random forest method is used for classification, regression, clustering, and relationship identification. Numerous sectors have widely adopted this method, demonstrating its increased effectiveness. Because of the selectivity and significant volatility, a choice based on a single tree yields a very poor categorization (Wang et al., 2022). However, because the RF model employs ensemble trees, it is capable of resolving these issues. The RF model uses several arbitrary binary trees to create a forest (Gupta & Sihag, 2022). The CART (classification and regression trees) technique creates each tree based on a random subset using randomly selected and chosen attributes at each node. When making predictions, the random forest model took into account the majority of the approximate (unweighted) majority class while addressing the problem of clusters.

Deep Learning Neural Network (DLNN)

DLNNs are ANN methods with numerous (deep) concealed layers (depending on the characteristics) and a spine training strategy that uses a feed-forward connection (Kaur & Gandhi, 2020). The method can reveal important nonlinear and complicated phenomena like landslides due to the use of several concealed layers (Mohan et al.,

2021). In this study, we adjusted the concealed layer to three to achieve greater feature extraction, drawing on past research. Every neuron in the concealed layers of the DLNN implements the sigmoid operation, conducting the spine and weighted linear framework. The sigmoid method f(x) is defined as (1 + ex). ReLU is being employed to prevent scattering slope since it is trained using a horizontal stripe approach with backpropagation.

Validation Techniques

Without comprehensive validation using sound methods, machine learning and deep learning models are useless (Saha et al.,2021a; Arabameri et al., 2021). In the present study, for testing the applied models, receiver operating characteristic (ROC), accuracy, F1-measure, Jaccard, MCC, MAE, and ROC were used (Table 2.3). The ROC curve is a widely used technique for model validation. Based on the area under the curve (AUC) of the ROC performances of the models, they are classified as poor to excellent. True positive (TP), false positive (FP), true negative (TN), and false negative (FN), which indicate the correct or incorrect classification of landslide location in the used models, are used for calculating the value of AUC, accuracy, F1_ measures, Jaccard, and MCC. On the other hand, the statistical methods MAE and RMSE measure the dissimilarity between the actual value and the predicted value of a landslide occurrence.

Results

Multi-collinearity Analysis

The VIF value is the maximum in cases of road density (2.224) and rainfall, as shown in the table (Table 2.4) of multicollinearity (2.003). Because their VIF values are high, road density and rainfall have the greatest impact on landslide susceptibility, followed by LULC (1.966), elevation (1.711), curvature (1.654), soil classes (1.639), and aspect (1.628). On the other hand, the topographical position index (1.076), lineament density (1.117), and drainage density (1.155), on the other hand, have much less impact on the risk of landslides since their values are closer to 1. On the other hand, road density (0.452) has the greatest impact on land slide susceptibility, followed by rainfall (0.489) and LULC (0.509), which are all dependent on tolerance levels. On the other hand, drainage density and topographical position index have the least impact on landslide susceptibility because their tolerance values of 0.938 and 0.930, respectively, are the highest among all components.

Table 2.3 Formulas with suitable references of the validation method applied in the present study

Methods	Formula	References
Accuracy	$Accuracy = \frac{TP+TN}{N}$ where true positives = (TP) and true negatives = (TN) represent the number of pixels that are correctly classified	Azmoon et al. (2021)
F1-measures	$F_1 - measure = 2 \times \left[\left[\frac{\left(\frac{TP}{TP+FP}\right)\times\left(\frac{TP}{TP+FN}\right)}{\left(\frac{TP}{TP+FP}\right)+\left(\frac{TP}{TP+FN}\right)}\right]\right]$ Where, true positive (TP), false positive (FP), true negative (TN), and false negative (FN)	Kumar et al. (2021)
Jaccard	$Jaccard(A, B) = \frac{\lvert A\cap B\rvert}{\lvert A\cup B\rvert}$	Zhang et al. (2021)
MCC (Matthews correlation co-efficient)	$MCC = \frac{(TP\times TN)-(FP\times FN)}{\sqrt{[(TP+FP)(TP+FN)(TN+FN)]-(TN\times FN)}}$	Wang et al. (2021)
RMSE	$RMSE = \sqrt{\frac{1}{n}\sum_{i=1}^{n}(Lpredicted - Lactual)^2}$ where, $L_{predicted}$ and L_{actual} = Predicted and actual values of landslides and, n = Sample size for training or testing subsets	Alqadhi et al. (2021)
MAE	$MAE = \frac{1}{n}\sum_{i=1}^{n}\lvert Lpredicted - Lactual\rvert$ where, $L_{predicted}$ and L_{actual} = Predicted and actual values of landslides and, n = Sample size for training or testing subsets	Alqadhi et al. (2021)

Table 2.4 Result of Multi-collinearity test of used LCFs

Factors	Tolerance	VIF
Elevation	0.584	1.711
Slope	0.830	1.196
Curvature	0.603	1.654
Aspect	0.614	1.628
Drainage density	0.938	1.155
Sediment transport index (STI)	0.743	1.345
Topographical wetness index (TWI)	0.845	1.198
Topographical position index (TPI)	0.930	1.076
Geology	0.856	1.158
Lineament density	0.795	1.117
Soil classes	0.610	1.639
Road density	0.452	2.224
Rainfall	0.489	2.003
LULC	0.509	1.966

Future Rainfall Pattern

With the ANN model, we have predicted the future rainfall map of our study region. Here we have forecasted the rainfall maps of 2050 and 2080 based on the rainfall maps of 1990 and 2020 (Fig. 2.7). Here, we can see that, in the case of the map of 1990, there is a gradual increase in the amount of rainfall from north to south. On average, about 45% of the area in the southern half of the study region has experienced a very high amount of rainfall (>3500). Above this very high zone of rainfall, the amount of rainfall has been decreasing in the northern part. On average, about 20.11% of the study region occupies a relatively small amount of the rainfall zone. This sequence of rainfall zones is almost the same for the next three years as well (2020, 2050, and 2080). In comparison to the 1990 rainfall map, the area of very low rainfall (1500) has decreased in the 2020 rainfall map. The 2020 map reveals a slight increase and decrease in the very high and very low rainfall category zones compared to the 1990 map, covering approximately 49.21 and 18.76% of our study region, respectively. The predicted rainfall map for 2050 identifies the same gradual changes in the rainfall category zones. Compared to the previous map (2020), there has been a slight decrease and increase in the very high and very low amounts of rainfall category zones, covering approximately 33.08 and 30.27% of our study region, respectively. Finally, the predicted rainfall map of 2080 shows a slight increase and decrease in the very high and very low amounts of rainfall category zones compared to the previous map of 2050, covering approximately 36.25 and 26.46% of our study region, respectively (Table 2.5).

Future LULC Pattern

With the Marcove Chain model here, we have predicted the future LULC map of our study region. Here we have forecasted the LULC maps of 2050 and 2080 by using the LULC maps of 1990 and 2020. Here, we can see that over the past few years, the settlement area of agricultural land has increased significantly. On the other hand, arable land, sparse forest, shrubs, dense forest, and rocky land have decreased significantly. In the years 1990, 2020, 2050, and 2080, the percentage of settlement area was 4.18, 5.56, 7.61, and 8.62%, respectively. This zone occupies the central to northern part of the map. In the case of agricultural land, 15.04, 18.54, 22.07, and 23.78% of the area come under the years of 1990, 2020, 2050, and 2080 respectively. The majority of this zone is located in the south. In the years of 1990, 2020, 2050, and 2080, the percentage of arable land was 5.76, 6.67, 5.69, and 4.93%, respectively. In the case of sparse forest, 15.78, 14.26, 13.06, and 12.36% of the area fall under the years 1990, 2020, 2050, and 2080, respectively. In For dense forest, the areas falling under these categories in 1990, 2020, 2050, and 2080 are 46.55, 42.14, 39.16, and 37.89%, respectively. The case of shrubs, the areas is 10.57, 10.38, 10.09, and 9.72% for the years 1990, 2020, 2050, and 2080 respectively (Fig. 2.8). On the other hand,

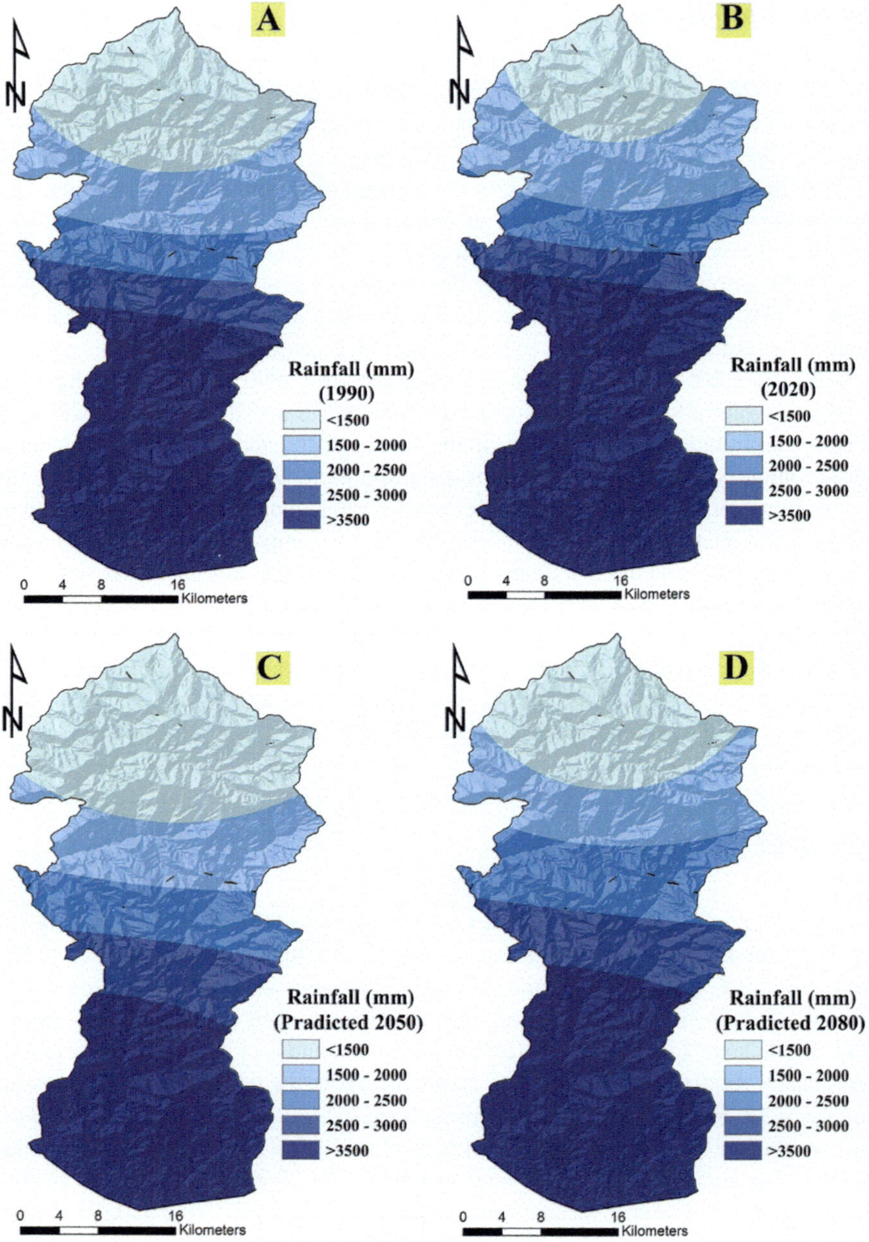

Fig. 2.7 Present and projected rainfall maps of **a** 1990, **b** 2020, **c** 2050 and **d** 2080

Table 2.5 Rainfall trends scenarios from 1990 to 2020 & 2050 predicted; and 2080 predicted using ANN model

Rainfall (mm)	1990		2020		2050		2080	
	%	Area	%	Area	%	Area	%	Area
Very low (<1500)	20.11	215.00	18.76	200.57	30.27	323.63	26.46	282.89
Low (1500–2000)	10.46	111.83	13.24	141.55	14.28	152.67	14.14	151.18
Moderate (2000–2500)	13.03	139.31	9.13	97.61	8.85	94.62	9.98	106.70
High (2500–3000)	15.28	163.36	9.76	104.35	13.52	144.55	13.17	140.80
Very high (<3000)	41.12	439.63	49.21	526.12	33.08	353.67	36.25	387.56

for the category of rocky land, the areas have been found to be 2.66, 2.45, 2.31, and 2.25% for the years 1990, 2020, 2050, and 2080 respectively (Table 2.6).

Past and Present Landslide Susceptibility Assessment

For the ANN model, In Fig. 2.8 and Table 2.7, we can see that the very low landslide-prone area has increased to some extent from the past to the present (from 1990 to 2020). In the 1990 map, the very susceptible zone covered 30.24% of the study region, a figure that has now increased to 31.52% in 2020, primarily on the north-eastern side. On the other hand, extremely vulnerable zones have grown from 1990 to the present (1990–2020). This zone covered 7.22% of the total area in 1990 and increased to 8.67% in 2020. Over time, patches of very high and highly landslide-prone zones have expanded, now haphazardly spreading throughout the region. The RF model exhibits an increasing and decreasing trend for very high and very low susceptible zones, respectively. We also see a decrease in very low-landslide-susceptible areas over time, specifically from 1990 to 2020. From 1990 to 2020, there has been a continuous increase in the very high susceptible zone. The area that falls under the very high susceptible zone has increased from 73.46503 km^2 in 1990 to 83.62 km^2 in 2020. On the other hand, conversely, the highly susceptible zone occupied a total area of 133.4561 km^2 in 1990 and 146.40 km^2 in 2020. In the 1990 susceptibility map, the very low susceptible zone covered a total area of 351.4986 km^2 in the study region, whereas in 2020, it expanded to 317.386 km^2, primarily on the north-eastern side. In the DLNN model, very low susceptibility zones occupied 31.36 and 28.34% of the time in 1990 and 2020, respectively. Over the period 1990–2020, we have found a decreasing trend in the case of very susceptible zones. On the other hand, in the case of the very high and high landslide susceptibility zones, there is an increasing trend over time. The area under the high and very high susceptibility zones was 11.36 and 8.16%, respectively, in 1990, and 12.95 and 8.46% in 2020.

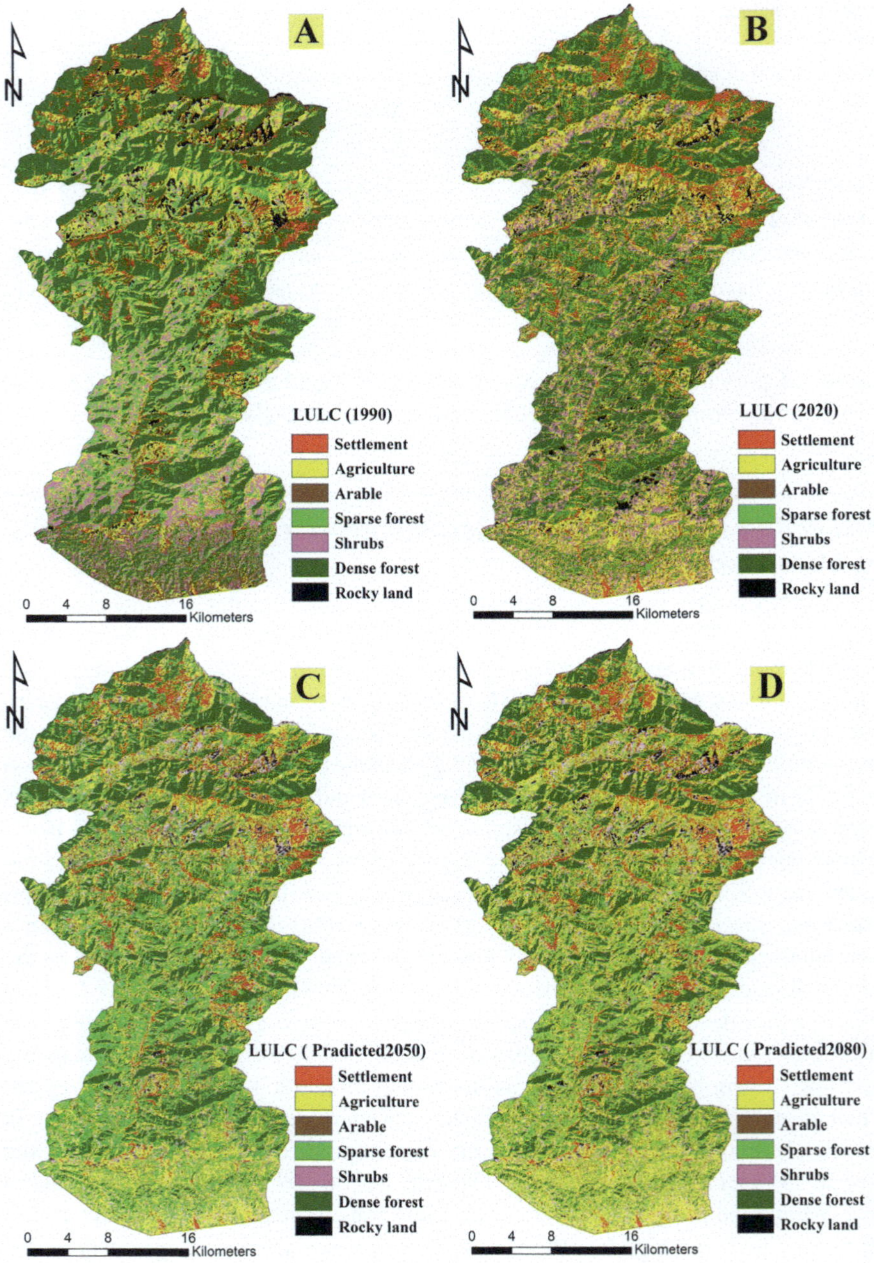

Fig. 2.8 Present and projected LULC maps of **a** 1990, **b** 2020, **c** 2050 and **d** 2080

Table 2.6 Land use and Land-cover change scenarios from 1990 to 2080 predicted using MC model

LULC	1990		2020		2050		2080	
	%	Area	%	Area	%	Area	%	Area
Settlement	4.18	44.69	5.56	59.44	7.61	81.36	8.62	92.16
Agriculture	15.04	160.80	18.54	198.22	22.07	235.96	23.78	254.24
Arable	5.76	61.58	6.67	71.31	5.69	60.83	4.93	52.71
Sparse forest	15.78	168.71	14.26	152.46	13.06	139.63	12.36	132.14
Shrubs	10.57	113.00	10.38	110.98	10.09	107.88	9.72	103.92
Dense forest	46.55	497.68	42.14	450.53	39.16	418.67	37.89	405.09
Rocky land	2.66	28.44	2.45	26.19	2.31	24.70	2.25	24.05

Table 2.7 Areal distribution of ANN, RF and DLNN approach based landslide susceptibility maps for year 1990, 2020, predicted 2050 & predicted 2080 (LSMs)

Susceptibility class	1990		2020		2050		2080	
	km^2	%	km^2	%	km^2	%	km^2	%
ANN								
Very low	323.37	30.24	337.06	31.52	286.79	26.82	293.89	27.48
Low	334.39	31.27	270.01	25.25	304.93	28.52	296.79	27.75
Moderate	212.70	19.89	252.69	23.63	264.34	24.72	236.43	22.11
High	122.76	11.48	116.77	10.92	114.6	10.72	130.51	12.20
Very high	77 21	7.22	92.71	8.67	97.54	9.12	106.27	9.94
RF								
Very low	351.50	32.87	317.39	29.68	283.04	26.47	291.04	27.22
Low	292.90	27.39	289.69	27.09	342.71	32.05	316.71	29.62
Moderate	217.94	20.38	237.83	22.24	196.66	18.39	206.6	19.32
High	133.46	12.48	146.40	13.69	147.09	13.75	159.09	14.88
Very high	73.47	6.87	83.62	7.82	93.76	8.77	97.54	9.12
DLNN								
Very low	335.35	31.36	303.06	28.34	283.04	26.47	275.06	25.72
Low	287.34	26.87	292.58	27.36	348.71	32.61	258.99	24.22
Moderate	237.93	22.25	244.56	22.87	196.65	18.39	265.89	24.86
High	121.48	11.36	138.48	12.95	147.09	13.75	165.24	15.45
Very high	87.26	8.16	90.47	8.46	93.79	8.77	102.49	9.58

Future Landslide Susceptibility Assessment

For the ANN model, in Fig. 2.9 and Table 2.7, we can see that the very low landslide susceptible areas have decreased from present to future in the predicted map of landslide susceptibility (2050 and 2080). In the 2050 map, the very low susceptible zone covered 26.81% of the study region, whereas in the 2080 map, it increased to 27.48%, primarily concentrated on the north-eastern side. Conversely, the future has seen a continuous increase in the very high susceptible zone compared to the present. This zone covers a total area of 8.67% in 2020, 9.12% in 2050, and 9.93% in 2080. Over time, the patches of very high and highly landslide-prone zones have expanded, spreading haphazardly throughout the region. The RF model shows the same increasing and decreasing trends for very high and very low susceptible zones, respectively (Fig. 2.10). We also observe a decrease in very low landslide-susceptible areas from the present to the future, specifically from 2020 to 2080. Similar to the ANN model, there has been a continuous increase in the very high and highly susceptible zones from the present to the future (2020–2080). The total area that has fallen under the very high susceptible zone is 83.62 km^2 in 2020, 93.76 km^2 in 2050, and 97.54 km^2 in 2080. Conversely, the high susceptible zone occupies a total area of 146.3954 km^2 in 2020, 147.09 km^2 in 2050, and 97.54 km^2 in 2080. In the case of the susceptibility map of 2020, the total area covered by the very low susceptible zone was 317.386 km^2, which has been found to be 291.04 km^2 in the case of 2080, which is mainly concentrated on the north-eastern side of the region. In the DLNN model, a very low susceptibility zone occupies 28.34% in 2020, compared to 26.46% and 25.72.5% in 2050 and 2080, respectively (Fig. 2.11). We found a decreasing trend in the case of very susceptible zones over the period (2020–2080). On the other hand, in the case of the very high and high landslide susceptibility zones, there is an increasing trend over time. In terms of high and very high susceptibility zones, they are 12.95 and 8.46% in 2020, 13.75 and 8.77% in 2050, and 15.45 and 9.58% in 2050.

Assessment of Importance of Landslide Conditioning Factors

From Table 2.8, it can be seen that slope (29.245) and elevation (25.33) have the highest contribution among all the conditioning factors of a landslide, as their mean decrease Gini value is the highest among all the parameters, followed by rainfall (2020) (14.827), LULC (2020) (13.879), rainfall (1990) (13.285), LULC (1990) (12.547), and curvature (10.684). All these factors have a huge influence on the landslide events in our study region. On the contrary, STI (2.541), lineament density (2.877), soil classes (3.447), and TPI (3.911) have very little influence on the landslide event of the study area, as their mean decrease Gini value is the lowest among all the parameters. Besides that, geology (9.694), road density (7.878), drainage density (7.244), aspect (5.247), and TWI (3.911) have a moderate influence on the land susceptibility.

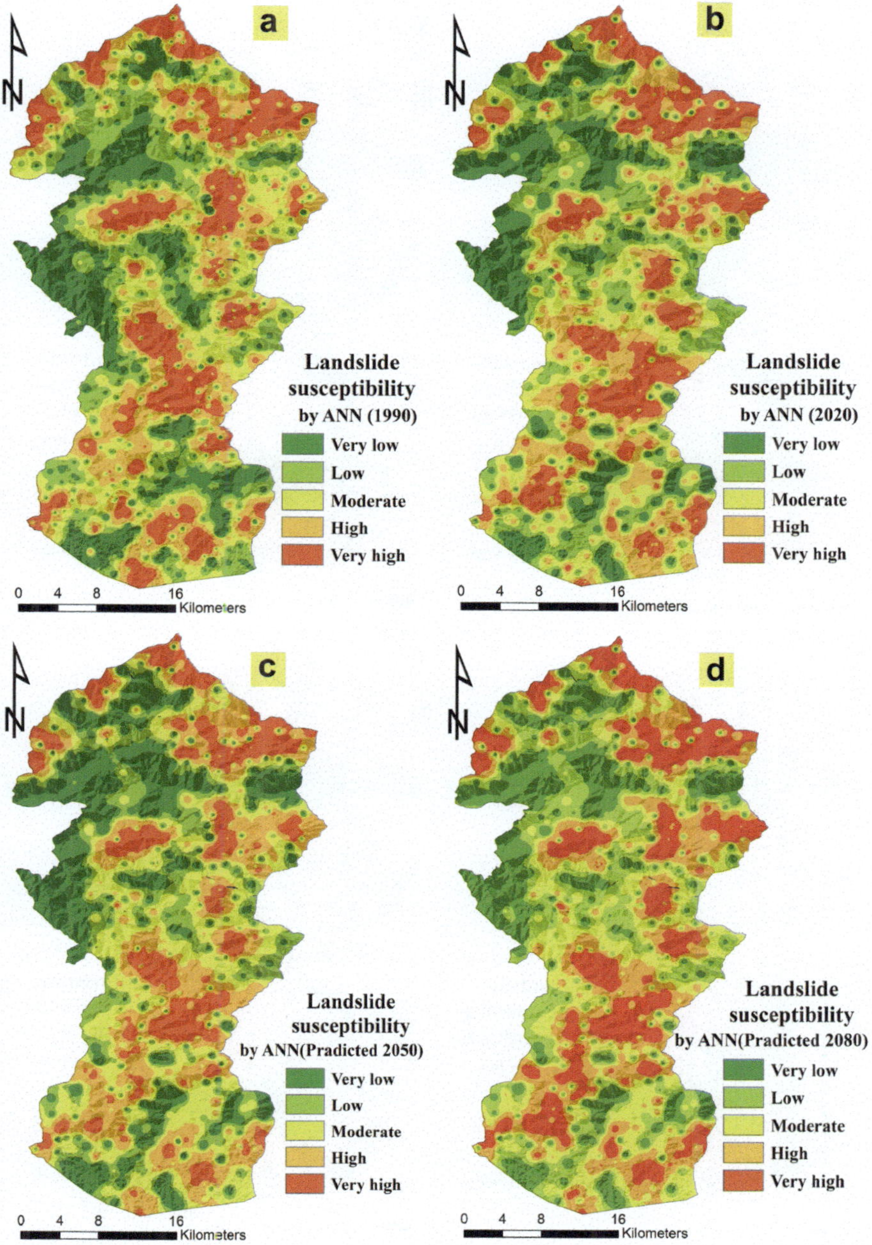

Fig. 2.9 Landslide susceptibility and future projection using ANN: **a** 1990, **b** 2020, **c** 2050 and **d** 2080

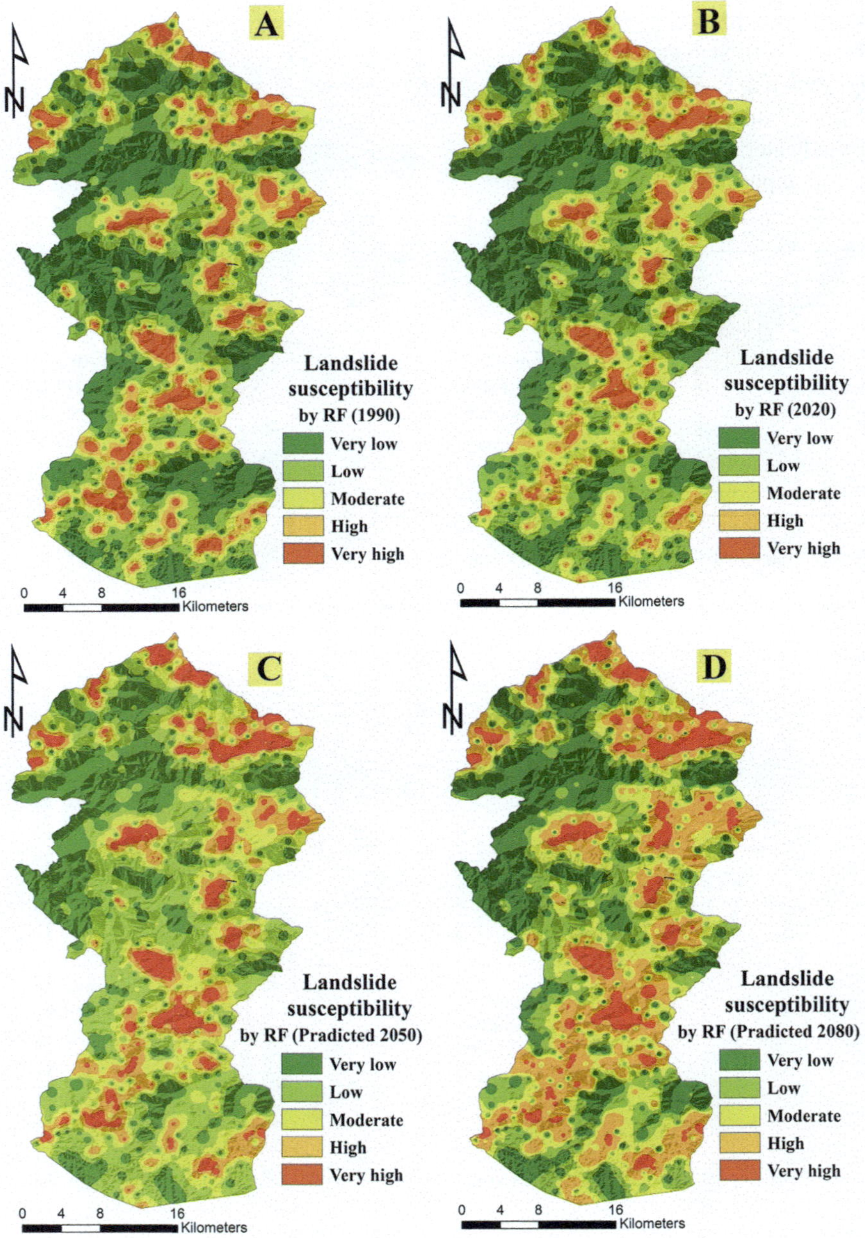

Fig. 2.10 Landslide susceptibility and future projection using RF: **a** 1990, **b** 2020, **c** 2050 and **d** 2080

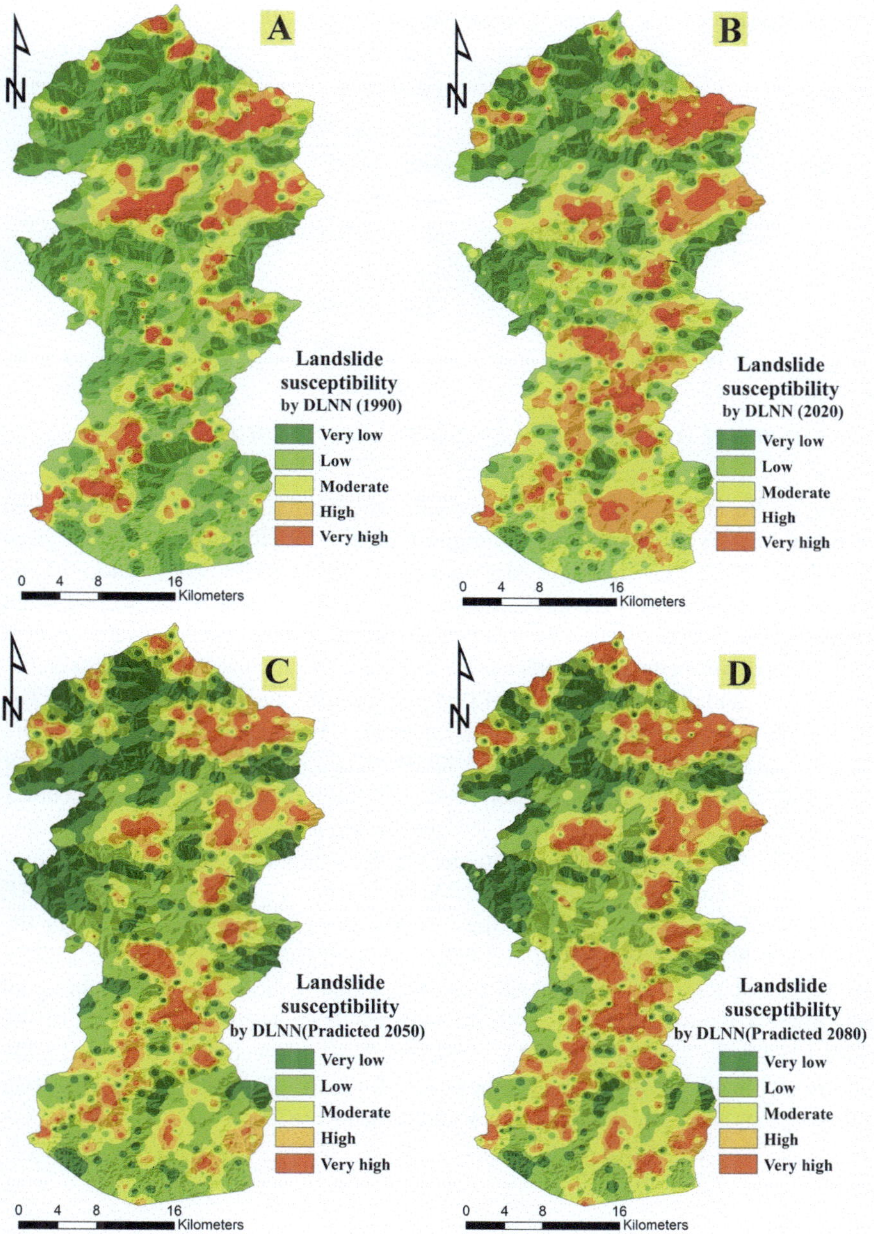

Fig. 2.11 Landslide susceptibility and future projection using DLNN: **a** 1990, **b** 2020, **c** 2050 and **d** 2080

Table 2.8 Landslide factor contribution analysis by random forest (RF)

Influencing factor	Mean decrease Gini	Influencing factor	Mean decrease Gini
Elevation	25.332	Geology	9.694
Slope	29.245	Lineament density	2.877
Curvature	10.684	Soil classes	3.447
Aspect	5.247	Road density	7.878
Drainage density	7.244	Rainfall (1990)	13.285
STI	2.541	LULC (1990)	12.547
TWI	5.142	Rainfall (2020)	14.827
TPI	3.911	LULC (2020)	13.879

Assessment of Accuracy of Landslide Susceptibility Maps

Among all the models of landslide susceptibility, the DLNN model is best suited for the study. As the AUC value is highest for the DLNN model, For the years 1990, 2020, 2050, and 2080, the AUC value was 88.29, 89.96, 91.72, and 90.57%, respectively (Table 2.9). We can conclude that random forest is the second fitted model for landslide susceptibility, as it yielded the second highest AUC value against the RF model, following the DLNN model. The AUC values have been found at 86.03, 86.63, 89.18, and 88.27% for the years 1990, 2020, 2050, and 2080, respectively. The DLNN model yielded the highest accuracy measures. For the years 1990, 2020, 2050, and 2080, the accuracy values are 88.29, 89.96, 91.72, and 90.57%, respectively, for the testing data set (Fig. 2.12). Conversely, the training data set shows accuracy values of 87.93, 88.37, 91.09, and 89.19% for the years 1990, 2020, 2050, and 2080, respectively. Conversely, the DLNN model yields the highest value for the measures of specificity, sensitivity, jacard, and F1. Measuring the error of the models by MAE and RMSE, it has been found that the lowest error value has been found for the model of DLNN, such as 0.191, 0.112, 0.123, and 0.121 values for 1990, 2020, 2050, and 2080, respectively, for the MAE measure, which is the lowest value among all three models. Conversely, the RMSE measure yields values of 0.378, 0.245, 0.195, and 0.188 for the years 1990, 2020, 2050, and 2080, respectively. We have finally determined that the DLNN model is the best fit for landslide susceptibility modeling in this region, based on all the accuracy and error measures.

Discussion

The intensity of landslides has increased in the last decade, not only in Bhutan but also all over the world's mountainous regions, due to climate and land use changes. In the present study, past as well as future landslide susceptibilities were assessed keeping in mind the changing patterns of rainfall and land uses, and data from an

Table 2.9 Results of statistical measures used for validating the models

Model	Dataset	Sensitivity	Specificity	AUC (%)	F1 measure	Jaccard	MCC	MAE	RMSE
ANN (1990)	Training	0.791	0.807	80.93	0.725	0.811	0.798	0.338	0.259
	Testing	0.806	0.796	82.56	0.812	0.826	0.812	0.288	0.267
ANN (2020)	Training	0.803	0.811	84.37	0.814	0.838	0.821	0.042	0.137
	Testing	0.811	0.824	86.41	0.819	0.844	0.838	0.055	0.196
ANN (2050)	Training	0.862	0.798	83.87	0.872	0.887	0.862	0.239	0.206
	Testing	0.866	0.715	85.24	0.878	0.835	0.868	0.164	0.348
ANN (2080)	Training	0.865	0.798	83.34	0.862	0.881	0.865	0.222	0.324
	Testing	0.874	0.684	85.82	0.875	0.829	0.845	0.184	0.267
RF (1990)	Training	0.856	0.791	85.31	0.792	0.855	0.814	0.215	0.248
	Testing	0.864	0.832	86.03	0.854	0.822	0.839	0.207	0.199
RF (2020)	Training	0.856	0.848	85.86	0.871	0.896	0.841	0.264	0.198
	Testing	0.895	0.877	86.63	0.864	0.856	0.884	0.166	0.178
RF (2050)	Training	0.894	0.881	88.19	0.938	0.861	0.887	0.129	0.231
	Testing	0.885	0.881	89.18	0.956	0.879	0.884	0.122	0.245
RF (2080)	Training	0.866	0.785	88.02	0.938	0.889	0.869	0.178	0.314
	Testing	0.889	0.784	88.27	0.895	0.875	0.874	0.183	0.328
DLNN (1990)	Training	0.889	0.894	87.93	0.91	0.899	0.847	0.175	0.288
	Testing	0.876	0.912	88.29	0.922	0.875	0.884	0.191	0.378
DLNN (2020)	Training	0.94	0.896	88.37	0.935	0.911	0.937	0.049	0.211
	Testing	0.882	0.901	89.96	0.944	0.939	0.884	0.112	0.245
DLNN (2050)	Training	0.926	0.885	91.06	0.952	0.901	0.919	0.122	0.193
	Testing	0.913	0.984	91.72	0.948	0.895	0.911	0.123	0.195
DLNN (2080)	Training	0.889	0.954	89.19	0.945	0.909	0.914	0.125	0.198
	Testing	0.876	0.932	90.57	0.956	0.925	0.898	0.121	0.188

electric resistivity meter was also used to validate the results. First, the First, the tolerance and VIF values for LCFs were determined to improve the forecasting accuracy of the produced landslide susceptibility models (Table 2.4). These have recently become recurrent phenomena in our study area, as is clearly visible from Table 2.1, which shows the frequency of young landslides in the years 2020 and 2021 and their locations. In their research on a landslide occurrence, Chowdhuri et al. (2021) identified rainfall and LULC as important factors. As a result, rainfall and LULC have been recognized as predicting predictor Bhutan's LSMs (Saha et al., 2021a, 2021b, 2022; Gupta et al., 2022). In the ongoing work, a model of ANN was utilized to estimate rainfall events for the years 2050 and 2080 in order to generate the susceptibility map of landslides in Bhutan. On the other hand, the Markov Chain Model was employed to estimate the LULC map for the years 2050 and 2080 in Bhutan. The rainfall pattern predicted from the ANN model shows that the very low

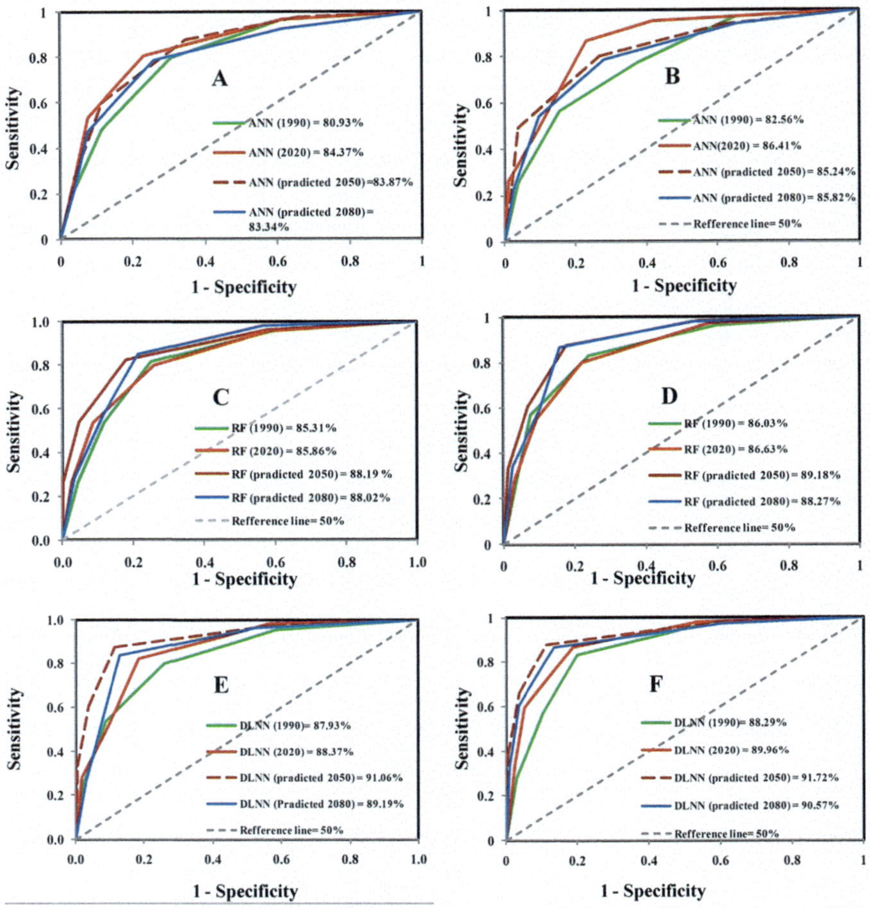

Fig. 2.12 ROC curves used for validating the landslide susceptibility models: A. ROC curves of ANN model using training data, B. ROC curves of ANN model using testing data, A. ROC curves of RF model using training data, B. ROC curves of RF model using testing data, A. ROC curves of DLNN model using training data & B. ROC curves of DLNN model using testing data

to low rainfall area will increase by 2080, whereas the very high to high rainfall area will decrease during the same period (Table 2.5). The LULC future pattern shows that the percentage of area for settlement will increase, and thus the share of agricultural area will increase too, but on the cost of forest and arable land, which show a declining trend (Table 2.6).

According to several scholars (Zhao et al., 2022), DLNN is the most competent deep learning technique specifically for modeling landslide vulnerability. A widely acknowledged machine learning model, ANN and RF, is also commonly used in the mapping of landslide susceptibility (Meghanadh et al., 2022). The DLNN method used here improves accuracy and reduces human errors by employing multiple layers (Saha et al., 2022). The ANN model can split and restructure multiple data sets within a large data range and thus be applied to non-linear and complex relationships (Saha et al., 2022). The random forest model carries out both regression and classification tasks, which produce accurate predictions (Breiman, 2001). The AUC-ROC analysis of the results of several models revealed that the DLNN model has the greatest accomplishment rate (the AUC values for 1990, 2020, 2050, and 2080 are 88.29, 89.96, 91.72, and 90.57%, respectively). The FR model came in second (88.36% AUC for training and 89.92% AUC for testing points), followed by 86.03%, 86.63, 89.18, and 88.27% for the years 1990, 2020, 2050, and 2080, respectively. The DLNN model has the best accuracy value that has been found (for the years 1990, 2020, 2050, and 2080, the accuracy values are 88.29, 89.96, 91.72, and 90.57%, respectively). In addition, the RF model has demonstrated high accuracy. These results show that using the right ensemble approach may improve the prediction outcome to a certain extent. DLNN, on the other hand, has independently outperformed the models in the current evaluation. As shown in Table 2.9, the other assessment techniques utilized revealed identical findings in the case of ROC-AUC for each model utilized. Other research, such as Saha et al. (2021a, 2021b) and Pal et al. (2022), has confirmed the exceptional performance of DLNN in this investigation. The results of all of the models' analyses revealed that, with the exception of the extreme western part, the high and very high landslide susceptibility zones have spread haphazardly throughout the region, with rainfall ranging from 1500 to 3500 cm, elevations ranging from 2032 to 3927m, and slopes ranging from 15 to 72.16°. Numerous studies on landslide susceptibility have also revealed that the surface's characteristics are substantially sensitive to landslide susceptibility. very high class shows an increasing rate of landslide susceptibility in different periods, and the overall change from 1990 to 2080 is 2.72, 2.25, and 1.42% according to ANN, RF, and DLNN, respectively, whereas the very low class shows a decreasing trend from 1990 to 2080, which is −2.76, −5.65, and −5.64% according to ANN, RF, and DLNN, respectively (Table 2.10). ANN and DLNN show an increasing level of landslide susceptibility in 2020–2050 and 2050–2080, which is 0.45% and 0.31 to 0.82 and 0.81%, respectively (Table 2.10). The different factors selected for this study include elevation, slope, curvature, aspect, drainage density, STI, TWI, TPI, geology, distance from lineament, soil, road density, LULC, and rainfall. Out of which, it has been found that road density, rainfall, and LULC are the most dominant factors in these study areas, with the highest VIF values of 2.22, 2.00, and 1.97, respectively. The same factors were found to be dominant in other literature (Sonker et al., 2022; Pasang et al., 2020). The least impacting factors include drainage density and topographical position index, with VIF values of 0.94 and 0.93, respectively. Landslide probability is importantly determined by soil resistivity. To compare the soil resistivity between very high landslide susceptibility and very low landslide susceptibility, an electric resistivity meter has been used. The vertical soil resistivity Fig. 2.13a–e represent the areas of low soil resistivity and thus high landslide susceptibility. which are Narphu, Narphu to Nagzor roads, Khaling, Monger Trashingang, and Dangling Tsho, respectively, whereas Fig. 2.13f–h represent the areas of high soil resistivity and thus low landslide susceptibility, which are Tsangpo, Yonphula Lhakhang, and Baekha Woesel Choeling Yogi Monastery, respectively. So, in the study area, low soil resistivity is also an important cause of landslide occurrence. The present study evaluated landslide susceptibility by employing a total of 14 relative parameters. A few more relative parameters could have been taken into account for better analysis of landslide susceptibility on the basis of studies of other literature. This study can help researchers and scholars in the future with their studies and analyses. This landslide susceptibility analysis, if combined with landslide vulnerability, would help to assess the risk analysis in the future. This study could help in taking preventive measures in the area. This analysis of landslide susceptibility will be very useful to engineers and planners in selectively locating their skills and resources to minimize the evil effects of landslides in Bhutan.

Conclusion

Landslides have been a common occurrence in eastern Bhutan. A valuable landslide susceptibility map may play a significant role in the management of land-use development, reservoirs, reservoir deposition, and other lengthy time periods. The most precise and optimal way of creating LSM has yet to be discovered. As a result, a number of measures have been taken to improve the findings. In order to detect the link between the landslide and associated influencing elements and develop landslide susceptibility mapping for the current and distant future, a model of an artificial neural network, a random forest, and a deep learning neural network were used. The main influencing elements for the landslide occurrence in this location, according to this study, are LULC and rainfall. As a result of these two elements, we have projected future landslide susceptibility maps using the Markov Chain and ANN Model models. The results show that the DLNN model is performing better. The completed maps reveal that approximately 19.52% (1990), 21.41% (2020), 22.52% (2050), and 25.03% (2080) of our study territory, namely the eastern and south-eastern portions of the region, have slipped into a highly to extremely vulnerable land slide area. Researchers have also found that the level of soil resistivity is significantly lower in areas with high and high landslide susceptibility compared to areas with low landslide proneness. However, compared to other models used in this research, DLNN has produced the most advanced results and is best suited for creating the LSMs map. Several researchers have appropriately validated these models in numerous landslide-related publications. Furthermore, because the Eastern Bhutan region and its adjacent environment are valuable assets of the nation, the present study might be incredibly useful for planners and developers in designing any short-to-long-term initiative in this zone. As a result, for a long-term view in Eastern Bhutan, meticulous preparation is essential.

Table 2.10 Showing the areal change of susceptibility classes from 1990–2020, 2020–2050, 2050–2080, 1990–2080 and 2020–2080

Susceptibility classes	Areal change 1990–2020		Areal change 2020–2050		Areal change 2050–2080		Areal change 1990–2080		Areal change 2020–2080	
	km^2	%	km^2	%	km^2	%	km^2	%	km^2	%
ANN										
Very low	13.69	1.28	−50.27	−4.7	7.1	0.66	−29.48	−2.76	−43.17	−4.04
Low	−64.38	−6.02	34.92	3.27	−8.14	−0.77	−37.6	−3.52	26.78	2.5
Moderate	39.99	3.74	11.65	1.09	−27.91	−2.61	23.73	2.22	−16.26	−1.52
High	−5.99	−0.56	−2.17	−0.2	15.91	1.48	7.75	0.72	13.74	1.28
Very high	15.5	1.45	4.83	0.45	8.73	0.82	29.06	2.72	13.56	1.27
RF										
Very low	−34.11	−3.19	−34.35	−3.21	8	0.75	−60.46	−5.65	−26.35	−2.46
Low	−3.21	−0.3	53.02	4.96	−26	−2.43	23.81	2.23	27.02	2.53
Moderate	19.89	1.86	−41.17	−3.85	9.94	0.93	−11.34	−1.06	−31.23	−2.92
High	12.94	1.21	0.69	0.06	12	1.13	25.63	2.4	12.69	1.19
Very high	10.15	0.95	10.14	0.95	3.78	0.35	24.07	2.25	13.92	1.3
DLNN										
Very low	−32.29	−3.02	−20.02	−1.87	−7.98	−0.75	−60.29	−5.64	−28	−2.62
Low	5.24	0.49	56.13	5.25	−89.72	−8.39	−28.35	−2.65	−33.59	−3.14
Moderate	6.63	0.62	−47.91	−4.48	69.24	6.47	27.96	2.61	21.33	1.99
High	17	1.59	8.61	0.8	18.15	1.7	43.76	4.09	26.76	2.5
Very high	3.21	0.3	3.32	0.31	8.7	0.81	15.23	1.42	12.02	1.12

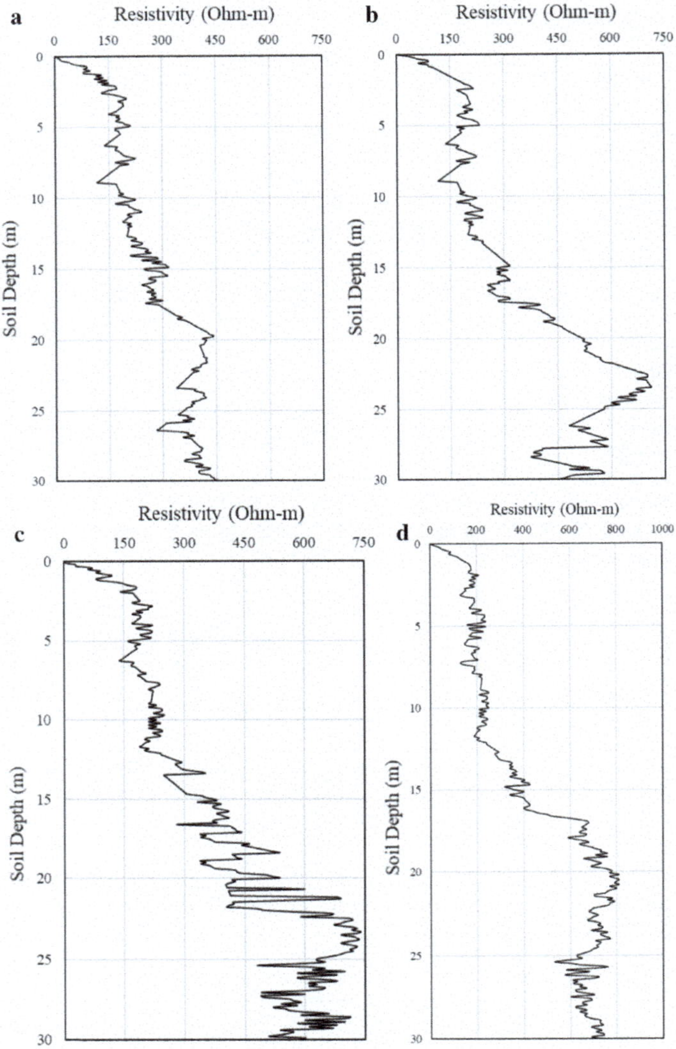

Fig. 2.13 Representation of vertical soil resistivity status of few selected places in very high and very low landslide susceptibility zones measured by electric resistivity meter: **a** Narphu (27°01′05"N, 91°30′25"E)-**VHSZ**, **b** Narphu to Nagzor roads (26°57′42"N, 91°32′24"E)-**VHSZ**, **c** Khaling (27°13′10"N, 91°35′50"E)-**VHSZ**, **d** Rongthng Monger Trashingang (27°16′48"N, 91°16′48"E)-**VHSZ**, **e** Dangling Tsho (27°14′46"N, 91°37′20"E)-**VHSZ**, **f** Tsangpo (27°07′01"N, 91°37′26"E)-**VLSZ**, **g** Yonphula Lhakhang (27°15′54"N, 91°30′45"E)-**VLSZ**, **h** Baekha Woesel Choeling Yogi Monastery (27°09′53"N, 91°38′16"E)-**VLSZ**

Fig. 2.13 (continued)

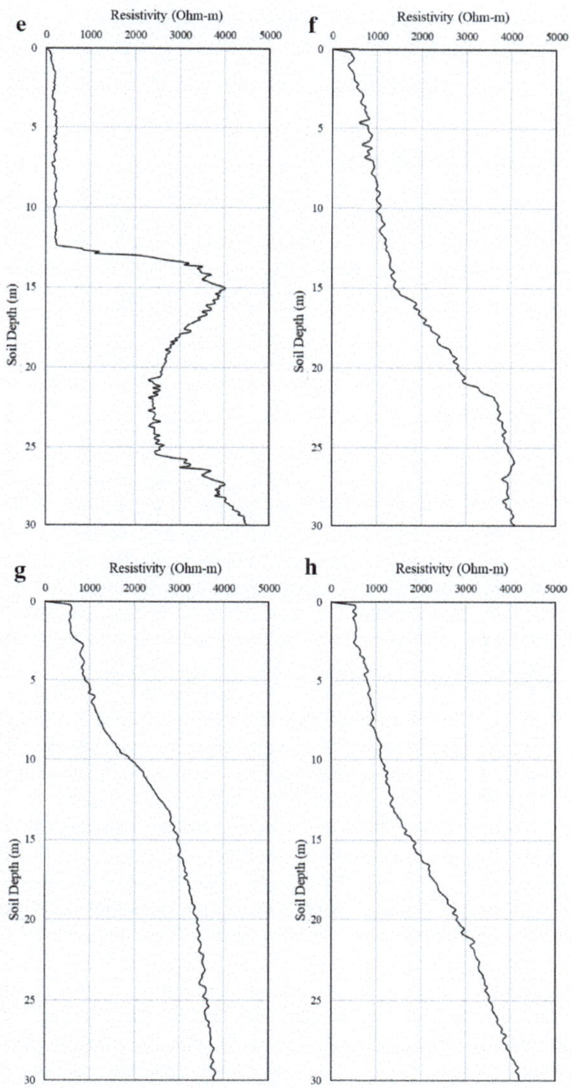

Acknowledgements We would like to extend hearty acknowledgement to the local dwellers for their cooperation during field survey.

Funding No funding was received for this work.

Availability of Data and Materials The datasets used during the current study are available from the corresponding author on reasonable request.

Ethical approval Not applicable.

Conflict of Interest: The authors declare that they have no competing interests.

Consent to Participate All participants were provided their consent for participating into the interview.

Consent to Publish Not applicable

Author Credit statement SS and AS decided the Methodological framework; SS, AS, RS and PG did the Formal analysis; SS and AS completed the Presentation; SS, AS, RS, BK and PG wrote the original draft; SS, AS, RS, BK, SA and PG Reviewed the manuscript; SS, AS and RS handle the Software; SA collected the Data; SA Investigated the field.

References

Abuzied, S. M., & Pradhan, B. (2021). Hydro-geomorphic assessment of erosion intensity and sediment yield initiated debris-flow hazards at Wadi Dahab Watershed, Egypt. *Georisk: Assessment and Management of Risk for Engineered Systems and Geohazards, 15*(3), 221–246.

Achour, Y., Saidani, Z., Touati, R., Pham, Q. B., Pal, S. C., Mustafa, F., & Balik Sanli, F. (2021). Assessing landslide susceptibility using a machine learning-based approach to achieving land degradation neutrality. *Environmental Earth Sciences, 80*(17), 1–20.

Ahmadi, H., & Pekkan, E. (2021). Fault-based geological lineaments extraction using remote sensing and GIS—A review. *Geosciences, 11*(5), 183.

Ali, S. A., Parvin, F., Vojteková, J., Costache, R., Linh, N. T. T., Pham, Q. B., & Ghorbani, M. A. (2021). GIS-based landslide susceptibility modeling: A comparison between fuzzy multi-criteria and machine learning algorithms. *Geoscience Frontiers, 12*(2), 857–876.

Alqadhi, S., Mallick, J., Talukdar, S., Bindajam, A. A., Saha, T. K., Ahmed, M., & Khan, R. A. (2021). Combining Logistic Regression-based hybrid optimized machine learning algorithms with sensitivity analysis to achieve robust landslide susceptibility mapping. *Geocarto International* (just-accepted), 1–25.

Arabameri, A., Santosh, M., Saha, S., Ghorbanzadeh, O., Roy, J., Tiefenbacher, J.P., Moayedi, H. & Costache, R., (2021). Spatial prediction of shallow landslide: Application of novel rotational forest-based reduced error pruning tree. *Geomatics, Natural Hazards and Risk*, *12*(1), pp.1343-1370.

Azmoon, B., Biniyaz, A., Liu, Z., & Sun, Y. (2021). Image-data-driven slope stability analysis for preventing landslides using deep learning. *IEEE Access, 9*, 150623–150636.

Breiman, L. (2001). Random forests. *Machine learning*, *45*(1), 5-32.

Chen, W., & Zhang, S. (2021). GIS-based comparative study of Bayes network, Hoeffding tree and logistic model tree for landslide susceptibility modeling. *Catena, 203*, Article 105344.

Chen, H., Qin, S., Xue, L., Xu, C. (2021a). Why the Xintan landslide was not triggered by the heaviest historical rainfall: Mechanism and review. *Engineering Geology, 294*, 106379.

Chen, Z., Huang, M., Zhu, D., & Altan, O. (2021b). Integrating remote sensing and a Markov-FLUS model to simulate future land use changes in Hokkaido, Japan. *Remote Sensing, 13*(13), 2621.

Chiarelli, D. D., D'Odorico, P., Davis, K. F., Rosso, R., & Rulli, M. C. (2021). Large scale land acquisition as a potential driver of slope instability. *Land Degradation & Development, 32*(4), 1773–1785.

Chowdhuri, I., Pal, S. C., Chakrabortty, R., Malik, S., Das, B., Roy, P., & Sen, K. (2021). Spatial prediction of landslide susceptibility using projected storm rainfall and land use in Himalayan region. *Bulletin of Engineering Geology and the Environment, 80*(7), 5237–5258.

Dada, E. G., Yakubu, H. J., & Oyewola, D. O. (2021). Artificial neural network models for rainfall prediction. *European Journal of Electrical Engineering and Computer Science, 5*(2), 30–35.

Das, S., Shit, P. K., & Patel, P. P. (2021). Ecosystem services value assessment and forecasting using integrated machine learning algorithm and CA-Markov model: An empirical investigation of an Asian megacity. *Geocarto International*, 1–23.

Davies, T., & Moretti, D. (2022). Geomorphic precursors of large landslides: Seismic preconditioning and slope-top benches. In *Landslide Hazards, Risks, and Disasters* (pp. 641–666). Elsevier.

Dejene, M. (2021). *Urban land use land cover change detection: A case study of Adama City, Ethiopia*. Doctoral dissertation, ASTU.

Depicker, A., Jacobs, L., Mboga, N., Smets, B., Van Rompaey, A., Lennert, M., … & Govers, G. (2021). Historical dynamics of landslide risk from population and forest-cover changes in the Kivu Rift. *Nature Sustainability*, *4*(11), 965–974.

Dikshit, A., Sarkar, R., Pradhan, B., Jena, R., Drukpa, D., & Alamri, A. M. (2020). Temporal probability assessment and its use in landslide susceptibility mapping for eastern Bhutan. *Water, 12*(1), 267.

Dorji, T., Ona, B. J. L., & Raghavan, S. V. (2021). *Statistical analyses on the seasonal rainfall trend and annual rainfall variability in Bhutan*. SOLA.

Evelpidou, N., Tzouxanioti, M., Spyrou, E., Petropoulos, A., Karkani, A., Saitis, G., & Margaritis, M. (2022). GIS-based assessment of fire effects on flash flood hazard: The case of the summer 2021 forest fires in Greece. *GeoHazards*, *4*(1), 1-22.

Erasu Tufa, D., & Lika Megento, T. (2022). Conversion of farmland to non-agricultural land uses in peri-urban areas of Addis Ababa Metropolitan city, Central Ethiopia. *GeoJournal*, *87*(6), 5101-5115.

Farooq, S., & Akram, M. S. (2021). Comparison of data-driven landslide susceptibility assessment using weight of evidence, information value, frequency ratio and certainty factor methods. *Acta Geodynamica et Geromaterialia, 18*(3), 301–318.

Fuller, I. C., & Rutherfurd, I. D. (2021). Geomorphic responses to anthropogenic land-cover change in Australia and New Zealand.

Gaidzik, K., & Ramírez-Herrera, M. T. (2021). The importance of input data on landslide susceptibility mapping. *Scientific Reports, 11*(1), 1–14.

Ghamariadyan, M., & Imteaz, M. A. (2021). A wavelet artificial neural network method for medium term rainfall prediction in Queensland (Australia) and the comparisons with conventional methods. *International Journal of Climatology, 41*, E1396–E1416.

Gholami, V., & Sahour, H. (2022). Simulation of rainfall-runoff process using an artificial neural network (ANN) and field plots data. *Theoretical and Applied Climatology, 147*(1), 87–98.

Gupta, N., Pal, S. K., & Das, J. (2022). GIS-based evolution and comparisons of landslide susceptibility mapping of the East Sikkim Himalaya. *Annals of GIS*, 1–26.

Gupta, S., & Sihag, P. (2022). Prediction of the compressive strength of concrete using various predictive modeling techniques. *Neural Computing and Applications*, *34*(8), 6535-6545.

Jakob, M. (2022). Landslides in a changing climate. In *Landslide hazards, risks, and disasters* (pp. 505–579). Elsevier.

Jakob, M., & Owen, T. (2021). Projected effects of climate change on shallow landslides, North Shore Mountains, Vancouver, Canada. *Geomorphology, 393*, Article 107921.

Jones, J. N., Boulton, S. J., Bennett, G. L., Stokes, M., & Whitworth, M. R. (2021). Temporal variations in landslide distributions following extreme events: Implications for landslide susceptibility modeling. *Journal of Geophysical Research: Earth Surface, 126*(7), e2021JF006067.

Kaur, T., & Gandhi, T. K. (2020). Deep convolutional neural networks with transfer learning for automated brain image classification. *Machine vision and applications*, *31*(3), 20.

Khaidem, S., & Singh, K. (2021, November). Landslide susceptibility mapping along Manipur-Assam NH-37. In *IOP Conference Series: Earth and Environmental Science* (Vol. 889, No. 1, p. 012002). IOP Publishing.

Khelifa, R., Mahdjoub, H., & Samways, M. J. (2022). Combined climatic and anthropogenic stress threaten resilience of important wetland sites in an arid region. *Science of the Total Environment*, *806*, 150806.

Koley, B., Nath, A., Saraswati, S., Chatterjee, U., Bandyopadhyay, K., Bhatta, B., & Ray, B. C. (2022). Assessment of spatial distribution of rain-induced and earthquake-triggered landslides using geospatial techniques along North Sikkim Road Corridor in Sikkim Himalayas, India. *GeoJournal*, 1–39.

Kumar, R., Singh, M. P., Roy, B., & Shahid, A. H. (2021). A comparative assessment of metaheuristic optimized extreme learning machine and deep neural network in multi-step-ahead long-term rainfall prediction for All-Indian regions. *Water Resources Management, 35*(6), 1927–1960.

Leal Filho, W., & Manolas, E. (2022). Addressing climate change in the Mediterranean Basin: Some thoughts for the future. In *Climate change in the Mediterranean and Middle Eastern Region* (pp. 521–527). Springer.

Lin, C. Y., Wu, J. P., Wu, S. W., Lin, J. Y., & Chen, T. Y. (2022). Watershed landslide evolution combined with topographic factors to delineate the spatio-temporal distribution of sediment sources in central Taiwan. *CATENA, 211*, Article 105978.

Majeed, M., Tariq, A., Anwar, M. M., Khan, A. M., Arshad, F., Mumtaz, F., … & Shaukat, S. (2021). Monitoring of land use–land cover change and potential causal factors of climate change in Jhelum district, Punjab, Pakistan, through GIS and multi-temporal satellite data. *Land, 10*(10), 1026.

Mallick, J., Alqadhi, S., Talukdar, S., AlSubih, M., Ahmed, M., Khan, R. A., … & Abutayeh, S. M. (2021). Risk assessment of resources exposed to rainfall induced landslide with the development of GIS and RS based ensemble metaheuristic machine learning algorithms. *Sustainability, 13*(2), 457.

Marzen, M., Porten, M., & Ries, J. B. (2022). Quantification of dust emissions during tillage operations in steep slope vineyards in the Moselle Area. *Agriculture, 12*(1), 100.

Makwinja, R., Kaunda, E., Mengistou, S., & Alamirew, T. (2021). Impact of land use/land cover dynamics on ecosystem service value—A case from Lake Malombe, Southern Malawi. *Environmental Monitoring and Assessment, 193*(8), 492.

McColl, S. T. (2022). Landslide causes and triggers. In *Landslide hazards, risks, and disasters* (pp. 13–41). Elsevier.

Meghanadh, D., Maurya, V. K., Tiwari, A., & Dwivedi, R. (2022). A multi-criteria landslide susceptibility mapping using deep multi-layer perceptron network: A case study of Srinagar-Rudraprayag region (India). *Advances in Space Research, 69*(4), 1883–1893.

Mkuzi, H. T. (2020). *Assessment of land tenure, land use and land cover changes in Taita Hills forest fragments: A case study of Ngerenyi Forest Fragments in Taita Taveta County, Kenya.* Doctoral dissertation, Pwani University.

Mohan, A., Singh, A. K., Kumar, B., & Dwivedi, R. (2021). Review on remote sensing methods for landslide detection using machine and deep learning. *Transactions on Emerging Telecommunications Technologies, 32*(7), Article e3998.

Mondini, A. C., Guzzetti, F., Chang, K. T., Monserrat, O., Martha, T. R., & Manconi, A. (2021). Landslide failures detection and mapping using Synthetic Aperture Radar: Past, present and future. *Earth-Science Reviews*, 103574.

Mouloodi, S., Rahmanpanah, H., Gohari, S., Burvill, C., Tse, K. M., & Davies, H. M. (2021). What can artificial intelligence and machine learning tell us? A review of applications to equine biomechanical research. *Journal of the Mechanical Behavior of Biomedical Materials, 123*, Article 104728.

Nanda, A. M., Ahmed, P., & Kanth, T. A. (2021). Landslide susceptibility assessment of national highway 1D from Sonamarg to Kargil, Jammu and Kashmir, India using frequency ratio method. *GeoJournal, 86*(6), 2945–2956.

Nath, R. R., Pareek, N., & Sharma, M. L. (2022). Implications and inclusion of size-dependent scenario earthquakes on landslide hazard zonation: A case study of the Indian Himalayas. *CATENA, 212*, Article 106027.

Ogila, W. A. M. (2021). Analysis and assessment of slope instability along international mountainous road in North Africa. *Natural Hazards, 106*(3), 2479–2517.

Pasang, S., & Kubíček, P. (2020). Landslide susceptibility mapping using statistical methods along the Asian Highway, Bhutan. *Geosciences, 10*(11), 430.
Pal, S. C., Chakrabortty, R., Saha, A., Bozchaloei, S. K., Pham, Q. B., Linh, N. T. T., … & Ahmadi, K. (2022). Evaluation of debris flow and landslide hazards using ensemble framework of Bayesian- and tree-based models. *Bulletin of Engineering Geology and the Environment, 81*(1), 1–25.
Peiris, H. M. P., & Gunarathne, N. (2022). The changing landscape of the plantation sector in the central highlands of Sri Lanka. In *Mountain landscapes in transition* (pp. 539–554). Springer.
Pulvermüller, F., Tomasello, R., Henningsen-Schomers, M. R., & Wennekers, T. (2021). Biological constraints on neural network models of cognitive function. *Nature Reviews Neuroscience, 22*(8), 488-502.
Qi, T., Meng, X., Qing, F., Zhao, Y., Shi, W., Chen, G., … & Dijkstra, T. (2021). Distribution and characteristics of large landslides in a fault zone: A case study of the NE Qinghai-Tibet Plateau. *Geomorphology, 379*, 107592.
Saha, A., & Saha, S. (2020). Comparing the efficiency of weight of evidence, support vector machine and their ensemble approaches in landslide susceptibility modelling: A study on Kurseong region of Darjeeling Himalaya, India. *Remote Sensing Applications: Society and Environment, 19*, Article 100323.
Saha, S., Arabameri, A., Saha, A., Blaschke, T., Ngo, P. T. T., Nhu, V. H., & Band, S. S. (2021a). Prediction of landslide susceptibility in Rudraprayag, India using novel ensemble of conditional probability and boosted regression tree-based on cross-validation method. *Science of the Total Environment, 764*, 142928.
Saha, S., Sarkar, R., Roy, J., Hembram, T. K., Acharya, S., Thapa, G., & Drukpa, D. (2021). Measuring landslide vulnerability status of Chukha, Bhutan using deep learning algorithms. *Scientific Reports, 11*(1), 1–23.
Saha, S., Sarkar, R., Roy, J., Bayen, B., Bhardwaj, D., & Wangchuk, T. (2022). Application of RBF and MLP Neural Networks integrating with rotation forest in modeling landslide susceptibility of Sampheling, Bhutan. In *Impact of climate change, land use and land cover, and socio-economic dynamics on landslides* (pp. 221–245). Springer.
Saleh, A., Yuzir, A., Sabtu, N., Abujayyab, S. K., Mudashiru, R. B., & Pham, Q. B. (2022). Flash flood susceptibility mapping in urban area using genetic algorithm and ensemble method. *Geocarto International (just-accepted)*, 1–27.
Sankarrao, L., Ghose, D. K., & Rathinsamy, M. (2021). Predicting land-use change: Intercomparison of different hybrid machine learning models. *Environmental Modelling & Software, 145*, Article 105207.
Shu, H., Guo, Z., Qi, S., Song, D., Pourghasemi, H. R., & Ma, J. (2021). Integrating landslide typology with weighted frequency ratio model for landslide susceptibility mapping: A case study from Lanzhou City of Northwestern China. *Remote Sensing, 13*(18), 3623.
Sonker, I., & Tripathi, J. N. (2022). Remote sensing and GIS-based landslide susceptibility mapping using frequency ratio method in Sikkim Himalaya. *Quaternary Science Advances, 8*, 100067.
Tanyu, B. F., Abbaspour, A., Alimohammadlou, Y., & Tecuci, G. (2021). Landslide susceptibility analyses using Random Forest, C4. 5, and C5. 0 with balanced and unbalanced datasets. *Catena, 203*, 105355.
Verma, N., Patel, R. K., & Choudhari, P. (2022). Watershed prioritization for soil conservation in a drought prone watershed of Eastern India: Tel River Basin, Odisha. *Geology, Ecology, and Landscapes*, 1–14.
Wang, D., Yuan, Z., Cai, Y., Jing, D., Liu, F., Tang, Y., … & Fu, X. (2021). Characterisation of soil erosion and overland flow on vegetation-growing slopes in fragile ecological regions: A review. *Journal of Environmental Management, 285*, 112165.
Wang, M., Feng, D., Li, D., & Wang, J. (2022). Reservoir parameter prediction based on the neural random forest model. *Frontiers in Earth Science, 10*, 888933.
Zhang, M., Li, J., & Li, N. (2021). Fine-scale characteristics of summer precipitation over Cang Mountain. *Journal of Applied Meteorology and Climatology, 60*(9), 1285–1300.

Zhao, P., Masoumi, Z., Kalantari, M., Aflaki, M., & Mansourian, A. (2022). A GIS-based landslide susceptibility mapping and variable importance analysis using artificial intelligent training-based methods. *Remote Sensing, 14*(1), 211.

Chapter 3
Hydrologic Sensitivity of Himalayan Slopes: Pre- and Post-rainfall LEM (Morgenstern–Price) Evaluation in Uttarakhand

Gaurav Bharti, Vaishnavi Bansal, and B. Bharath

Abstract This study evaluates rainfall-induced changes in slope stability across representative corridors of Uttarakhand—Rishikesh, Chamoli, Kathgodam, Karnaprayag, and Bageshwar—using SLOPE/W with the Morgenstern–Price limit-equilibrium formulation. Fourteen field-mapped cross-sections were modeled in 2-D with site-specific geometry (height, dip direction/angle) and Mohr–Coulomb strength parameters (unit weight, cohesion, friction angle). Two hydrologic scenarios were analysed for each section: a pre-rainfall ("dry") case with low pore-water pressures and a post-rainfall ("wet") case represented by a raised phreatic surface or equivalent pore-pressure increase. Factors of Safety (FoS) computed by a dense entry–exit slip-surface search show strong spatial contrasts and varied hydrologic sensitivity. Most Rishikesh sections are persistently unstable (FoS < 1.0) except one stable site, Chamoli exhibits transitional behavior with one slope (S5) tipping from stable (FoS $\approx$ 1.000) to unstable (FoS $\approx$ 0.999) under saturation, Kathgodam remains robustly stable in both states, and Karnaprayag as well as Bageshwar are consistently below unity regardless of rainfall. Although FoS shifts between dry and wet conditions are generally modest, small changes near the threshold are decisive for classification and risk.

Keywords Slope stability · Geo-studio · Factor of safety · Limit-equilibrium · Rainfall and SLOPE/W

G. Bharti · V. Bansal (✉) · B. Bharath
Department of Civil Engineering, JSS Academy of Technical Education, Noida, India
e-mail: vaishnavibansal@jssaten.ac.in

G. Bharti
e-mail: gauravbharti000@gmail.com

B. Bharath
e-mail: bharathb@jssaten.ac.in

R. Sarkar et al. (eds.), *Climate Change, Environmental Hazards and Community-Based Resilience*, Disaster Risk Reduction, https://doi.org/10.1007/978-981-95-9485-6_3

Introduction

The Himalayan arc hosts some of the world's most active, complex, and hazard-prone mountain belts, where rapid tectonic uplift, steep relief, heterogeneous lithologies, and intense monsoonal forcing combine to drive pervasive slope instability (Bharti et al., 2025a; Shukla et al., 2024). Within the Lesser Himalaya of Uttarakhand and particularly across the Kumaun sector, this hazard intersects dense networks of settlements, highways, hydropower corridors, and vital lifelines (Bansal & Shukla, 2022; Rawat et al., 2024; Shukla et al., 2023a). The result is a chronic pattern of landslides—ranging from shallow translational failures and debris flows to deep-seated rotational slides and rock mass instabilities—that degrade infrastructure serviceability, threaten lives and livelihoods, and impose recurrent repair and maintenance costs on agencies (Shukla et al., 2023b, 2025a). The region's geologic architecture folded, faulted, and thrusted stratigraphic sequences from Proterozoic to Cenozoic age creates abrupt spatial transitions in rock mass quality, weathering profiles, and discontinuity sets, while the orographic gradient of monsoonal rainfall frequently elevates pore pressures to destabilizing levels (Bharti et al., 2025b; Chamyal, 1991; Merh, 1977; Valdiya, 1980, 1983; Yin, 2006).Engineering practice in such settings needs robust, site-specific stability assessments that integrate geometric characterization, geotechnical property determination, and realistic hydrologic load cases (Gowda et al., 2025; Shukla et al., 2025b). Traditional limit equilibrium methods (LEMs) remain indispensable screening tools because they are transparent, computationally efficient, and able to explore multiple slip surfaces to estimate the factor of safety (FoS). Yet, their assumptions—e.g., predefined failure mechanisms, simplified interslice force representations—can limit realism for complex geometries or strongly heterogeneous rock/soil masses (Duncan, 1996; Matthews et al., 2014). Finite element methods (FEM) and shear strength reduction (SSR) analysis provide a complementary pathway: they solve for the stress–strain field and "discover" failure mechanisms emergently, while the SSR procedure progressively scales down strength parameters (e.g., (c), (\phi)) until global non-convergence reveals the collapse state (Dawson et al., 1999; Griffiths & Lane, 1999; Matsui & San, 1992; Zheng et al., 2005). The FEM/SSR framework is particularly attractive in the Himalaya, where rock slopes with blocky structure, anisotropic jointing, and variable overburden can produce noncircular, composite, or structurally controlled failure modes (Eberhardt, 2003; Hammah et al., 2007; Pain et al., 2014; Awasthi et al., 2019).

From a geomechanics standpoint, the mobilizing and resisting forces that govern slope stability are strongly modulated by (i) rock/soil strength (cohesion and friction angle), (ii) unit weight and suction (for soils), (iii) joint orientation, persistence, and infill (for rock slopes), (iv) groundwater regime, and (v) geometry (slope height, angle, curvature, toe conditions, and reinforcing elements if present). While Mohr–Coulomb (MC) strength parameters dominate routine design and analysis, their interpretation for intact rock or jointed rock masses requires care: polyaxial stress states, confinement effects, and scale dependences can deviate from simple MC idealizations (Bharti et al., 2022; Hackston & Rutter, 2016; Labuz & Zang,

2012). In the Kumaun Himalaya, field campaigns commonly reveal a mosaic of weathered phyllites, quartzites, slates, and limestones juxtaposed by thrusts and shear zones. Discontinuity sets and altered interfaces often localize shear, while perched groundwater during prolonged or intense rainfall rapidly reduces effective stress, sometimes aided by clayey gouge or infill that reduces friction and cohesion. These realities argue for (a) rigorous field and laboratory characterization; (b) hydrologic scenarios spanning dry, wet, and saturated (transient) conditions; and (c) numerical analyses that can accommodate material and geometric complexity (Kainthola et al., 2011, 2012;, 2015; Mahanta et al., 2016; Solanki et al., 2019; Pande, 1992; Shukla et al., 2025c, 2025d). Hydroclimatic variability adds another layer of uncertainty. Rainfall over Uttarakhand exhibits strong seasonality and interannual variability, with extremes capable of rapidly elevating pore pressures and triggering widespread failures along road cuttings, valley flanks, and engineered slopes (Kumar, 2017). Seismotectonic transients further complicate stability: while the Himalayan seismic gap debate continues, historical records demonstrate that moderate to strong earthquakes can trigger landslides across broad swaths of the orogen and degrade slope conditions for subsequent rainfall seasons (Bansal & Sarkar, 2024a, 2024b; Keefer, 2002). The adoption of RS–GIS inventories and susceptibility maps at corridor scale can guide these decisions, but their value increases when coupled to field-verified, section-scale mechanistic models as deployed here (Saraswat et al. 2020; Tripathi, 2018; Pain et al., 2014).

Literature Review

Foundational stratigraphic and structural syntheses of the Kumaun Lesser Himalaya delineate a stacked and imbricated architecture shaped by the Himalayan orogen's Cenozoic evolution (Valdiya, 1980, 1983; Chamyal, 1991; Merh, 1977; Prakash et al., 1978; Yin, 2006). These works identify thrust-bounded slices of metasediments (phyllites, quartzites, slates, dolomites/limestones) with local mylonitic shear zones and cataclasites. The combined effects of folding, faulting, and metamorphic overprint generate spatially abrupt changes in rock mass fabric and strength, while weathering patterns and joint persistence vary along strike and with elevation. For slope engineering, this geologic context implies: (i) anisotropic and scale-dependent rock mass behavior, (ii) structurally controlled failures (planar, wedge, toppling), and (iii) sensitivity to groundwater along fault gouges and schistose planes. Uttarakhand's rainfall climatology features strong monsoonal concentration (June–September) with significant spatial gradients; statistical studies document both heavy-tailed extremes and interannual variability (Kumar, 2017). For the Kumaun sector, therefore, realistic design should consider saturated or near-saturated states as credible load cases, and—where relevant—co-seismic reductions in apparent friction or joint into Limit equilibrium methods (Bishop, Janbu, Spencer, Morgenstern–Price variants) have long underpinned slope analysis by balancing forces and/or moments along assumed slip surfaces (Duncan, 1996). Their strengths are computational efficiency, transparency, and mature practice. However, when failure mechanisms are composite or noncircular, or when material and stress path complexity matter, LEMs may under- or overestimate stability depending on the interslice function assumed (Matthews et al., 2014). FEM's ability to model spatial variability, staged construction, reinforcement, and complex boundary conditions gives it distinct advantages for Himalayan rock slopes, particularly where joints and benches, toe unloading, and variable confinement prevail (Eberhardt, 2003; Pain et al., 2014; Awasthi et al., 2019). The SSR technique, introduced and refined in geotechnical FEM, scales Mohr–Coulomb parameters by a reduction factor (F) until the model fails to converge, identifying a collapse multiplier whose reciprocal is the FoS (Matsui & San, 1992; Dawson et al., 1999; Griffiths & Lane, 1999; Zheng et al., 2005. For reinforced rock slopes, SSR facilitates quantifying the interaction of bolts, anchors, and facing with the deforming rock mass (Awasthi et al., 2019).

Mohr–Coulomb remains the workhorse criterion for soils and many rock masses, but intact rock deviates under polyaxial stress states, confinement changes, and scale effects (Hackston & Rutter, 2016; Labuz & Zang, 2012). In practice, Hoek–Brown (HB) parameters derived from rock mass quality (GSI, disturbance) are often converted to equivalent MC parameters over the stress range expected in a specific slope, for input to LEM/FEM tools (Bansal & Sarkar, 2023; Kainthola et al., 2012; Pain et al., 2014). Case studies from the Indian Himalaya emphasize careful field logging, joint set measurements, and laboratory testing to reduce epistemic uncertainty in (c), (phi), and (gamma), along with groundwater characterization (Kainthola et al., 2011, 2015; Mahanta et al., 2016; Solanki et al., 2019). Finite element-based probabilistic slope assessments (e.g., Monte Carlo with random fields or random variables) show that spatial variability can dominate stability outcomes (Griffiths & Fenton, 2004). When full Monte Carlo is computationally expensive, point-estimate methods (PEM) provide efficient numerical quadrature over uncertain variables to approximate moments of response metrics (Christian & Baecher, 1999). Commercial FEM platforms have integrated reliability modules, enabling parametric scans (Valley & Duff, 2011; Rocscience Inc., 2004. Studies from Maharashtra's basaltic terrains and multiple Himalayan corridors show that geometry, joint orientation, and saturation are recurrent determinants of failure (Kainthola et al., 2015; Mahanta et al., 2016; Singh et al., 2013; Kainthola et al., 2012). In the Himalaya, FEM-based modeling has successfully captured structurally controlled failures and buttressing/drainage mitigation effects (Awasthi et al., 2019; Pain et al., 2014). Geological and geotechnical characterization specific to Kumaun landslides (e.g., Khotila, Dharchula region) highlight the importance of stratigraphic juxtaposition and groundwater along weak planes (Solanki et al., 2019). FEM accuracy hinges on sound discretization, appropriate elements, and robust nonlinear solvers. Foundational results on finite element discretization and parameter-dependent problems clarify stability and convergence properties (Arnold, 1981). Iterative algorithms for nonlinear FEM—e.g., modified Newton, line search, arc-length controls—are central to tracing pre- and post-peak response under SSR (Laouafa & Royis, 2004). In geotechnical contexts, mesh refinement in anticipated shear zones, stabilization of nearly incompressible formulations (where relevant), and careful control of step size in SSR are practical necessities (Griffiths & Lane, 1999; Zheng et al., 2005; Rocscience Inc., 2004).

Study Region

The study area spans multiple districts in Uttarakhand, India principally Rishikesh, Chamoli, Kathgodam, Karnaprayag, and Bageshwar as shown in Fig. 3.1. where fourteen representative slopes (S1–S14) were investigated to capture contrasting geomorphic and hydrological settings along transport corridors and valley flanks. For each site, the report documents geographic coordinates (latitude–longitude) alongside slope geometry (height, dip direction, dip angle), which together provide the spatial and kinematic context for stability analysis. These inputs are listed in the dataset description and the "Investigated slope and overall geometry" is seen in Table 3.1., which specify, that Rishikesh slopes include S1 (25 m height, dip direction ~105°, dip angle ~55°), S2 (21 m, ~310°, ~64°), S3 (29 m, ~54°, ~80°), and S4 (25 m, ~95°, ~76°) at the recorded coordinates, while Chamoli slopes include S5 (20 m, ~90°, ~64°), S6 (16 m, ~270°, ~67°), and S7 (16 m, ~350°, ~74°). Kathgodam features S8 (15 m, ~95°, ~60°), Karnaprayag includes S9 (20 m, ~60°, ~65°) and S10, and Bageshwar includes S11; the table provides the corresponding coordinates for each (e.g., S8 at ~30°22′30″ N, 79°15′00″ E) so the cross-sections can be located precisely in the field and in GIS.

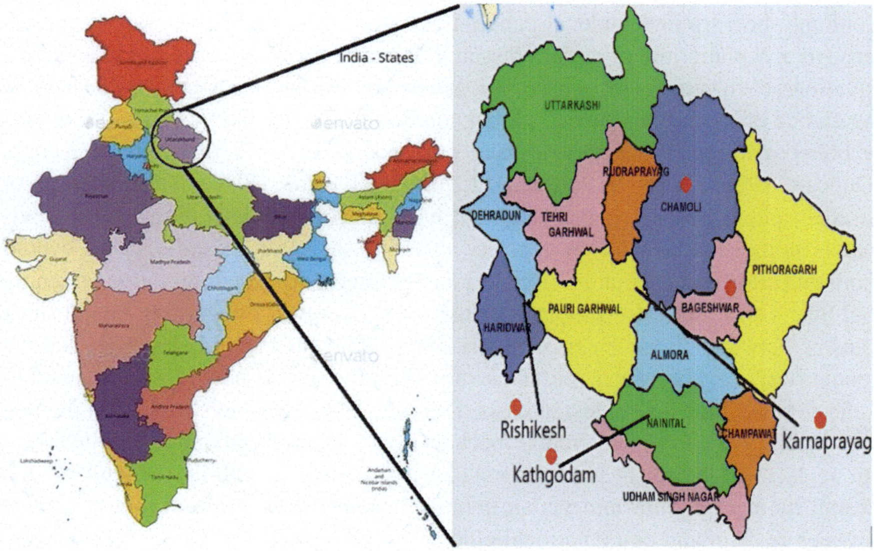

Fig. 3.1 Study region

Table 3.1 Specification of sites

S.NO	Slope	Site Latitude	Site Longitude	Area	Height (m)	Dip direction (°)	Dip angle (°)
1	S1	30°7′39.64″N	78°21′22.63″E	Rishikesh	25	105	55
2	S2	30°7′33.32″N	78°21′32.76″E	Rishikesh	21	310	64
3	S3	30°7′23.03″N	78°22′53.94″E	Rishikesh	29	54	80
4	S4	30°7′24.34″N	78°23′11.23″E	Rishikesh	25	95	76
5	S5	30°6′08.83″N	78°26′07.87″E	Chamoli	20	90	64
6	S6	30°4′08.22″N	78°27′33.93″E	Chamoli	16	270	67
7	S7	30°4′06.76″N	78°27′41.13″E	Chamoli	16	350	74
8	S8	30°22′30″N	79°15′00″E	Kathgodam	15	95	60
9	S9	29°16′00″N	79°33′00″E	Karnaprayag	20	60	65
10	S10	29°19′00″N	79°38′00″E	Karnaprayag	24	120	70
11	S11	30.75°00′N	79.908°00′00″E	Bageshwar	16	75	55
12	S12	30°40′00″N	79°50′00″E	Kathgodam	17	150	58
13	S13	30°7′39.66″N	78°21′22.64″E	Rishikesh	21	110	60
14	S14	30°7′33.33″N	78°21′32.75″E	Rishikesh	25	165	80

Methods

Limit-equilibrium methods (LEM) estimate the factor of safety (FoS) by enforcing equilibrium of a discretized sliding mass at limiting conditions, using Mohr–Coulomb shear strength under effective stress while prescribing pore-water pressures through a phreatic line or nodal pressures. Because there are more unknowns than equations, each variant closes the problem with an assumption about interslice forces; the choice of closure and which global conditions are enforced—forces, moments, or both—distinguish the methods and influence accuracy in different geometries (Duncan, 1996; Labuz & Zang, 2012; Matthews et al., 2014). The Fellenius or Swedish circle method is the historical baseline for circular failures in approximately homogeneous soils. It neglects interslice shear forces and satisfies overall moment equilibrium about the circle center, yielding a quick, conservative screening tool that is still useful for simple embankments. However, by ignoring lateral shear transfer between slices it can misrepresent stability in layered slopes, bench-cut geometries, or where water pressures vary laterally, so it is now mostly retained as a check rather than a sole design basis (Duncan, 1996). Bishop's simplified method remains the workhorse for circular mechanisms. It includes interslice normal forces but neglects interslice shear, solving slice vertical force equilibrium in tandem with overall moment equilibrium via an iterative scheme. This strikes a robust balance between realism and computational efficiency and typically produces higher, more representative FoS values than Fellenius for homogeneous or gently stratified soil slopes. Its principal limitation is kinematic: because it assumes a circular slip and does not enforce both force and moment equilibrium simultaneously, it can underperform when failure is noncircular or structurally controlled by weak seams or benches (Duncan, 1996).To address noncircular mechanisms common in stratified or partially rock-dominated hillslopes, Janbu methods extend the slice framework.

Spencer's method and the general limit-equilibrium formulation of Morgenstern–Price are widely regarded as the most reliable within LEM. Spencer imposes a constant interslice force angle—equivalently, a constant ratio of interslice shear to normal across slices—and thereby satisfies both global force and moment equilibrium for circular or noncircular surfaces. Morgenstern–Price generalizes this by prescribing an interslice function (often half-sine) scaled by a solved parameter so that both equilibrium sets are met without the constant-angle constraint. In modern software these two methods usually return closely aligned FoS when the same pore-pressure field and slip search are used, with Morgenstern–Price offering additional flexibility for layered, bench-cut, or weak-seam–controlled geometries typical of Himalayan corridors (Duncan, 1996; Matthews et al., 2014).For most engineering reports in complex, layered terrain, Morgenstern–Price with a half-sine interslice function—cross-checked by Bishop for circular plausibility or Spencer for constant-angle closure—provides a defensible, reproducible basis for decision-making (Duncan, 1996; Matthews et al., 2014).

Methodology

The methodology for slope stability assessment was carried out using GeoStudio (SLOPE/W) software, which employs the Limit Equilibrium Method (LEM) to evaluate slope safety under both dry and saturated conditions. The study focused on fifteen slopes distributed across key regions of Uttarakhand—primarily Rishikesh, Chamoli, Kathgodam, and Karnaprayag. Each slope was defined by its geometric parameters (height, dip direction, and dip angle) and geotechnical characteristics (unit weight, cohesion, and internal friction angle) as shown in Table 3.2. These parameters were derived from field observations and laboratory testing, ensuring that the modeled conditions closely replicated actual site profiles. Material strength parameters were entered separately for dry and saturated conditions to represent pre-rainfall and post-rainfall scenarios, respectively. For each slope, the Morgenstern–Price method—a rigorous LEM formulation satisfying both force and moment equilibrium—was applied to compute the Factor of Safety (FoS). This method is widely recognized for its accuracy and flexibility in handling non-circular or composite slip surfaces, making it suitable for complex Himalayan terrain where the failure mechanism is not always rotational.. The pore-water pressure distribution was varied to simulate dry and saturated conditions. The dry scenario represented the condition prior to monsoon rainfall when pore pressures are minimal, whereas the saturated case simulated post-rainfall conditions with a raised phreatic surface to reflect infiltration-induced pore-pressure buildup. By comparing FoS values between these two states, the impact of rainfall on slope stability was quantitatively assessed.

Table 3.2 Properties of different sites

S. no	Slope	Unit weight γ (kN/m^3)—Dry	Unit weight γ (kN/m^3)—Saturated	Cohesion C (kPa)	Angle of friction φ (°)
1	S1	15	17	20	34.78
2	S2	18	20	25	36.75
3	S3	17.6	19.6	26.8	37.5
4	S4	18	20	25.6	36.59
5	S5	19.8	21.8	24.6	32.3
6	S6	19.5	21.5	19.4	35.09
7	S7	17.3	19.3	23.6	38.05
8	S8	16	19	25	37.5
9	S9	15	17	16	30
10	S10	19.56	21.56	22	35
11	S11	19.5	21.5	21	28
12	S12	25	27	15	33
13	S13	16	18	16.96	34
14	S14	18.27	20.27	25.8	34.78
15	S15	19.6	21.6	27	36.75

Overall, the slope stability assessment through this methodology highlights the reliability of GeoStudio's LEM-based framework in replicating complex field conditions and quantifying rainfall-induced stability degradation as shown in the form of flowchart in Fig. 3.2. The pre- and post-rainfall comparative analysis effectively demonstrates the hydrological sensitivity of Himalayan slopes, particularly in the Rishikesh region where FoS consistently remained below 1.0. The use of the Morgenstern–Price method in SLOPE/W, supported by field-calibrated parameters, provided a robust foundation for identifying vulnerable slopes and understanding the mechanics of rainfall-triggered failures in the Himalayan context.

Result

The hydrological sensitivity of these same sites by reporting Factors of Safety (FoS) under "Dry" (pre-rainfall) and "Saturated" (post-rainfall) conditions, together with the resulting qualitative stability class "Before Rainfall" and "After Rainfall." In Rishikesh, S1 (30°07′39.64″ N, 78°21′22.63″ E) is unstable in both states with FoS ≈ 0.93 (dry) and 0.95 (saturated), and S3 (30°07′23.03″ N, 78°22′53.94″ E) and S4 (30°07′24.34″ N, 78°23′11.23″ E) are likewise unstable in both states with FoS well below unity, whereas S2 (30°07′33.32″ N, 78°21′32.76″ E) remains stably above 1.3 in both conditions as done by slope stability analysis in Geostudio software and shown in Figs. 3.3, 3.4, 3.5, and 3.6 for Site 1, 2, 3 and 4 respectively. This pattern

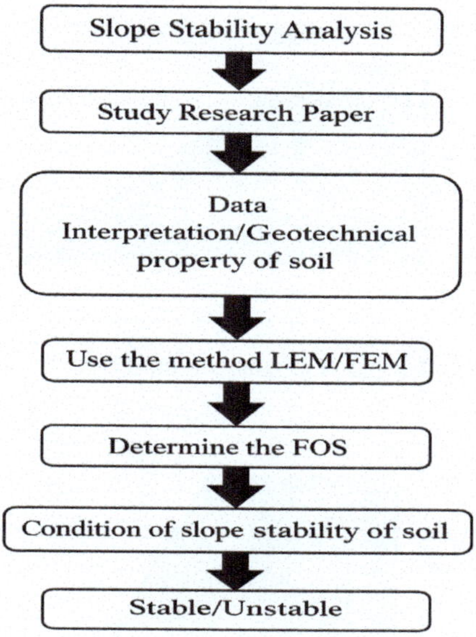

Fig. 3.2 Methodology flow chart

highlights strong intra-district variability in the Rishikesh corridor: despite similar relief and proximity, some cuts are persistently unstable while select reaches retain robust stability even under rainfall loading.

Chamoli exhibits transitional behavior across its sites. S5 (30°06′08.83″ N, 78°26′07.87″ E) sits on the stability threshold, shifting from "Stable" in dry conditions to "Unstable" under saturation (FoS changes from ~1.00 to ~0.999), directly illustrating rainfall sensitivity at this location. Nearby, S6 remains unstable in both conditions with FoS ≈ 0.99 (dry) and 0.96 (saturated), while S7 remains safely above unity in both states (≈ 1.106 dry, 1.081 saturated) as done in Geostudio software and shown in Figs. 3.7, 3.8 and 3.9 for Site 5, 6 and 7 respectively. The juxtaposition of S5–S7 underscores how modest differences in geometry, material condition, or groundwater response within the same district can flip the stability outcome during monsoon periods.

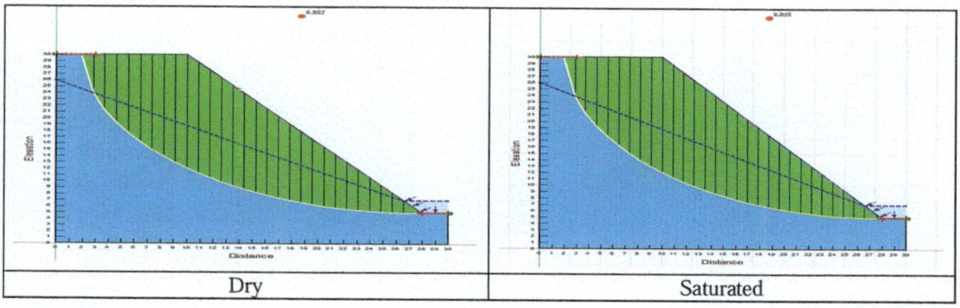

Fig. 3.3 Factor of safety for site 1

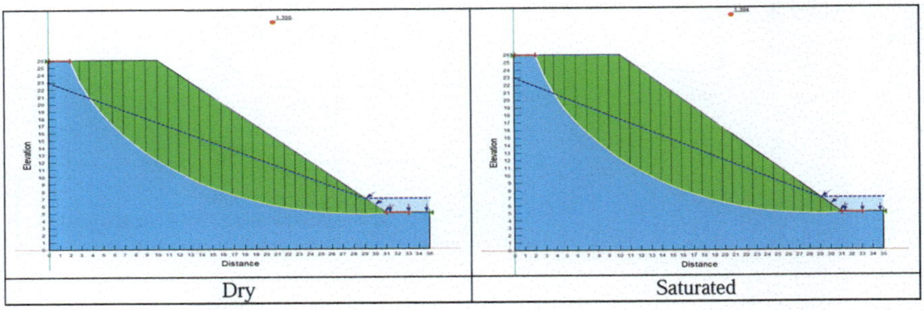

Fig. 3.4 Factor of safety for site 2

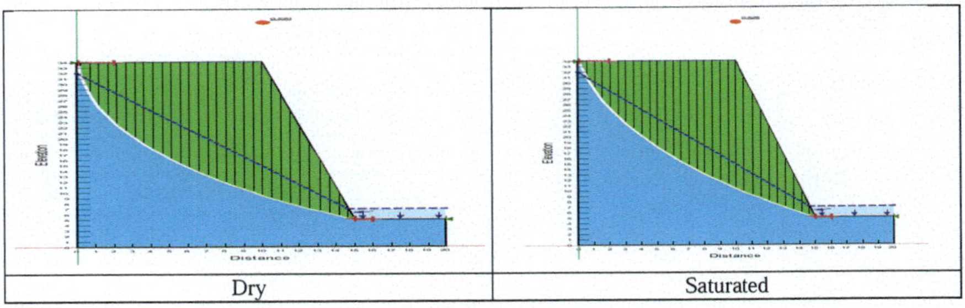

Fig. 3.5 Factor of safety for site 3

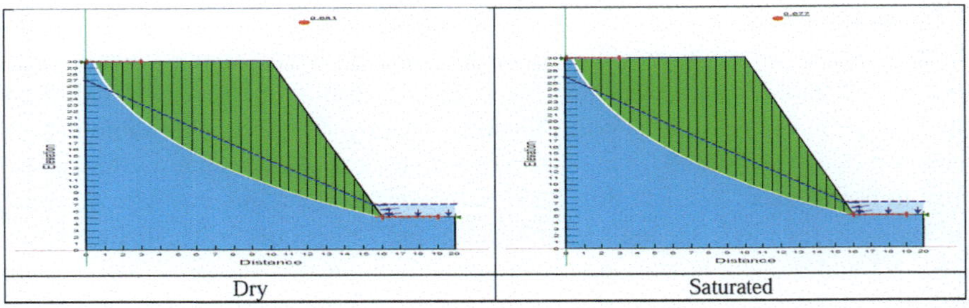

Fig. 3.6 Factor of safety for site 4

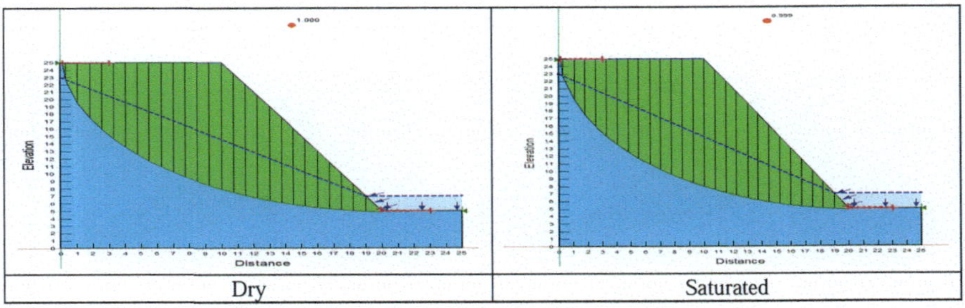

Fig. 3.7 Factor of safety for site 5

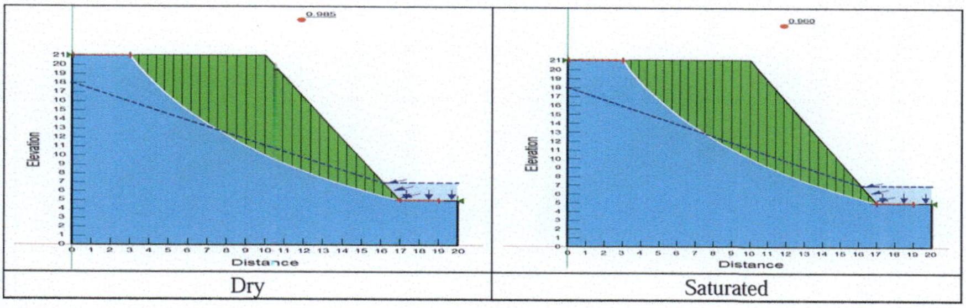

Fig. 3.8 Factor of safety for site 6

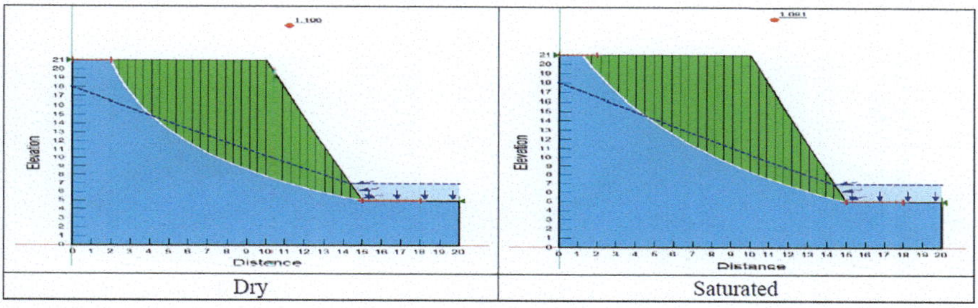

Fig. 3.9 Factor of safety for site 7

In the Kumaun foothill context of Kathgodam, the two profiled slopes (S8, S12) are comparatively robust, maintaining FoS comfortably above unity in both hydrologic scenarios (e.g., S8 ~1.37 dry, ~1.34 saturated; S12 ~ 1.07 dry, ~1.07 saturated) as shown in Figs. 3.10, 3.11, 3.12, and 3.13, reflecting more favorable combinations of geometry and material properties or better drainage response in these sections.

By contrast, both Karnaprayag sites (S9, S10) are persistently unstable across conditions (e.g., S9 ~ 0.85 dry, ~0.84 saturated; S10 ~ 0.58 dry, ~0.58 saturated), pointing to a pervasive risk regime along that reach; Bageshwar's S11 also remains

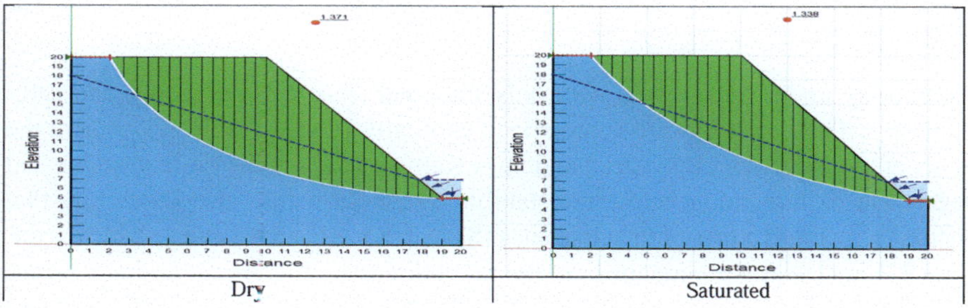

Fig. 3.10 Factor of safety for site 8

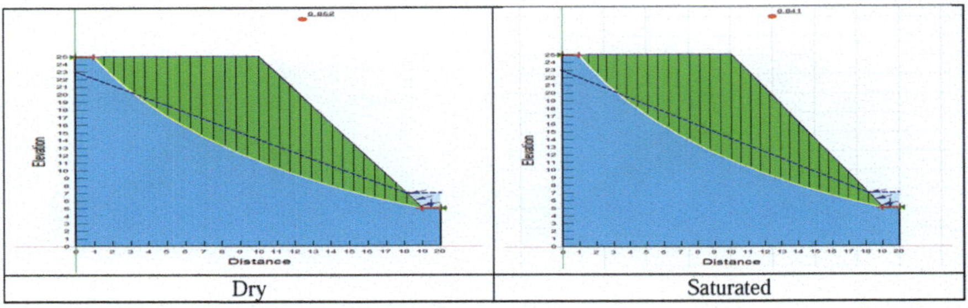

Fig. 3.11 Factor of safety for site 9

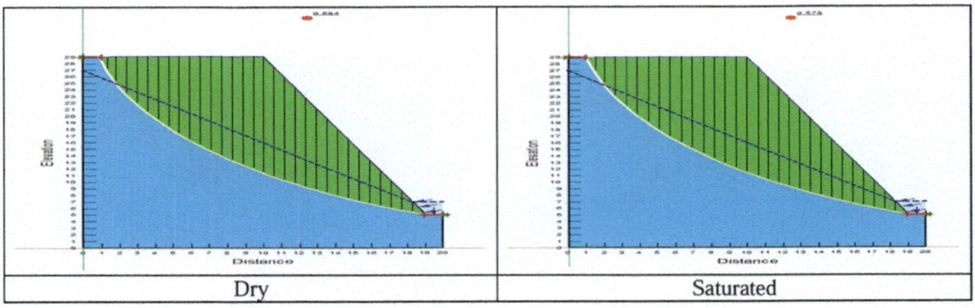

Fig. 3.12 Factor of safety for site 10

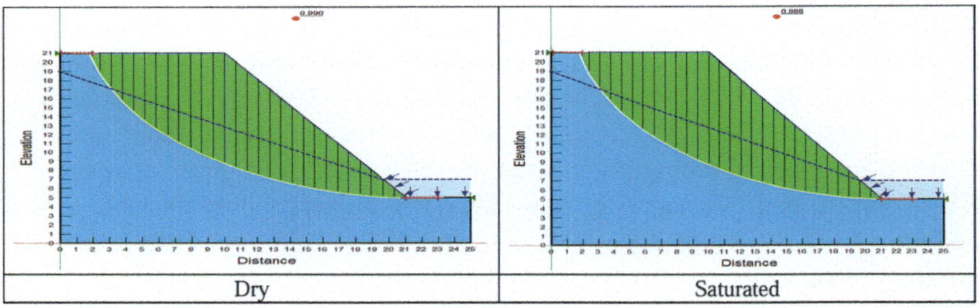

Fig. 3.13 Factor of safety for site 11

below unity under both states, with only marginal change between dry and saturated runs. These findings are summarized explicitly beneath Table 3.3., where the report groups outcomes by district and flags Karnaprayag as uniformly unstable and Bageshwar as remaining below the stability threshold, while also reiterating that Rishikesh includes a mix of unstable sites (S1, S3, S4, S13, S14) and a stable exception (S2), and that Kathgodam is stable in both cases. Together, the district-wise synthesis and the row-wise entries in Table 3.3.show that the most rainfall-sensitive transitions occur in Chamoli (notably S5), whereas structural or geometric controls keep Karnaprayag and several Rishikesh cuts below unity regardless of hydrologic state.

Table 3.3 Factor of Safety (F.O.S) before and after rainfall for different sites

S.n	Slope	Latitude	Longitude	Area	F.O.S (Dry)	F.O.S (Saturated)	Condition (Before Rainfall)	Condition (After Rainfall)
1	S1	30°7′39.64″N	78°21′22.63″E	Rishikesh	0.927	0.945	Unstable	Unstable
2	S2	30°7′33.32″N	78°21′32.76″E	Rishikesh	1.3	1.304	Stable	Stable
3	S3	30°7′23.03″N	78°22′53.94″E	Rishikesh	0.52	0.526	Unstable	Unstable
4	S4	30°7′24.34″N	78°23′11.23″E	Rishikesh	0.681	0.677	Unstable	Unstable
5	S5	30°6′08.83″N	78°26′07.87″E	Chamoli	1	0.999	Stable	Unstable
6	S6	30°4′08.22″N	78°27′33.93″E	Chamoli	0.985	0.96	Unstable	Unstable
7	S7	30°4′06.76″N	78°27′41.13″E	Chamoli	1.106	1.081	Stable	Stable
8	S8	30°22′30″N	79°15′00″E	Kathgodam	1.371	1.338	Stable	Stable
9	S9	29°16′00″N	79°33′00″E	Karnaprayag	0.852	0.841	Unstable	Unstable
10	S10	29°19′00″N	79°38′00″E	Karnaprayag	0.584	0.578	Unstable	Unstable
11	S11	30.75°00′N	79.908°00′00″E	Bageshwar	0.99	0.988	Unstable	Unstable
12	S12	30°40′00″N	79°50′00″E	Kathgodam	1.065	1.074	Stable	Stable
13	S13	30°7′39.66″N	78°21′22.64″E	Rishikesh	0.972	0.986	Unstable	Unstable
14	S14	30°7′33.33″N	78°21′32.75″E	Rishikesh	0.602	0.599	Unstable	Unstable

Methodologically, these regional patterns are produced from a common analysis framework in SLOPE/W in which each site's geometry and materials are entered from the field-derived dataset, then evaluated under dry and saturated pore-pressure states using a limit-equilibrium formulation (Morgenstern–Price) and a consistent slip-surface search, so that differences in FoS across the study region reflect local conditions rather than modeling artifacts. The report itself emphasizes that the Morgenstern–Price approach computes forces slice-by-slice and is well suited to complex slope situations where rotational and translational modes may coexist, and it notes that the results table is intended to enable district-level interpretation of hydrologic impacts on slope stability. In short, the study region is not only geographically diverse but also mechanically heterogeneous, and Table 3.3 captures this heterogeneity by pairing precise locations with comparable dry–wet stability metrics, allowing clear identification of rainfall-sensitive sites (e.g., Chamoli S5) versus persistently unstable corridors (e.g., Karnaprayag, several Rishikesh sites) and comparatively resilient areas (Kathgodam).

The factor-of-safety (FoS) mapping across all fourteen slopes reveals a distinct spatial pattern in stability and a generally weak sensitivity to saturation, with a few notable exceptions. In Rishikesh, three of the four primary sites modeled (S1, S3, S4) are persistently unstable in both hydrologic states, while only S2 remains stable throughout. Specifically, S1 yields FoS $\approx$ 0.927 (dry) and 0.945 (saturated), classified as unstable; S3 returns FoS $\approx$ 0.520 (dry) and 0.526 (saturated), also unstable; S4 shows FoS $\approx$ 0.681 (dry) and 0.677 (saturated), again unstable. By contrast, S2 maintains stability with FoS $\approx$ 1.300 (dry) and 1.304 (saturated). These values align with the tabulated "before rainfall/after rainfall" stability states, which remain "unstable" for S1, S3, and S4 and "stable" for S2. In Chamoli, the dataset captures a transitional response to rainfall. S5 is the clearest case of rainfall-induced loss of stability: it records FoS $\approx$ 1.000 (dry, "stable") that drops marginally to 0.999 when saturated, tipping the designation to "unstable," which underscores how close this slope sits to the failure threshold under wet conditions. Meanwhile, S6 remains unstable in both conditions with FoS $\approx$ 0.985 (dry) and 0.960 (saturated). S7 is comparatively robust, staying stable with FoS $\approx$ 1.106 (dry) and 1.081 (saturated). The report's narrative explicitly groups S5–S7 as "transitional," calling out S5's shift across the stability boundary during rainfall while noting S7's persistent stability. Kathgodam is consistently the most stable locality in your sample. S8 exhibits FoS $\approx$ 1.371 (dry) and 1.338 (saturated), while S12 records FoS $\approx$ 1.065 (dry) and 1.074 (saturated), leading to "stable" designations before and after rainfall for both sites as shown in Fig. 3.14. Table 3.3 emphasizes that both Kathgodam sites show FoS "well above 1.0" across conditions, which is consistent with the tabulated numbers.

Karnaprayag emerges as pervasively at risk. Both sites (S9 and S10) are unstable in dry and saturated states, with FoS for S9 $\approx$ 0.852 (dry) and 0.841 (saturated) and for S10 $\approx$ 0.584 (dry) and 0.578 (saturated), which the report synthesizes as "FoS well below 1.0," indicating systematic susceptibility to failure. Bageshwar (S11) is also assessed as unstable irrespective of rainfall, with only marginal change under saturation: FoS $\approx$ 0.990 (dry) and 0.988 (saturated). The accompanying discussion

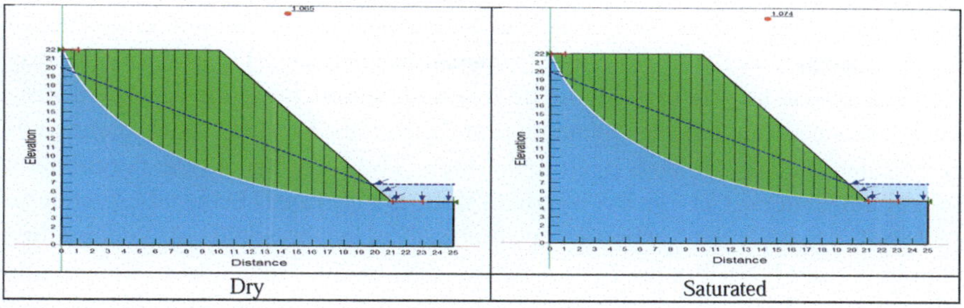

Fig. 3.14 Factor of safety for site 12

notes that S11 "remains unstable in both dry and saturated states," and that any saturated improvement is negligible—indeed here the saturated FoS is fractionally lower. Two additional Rishikesh slopes (S13 and S14) maintain the locality's overall instability signature. S13 remains unstable with FoS ≈ 0.972 (dry) and 0.966 (saturated). S14 is likewise unstable with FoS ≈ 0.602 (dry) and 0.599 (saturated). The narrative section that follows Table 3.3 groups S1–S4 along with S13 and S14 to conclude that "all the sites are unstable with FoS values less than 1.0," except S2, which stands out as stable. Slope failure envelopes for Site 13 and 14 are shown in Figs. 3.15 and 3.16 respectively.

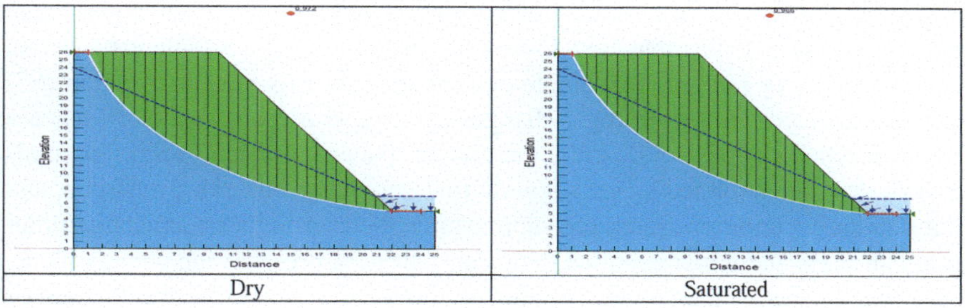

Fig. 3.15 Factor of safety for site 13

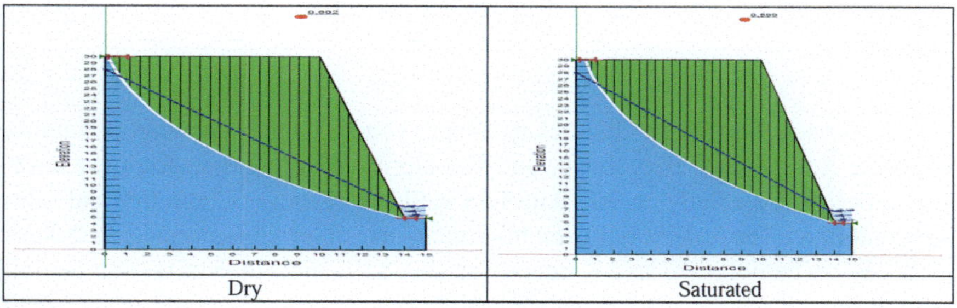

Fig. 3.16 Factor of safety for site 14

Taken together, the results depict four clear themes. First, spatial clustering of instability is evident: Rishikesh (apart from S2), Bageshwar (S11), and Karnaprayag (S9, S10) consistently show FoS < 1.0 even before rainfall, implying limited intrinsic shear resistance in the modeled geometries and materials. Second, Kathgodam stands out for robust stability in both hydrologic states. Third, the hydrologic perturbation from rainfall generally produces only small shifts in FoS values in the table, but those small shifts can be decisive when a slope is already at the threshold—as vividly illustrated by Chamoli S5 flipping from "stable" (dry) to "unstable" (saturated) despite a change from 1.000 to 0.999. Fourth, the sign of the saturated–dry difference is not uniform: some unstable slopes lose a bit more capacity when saturated (e.g., S6, S9, S10, S13, S14), while others change very little or even show tiny numerical increases; nevertheless, the categorical stability designations in Table 3.3 remain driven by whether FoS crosses 1.0. These four themes are stated or implied in the "Result" pages and the interpretive notes immediately after Table 3.3. From a site-specific perspective, the slopes with the most acute concern are those in Karnaprayag and several in Rishikesh, where FoS values are furthest below unity, implying greater mobilized driving forces relative to available shear strength and, therefore, higher probabilities of failure if conditions deteriorate further. The Bageshwar site's FoS, while close to one, remains below the stability threshold in both conditions, underscoring the need for mitigation in an already fragile setting. Conversely, Kathgodam's consistently high FoS suggests comparatively stronger materials, more favorable geometries, or more effective natural drainage, any of which could contribute to the observed resilience. In summary, demonstrates that most analyzed slopes are at least conditionally vulnerable, and several are unequivocally unstable even before the onset of heavy rainfall. The single case of stability loss under saturation (S5) illustrates how minimal hydrologic changes can trigger a categorical shift when a slope is perched at FoS $\approx$ 1.0, while the broader pattern of small FoS differentials between dry and saturated states suggests that geometry, material properties, and local boundary conditions (e.g., toe support) may be exerting a stronger first-order control in these particular sections than transient pore-pressure increments alone. These conclusions are documented around the Table 3.3 discussion and in the study's concluding remarks.

Conclusion

The report concludes that the Uttarakhand slopes you evaluated—spanning Chamoli, Rishikesh, Kathgodam, Bageshwar, and Karnaprayag—present widespread instability concerns, especially at sites with FoS well below unity in both dry and saturated conditions. The text explicitly highlights clusters of unstable slopes at Rishikesh (S1, S3, S4), Chamoli (S6), and Karnaprayag (S9, S10), and emphasizes that rainfall generally exacerbates instability, as observed in the S5 transition and in the small but adverse FoS shifts at several other sites. The conclusion section also reiterates the value of numerical methods (LEM with Morgenstern–Price in SLOPE/W) for diagnosing critical factors and framing mitigation concepts. In line with the body of the report and the literature reviewed within it, appropriate measures would include improved surface and subsurface drainage to manage infiltration, toe protection/ retaining systems where river undercutting or inadequate toe support is suspected, regrading to reduce driving moments on steep faces, and bio-engineering to enhance near-surface cohesion and reduce erosion—all in service of moving marginal slopes above an FoS of 1.0 in saturated conditions. These points are stated or directly implied in the concluding page of your document.

References

Arnold, D. N. (1981). Discretization by finite elements of a model parameter-dependent problem. *Numerische Mathematik, 37*, 405–421.

Awasthi, C. K., Jana, A., Dey, A., & Sreedeep, S. (2019). Stability assessment of reinforced rock slope based on two-dimensional finite element approach: A Himalayan case study. In R. Sundaram, J. Shahu, & V. Havanagi (Eds.), *Geotechnics for Transportation Infrastructure* (Lecture Notes in Civil Engineering, Vol. 28, pp. 639–650). Springer.

Bansal, V., & Sarkar, R. (2024a). Comparative analysis of slope stability for kalimpong region under dynamic loading using limit equilibrium method and machine benchmark learning classifiers. *Iranian Journal of Science Technology, Transactions of Civil Engineering, 48*, 2785–2807.

Bansal, V., & Sarkar, R. (2024b). Prophetical modeling using limit equilibrium method and novel machine learning ensemble for slope stability gauging in kalimpong. *Iranian Journal of Science Technology, Transactions of Civil Engineering, 48*, 411–430.

Bansal, V., & Sarkar, R. (2023). Numerical analysis and implications of fluctuating water levels on slope stability. *International Conference on Sustainable Infrastructure: Innovation, Opportunities and Challenges*. Springer, 407–419.

Bansal, V., & Shukla, B. K. (2022). Application of urea formaldehyde raisin in pavement subgrade stabilization at Hindon Barrage. *Journal of Physics: Conference Series, 2267*(1), 012150. IOP Publishing.

Bharti, G., Hurukadli, P., Shukla, B. K., Tripathi, A., Sharma, P. K., Singh, A., Singh, G., Patel, H., & Nigam, K. (2025a). Advanced geotechnical analysis and mitigation of slope stability in Dharchula, Pithoragarh, Uttarakhand, Using GeoStudio Software. In *Intelligent Infrastructure and Smart Materials: Sustainable Technologies for a Greener Future* (pp. 233–254). Cham: Springer Nature Switzerland. https://doi.org/10.1007/978-3-031-92421-7_14

Bharti, G., Shukla, B. K., Srinivasan, V., & Bansal, V. (2022). Experimental study on eccentrically loaded rectangular footing on sand for embedded condition to enhance serviceability of structures. *Materials Today: Proceedings, 61*, 517–522.

Bharti, G., Tripathi, A., Hurukadli, P., Bharath, B., Sharma, P. K., & Shukla, B. K. (2025b). Experimental analysis of fly ash and bagasse ash integration as sustainable stabilisers for improving the engineering properties of black cotton soil. In *Intelligent Infrastructure and Smart Materials: Sustainable Technologies for a Greener Future* (pp. 341–358). Cham: Springer Nature Switzerland. https://doi.org/10.1007/978-3-031-92421-7_20

Chamyal, L. S. (1991). Stratigraphy of the Lesser Himalayan rocks in Kumaun. *Proceedings of the Indian Academy of Sciences (Earth and Planetary Sciences), 100*, 293–306.

Christian, J. T., & Baecher, G. B. (1999). Point-estimate method as numerical quadrature. *Journal of Geotechnical and Geoenvironmental Engineering, 125*(9), 779–786.

Dawson, E. M., Roth, W. H., & Drescher, A. (1999). Slope stability analysis by strength reduction. *Géotechnique, 49*(6), 835–840.

Duncan, J. M. (1996). State of the art: Limit equilibrium and finite-element analysis of slopes. *Journal of Geotechnical Engineering, 122*(7), 577–596.

Eberhardt, E. (2003). *Rock slope stability analysis: Utilization of advanced numerical techniques* (Report, 41 pp.). University of British Columbia.

Gowda, S., Kumar, P. G., Suneetha, K. N., Kavya Sri, G., Gupta, A., & Shukla, B. K. (2025). Optimizing pavement condition index prediction using PSO-LSTM hybrid algorithm: A data-driven approach for road maintenance strategies. In *Intelligent Infrastructure and Smart Materials: Sustainable Technologies for a Greener Future* (pp. 29–53). Cham: Springer Nature Switzerland. https://doi.org/10.1007/978-3-031-92421-7_3

Griffiths, D. V., & Fenton, G. A. (2004). Probabilistic slope stability analysis by finite elements. *Journal of Geotechnical and Geoenvironmental Engineering, 130*(5), 507–518.

Griffiths, D. V., & Lane, P. A. (1999). Slope stability analysis by finite elements. *Géotechnique, 49*(3), 387–403.

Hackston, A., & Rutter, E. (2016). The Mohr-Coulomb criterion for intact rock strength and friction: A re-evaluation and consideration of failure under polyaxial stresses. *Solid Earth, 7*, 493–508.

Hammah, R. E., Yacoub, T. E., Corkum, B., Wibowo, F., & Curran, J. H. (2007). Analysis of blocky rock slopes with finite element shear strength reduction analysis. In *Proceedings of the 1st Canada–US Rock Mechanics Symposium* (pp. 329–334).

Kainthola, A., Singh, P. K., & Singh, T. N. (2015). Stability investigation of road cut slope in basaltic rock mass, Mahabaleshwar. *India. Geoscience Frontiers, 6*(6), 837–845.

Kainthola, A., Singh, P. K., Wasnik, A. B., Sazid, M., & Singh, T. N. (2012). Finite element analysis of road cut slopes using Hoek-Brown failure criterion. *International Journal of Earth Sciences and Engineering, 5*(5), 1100–1109.

Kainthola, A., Verma, D., Gupte, S. S., & Singh, T. N. (2011). A coal mine dump stability analysis: A case study. *Geomaterials, 1*(1), 1–13.

Keefer, D. K. (2002). Investigating landslides caused by earthquakes: A historical review. *Surveys in Geophysics, 23*, 473–510.

Kumar, V. (2017). Statistical distribution of rainfall in Uttarakhand, India. *Applied Water Science, 7*, 4765–4776.

Labuz, J. F., & Zang, A. (2012). Mohr-Coulomb failure criterion. *Rock Mechanics and Rock Engineering, 45*, 975–979.

Laouafa, F., & Royis, P. (2004). An iterative algorithm for finite element analysis. *Journal of Computational and Applied Mathematics, 164–165*, 469–491.

Mahanta, B., Singh, H. O., Singh, P. K., Kainthola, A., & Singh, T. N. (2016). Stability analysis of potential failure zones along NH-305. *India. Natural Hazards, 83*(3), 1341–1357.

Matsui, T., & San, K. C. (1992). Finite element slope stability analysis by shear strength reduction technique. *Soils and Foundations, 32*(1), 59–70.

Matthews, C., Farook, Z., & Helm, P. (2014). Slope stability analysis: Limit equilibrium or the finite element method. *Ground Engineering, 48*(5), 22–28.

Merh, S. S. (1977). Structural studies in parts of Kumaun Himalaya. *Himalayan Geology, 7*, 26–42.

Pain, A., Kanungo, D. P., & Sarkar, S. (2014). Rock slope stability assessment using finite element-based modelling: Examples from the Indian Himalayas. *Geomechanics and Geoengineering, 9*, 215–230.

Pande, R. K. (1992). Slope instability assessment in the Karmi area of Central Himalaya, India. *The Environmentalist, 12*, 211–221.

Prakash, G., Mehdi, S. H., & Kumar, G. (1978). Geology of the Chaukhtia–Bageshwar area, Almora district, Kumaun Himalaya, U.P. *Himalayan Geology, 8*(2), 1049–1063.

Rawat, S., Shukla, B. K., Sihag, P., Srivastav, Y., & Singh, H. K. (2024). Environmental effects and sustainable hydropower development: a review of deployment research, environmental valuation and IoT sensors integration. In *AIP Conference Proceedings* (Vol. 3050, No. 1, p. 040008). AIP Publishing LLC.

Rocscience Inc. (2004). *Application of the finite element method to slope stability*. https://www.rocscience.com/assets/resources/learning/papers/Application-of-the-Finite-ElementMethod-to-Slope-Stability.pdf

Saraswat, R., Parkash, S., & Pal, S. (2020). Landslide susceptibility assessment due to construction of buildings in garhwal area using GIS. In S. Ahmed, S. Abbas, & H. Zia (eds) *Smart Cities—Opportunities and Challenges*. Lecture Notes in Civil Engineering, vol. 58. Springer, Singapore. https://doi.org/10.1007/978-981-15-2545-2_70

Shukla, B. K., Bansal, V., Varadharajan, S., Bharti, G., Shenoy, A., Singh, U. K., & Kirthanashri, S. V. (2023a). Identification and analysis of factors of delay in road construction projects. *Lecture Notes in Civil Engineering, 280*, 505–511. https://doi.org/10.1007/978-981-19-4739-1_47

Shukla, B. K., Bharti, G., Nagavi, J. C., Hurukadli, P., Gupta, A., & Sharma, P. K. (2024). Enhancing the stability of black cotton soil through reinforcement with fly ash and bagasse ash mixtures: A comprehensive overview. *Lecture Notes in Civil Engineering, 508*, 55–74. https://doi.org/10.1007/978-981-97-3823-6_6

Shukla, B. K., Bharti, G., Parashar, B., Sharma, P. K., Abhimanyu, & Verma, S. (2025b). Variables Influencing the Performance of 3D Concrete Printing for Sustainable Construction: Current Status and Future Prospects. *Innovative Technologies for Smart and Sustainable Infrastructure*, 225–239. https://doi.org/10.1007/978-3-031-92801-7_13

Shukla, B. K., Gupta, A., Gowda, S., & Srivastav, Y. (2023b). Constructing a greener future: A comprehensive review on the sustainable use of fly ash in the construction industry and beyond. *Materials Today: Proceedings, 93*, 257–264.

Shukla, B. K., Gupta, L., Parashar, B., Sharma, P. K., Sihag, P., & Shukla, A. K. (2025a). Integrative assessment of surface water contamination using GIS, WQI, and machine learning in urban-industrial confluence zones surrounding the national capital territory of the Republic of India. *Water, 17*(7), 1076. https://doi.org/10.3390/w17071076

Shukla, B. K., Sharma, P. K., Singh, C., Kumari, N., Chaudhary, A., & Garg, A. (2025d). Integrated urban water management: optimizing rainwater harvesting and stormwater infrastructure for sustainable urban development. In *Geo-Data Revolution: Advances in Spatial Analysis and Natural Hazard Mapping* (pp. 103–118). Cham: Springer Nature Switzerland. https://doi.org/10.1007/978-3-031-94078-1_6

Shukla, B. K., Yadav, H., Singh, S., Verma, S., Shukla, A. K., & Sharma, C. (2025b). Sustainable engineering of recycled aggregate concrete: structural performance and environmental benefits under circular economy frameworks. *Construction Materials, 5*(3), 67.

Singh, P. K., Wasnik, A. B., Kainthola, A., Sazid, M., & Singh, T. N. (2013). The stability of road cut cliff face along SH-121: A case study. *Natural Hazards, 68*(2), 497–507.

Solanki, A., Gupta, V., Bhakuni, S. S., Ram, P., & Joshi, M. (2019). Geological and geotechnical characterization of the Khotila landslides in the Dharchula region, NE Kumaun Himalaya. *Journal of Earth System Science, 128*(4), 86.

Tripathi, D. K. (2018). Application of remote sensing and geographic information system in terrain analysis. *International Journal of Scientific Research in Computer Science Applications and Management Studies, 7*(5), 1–4.

Valdiya, K. S. (1980). *Geology of Kumaun Lesser Himalaya*. Wadia Institute of Himalayan Geology.

Valdiya, K. S. (1983). Lesser Himalayan geology: Crucial problems and controversies. *Current Science, 52*(18), 839–857.

Valley, B., & Duff, D. (2011). *Probabilistic analyses in Phase2 8.0*. https://www.rocscience.com/documents/pdfs/rocnews/spring2011/Phase2stat-Probabilistic-Analyses-in-Phase2.pdf

Yin, A. (2006). Cenozoic tectonic evolution of the Himalayan orogen as constrained by along-strike variation of structural geometry, exhumation history, and foreland sedimentation. *Earth-Science Reviews, 76*(1–2), 1–131.

Zheng, H., Liu, D. F., & Li, C. (2005). Slope stability analysis based on elasto-plastic finite element method. *International Journal for Numerical Methods in Engineering, 64*(14), 1871–1888.

Chapter 4
Geo-Environmental Assessment of Rainfall-Induced Slope Failure in Kalimpong Using Numerical Modelling

Vaishnavi Bansal, Gaurav Bharti, and Raju Sarkar

Abstract In the Indian Himalayas, slope instability remains a recurrent natural hazard, increasingly aggravated by the intensification of rainfall events linked to climate change. The growing frequency of extreme precipitation has amplified pore-water pressures and infiltration-induced failures, making it essential to evaluate slope stability under changing climatic conditions. This study employs numerical modelling approaches to assess the Factor of Safety (FOS) by considering the combined effects of soil properties, slope geometry, and rainfall infiltration. A detailed case study was conducted in Kalimpong, a highly landslide-prone region, focusing on a slope near Happy Villa that experienced rainfall-triggered failure. The analysis utilized SLOPE/W (Limit Equilibrium Method) and PLAXIS 2D (Finite Element Method) software to evaluate slope performance under unsaturated and saturated conditions, representing pre- and post-rainfall scenarios. In the limit equilibrium method, FOS values were 1.139 (unsaturated) and 0.449 (saturated), while in the finite element method, the corresponding values were 1.13 and 0.39. Both modelling approaches consistently revealed that steep slope gradients combined with rainfall-induced saturation significantly reduce slope stability, confirming rainfall infiltration as a critical triggering mechanism. The results highlight the growing influence of climate-induced hydrological variability on slope behavior and emphasize the need to incorporate climate change considerations into regional landslide hazard assessments and slope management strategies. Proactive adaptation through improved drainage, slope reinforcement, and climate-resilient planning is essential to mitigate future failure risks in Himalayan terrains.

V. Bansal (✉) · G. Bharti
Department of Civil Engineering, JSS Academy of Technical Education, Noida, India
e-mail: vaishnavibansal@jssaten.ac.in

G. Bharti
e-mail: gauravbharti000@gmail.com

R. Sarkar
Department of Civil Engineering, Delhi Technological University, New Delhi, India
e-mail: rajusarkar@dce.ac.in

R. Sarkar et al. (eds.), *Climate Change, Environmental Hazards and Community-Based Resilience*, Disaster Risk Reduction, https://doi.org/10.1007/978-981-95-9485-6_4

Keywords Climate change · Slope stability analysis · Numerical modelling · Kalimpong · Finite Element Method (FEM) · Limit Equilibrium Method (LEM)

Introduction

The increasing frequency and intensity of catastrophic natural events around the world have made slope failures a recurring phenomenon, posing significant threats to human lives, infrastructure, and the environment. Among these, rainfall-induced landslides represent one of the most common and destructive geohazards in mountainous terrains (Bharti et al., 2025a). In the Indian subcontinent, particularly in the geologically fragile Himalayan region, slope instability has emerged as a major concern due to a combination of adverse geomorphological, geological, and climatic factors (Bansal et al., 2025; Bharti et al., 2025b; Dar et al., 2023; Gowda et al., 2025). The Kalimpong district, located in the Darjeeling Himalayas, is particularly susceptible to landslides, being situated in a multi-hazard environment characterized by steep slopes, complex lithology, high seismicity, and intense monsoonal precipitation. The cumulative impact of these factors not only leads to recurrent slope failures but also severely disrupts socioeconomic development, damages infrastructure, and hampers disaster mitigation and rehabilitation efforts. Landslides in the Himalayan belt are predominantly triggered by rainfall, which contributes to slope saturation, increased pore water pressure, and reduction in shear strength of soil mass. Studies have shown that nearly 15% of India's landmass, approximately 0.49 million km^2, is prone to landslides, with the Himalayas contributing the majority of these occurrences (Sarma et al., 2019). Historical records reveal several devastating rainfall-induced slope failures, such as those at Malpa (1998), Mandakini Valley (1998), Agastyamuni (2005), Nilgiris (2009), Kedarnath (2013), Malin (2014), Balia Nala (2014), Kalimpong (2015), Shillong (2019), and Marappalam (Dey & Sengupta, 2018; Dikshit et al., 2019; Ering & Babu, 2016; Gupta et al., 2016; Khanduri et al., 2018; Pradhan et al., 2018; Senthilkumar et al., 2018; Umrao et al., 2017). These events underscore the need for detailed regional studies combining field data, geotechnical characterization, and numerical modelling to understand the mechanisms driving slope failures.

Kalimpong, due to its low lithological strength, rapid erosion by the Teesta River, and intense monsoon rainfall, frequently experiences slope instability (Bansal & Sarkar, 2024a, 2024b; Chatterjee, 2010). The area's geomorphology, comprising steep ridges and dissected valleys, is highly sensitive to infiltration-driven weakening during heavy rainfall episodes. Repeated failure events often isolate the district by damaging roads and lifelines, highlighting the urgency for robust slope stability assessments and predictive modelling. To address the complex interplay of factors causing slope failure, researchers employ numerical simulation methods such as the Limit Equilibrium Method (LEM), Finite Difference Method (FDM), and Finite Element Method (FEM) (Dey & Sengupta, 2018). Each method has its advantages and limitations. LEM, widely used due to its simplicity and efficiency, evaluates the equilibrium of forces along potential slip surfaces but neglects stress–strain behavior. In contrast, FEM provides a more comprehensive understanding by incorporating stress distribution, deformation, and pore pressure variations, offering a realistic representation of slope performance under varying environmental conditions. Advanced computational tools like TRIGRS, GeoStudio (SEEP/W and SLOPE/W), and PLAXIS 2D have facilitated precise analysis of rainfall infiltration and slope deformation (Chandrasekaran et al., 2013; Cho & Lee, 2002; Collins & Znidarcic, 2004; Deka et al., 2020; Kristo et al., 2017; Rahardjo et al., 2001, 2007; Raj & Sengupta, 2014). In particular, coupling seepage analysis with mechanical stability assessment helps in quantifying how rainfall-induced pore pressure changes affect the Factor of Safety (FOS) and overall slope behavior. These software packages allow for dynamic evaluation of unsaturated and saturated conditions, which is critical in regions like Kalimpong where intense short-duration rainfall often triggers failures.

In this context, the present study focuses on the Happy Villa landslide (October 19, 2021) in Kalimpong as a representative case. This site experienced catastrophic failure following continuous heavy rainfall, disrupting transportation and threatening residential areas. The event offers an opportunity to analyze the geo-environmental conditions, soil behavior, and slope response under varying saturation states using both SLOPE/W (LEM) and PLAXIS 2D (FEM) models. By comparing pre- and post-rainfall conditions, the study aims to assess how infiltration affects slope stability and to identify critical triggering factors. Slope instability in Kalimpong is a complex phenomenon influenced by several interrelated factors, including geological structure, lithology, hydrological conditions, and anthropogenic activities. The region forms part of the Darjeeling–Sikkim Himalayan belt, characterized by folded and faulted metamorphic rocks, schists, and phyllites that exhibit low shear strength and high weathering susceptibility (Ghosh et al., 2011; Kanungo et al., 2009, 2011). The steep slopes, combined with intense monsoonal precipitation averaging 250–300 cm annually (Sarkar et al., 2013; Soja & Starkel, 2007), create favorable conditions for rainfall-induced failures. The soil mantle, often composed of clayey and silty layers, exhibits high permeability and porosity, which facilitates rapid infiltration during rainfall. Consequently, transient pore pressure builds up, decreasing effective stress and reducing shear strength, ultimately leading to slope failure (Chatterjee, 2010; Collins & Znidarcic, 2004; Kristo et al., 2017). The hydrogeological setting of Kalimpong further amplifies the instability. The district is drained by numerous small tributaries and streams, locally termed kholas and jhoras, which evolve into larger channels feeding the Teesta River (Dikshit & Satyam, 2017; Mukherjee & Mitra, 2001). These channels act as natural pathways for infiltration and subsurface flow, saturating the slope materials during monsoon seasons. Additionally, the geomorphic processes of erosion and toe-cutting by the Teesta River intensify slope steepening and undercutting, destabilizing the overlying soil masses (Chatterjee, 2010). Therefore, a holistic understanding of the slope behavior requires integrating both hydrological and geomechanical analyses under varying rainfall intensities.

Recent studies have demonstrated that rainfall intensity, duration, and antecedent moisture conditions significantly influence slope response and failure initiation. For instance, the studies emphasized that continuous low-intensity rainfall may be as critical as high-intensity short-duration storms due to the progressive saturation of soil layers (Cho & Lee, 2002; Collins & Znidarcic, 2004). Also, they established that antecedent rainfall can maintain elevated pore pressures, predisposing slopes to failure even with moderate subsequent precipitation (Rahardjo et al., 2001). These insights underline the importance of transient analysis using numerical models that can simulate the time-dependent changes in pore pressure and shear strength. In the field of slope stability analysis, the Limit Equilibrium Method (LEM) remains a widely adopted approach owing to its simplicity and interpretability (Huang, 2014, Kaynia et al. 2011). It involves dividing the potential sliding mass into slices and computing the equilibrium of forces and moments along assumed slip surfaces. Although efficient, LEM neglects deformations and stress redistributions, often leading to approximate safety factors. On the other hand, the Finite Element Method (FEM) offers a more rigorous alternative by solving for the stress–strain response of the entire slope domain under given boundary and loading conditions (Moni & Sazzad, 2015). It accounts for nonlinear material behavior, seepage-induced pore pressure variations, and complex geometries, making it suitable for detailed geo-environmental studies. Comparisons between both methods have revealed that FEM tends to provide more conservative and realistic estimates of the Factor of Safety, particularly under saturated conditions.

Considering these aspects, the present study adopts a dual-method approach by employing both SLOPE/W (LEM) and PLAXIS 2D (FEM) to analyze the Happy Villa slope under unsaturated (pre-rainfall) and saturated (post-rainfall) conditions. The objective is to evaluate the influence of rainfall infiltration on slope stability and to quantify the variation in FOS due to pore pressure buildup. The Morgenstern-Price method implemented in SLOPE/W is used to compute FOS based on equilibrium conditions, while the Mohr–Coulomb model in PLAXIS simulates stress-deformation behavior and progressive failure mechanisms. This comparative assessment enhances understanding of the mechanical and hydraulic interactions within the slope system. Moreover, the study integrates laboratory geotechnical testing—including grain size distribution, Standard Proctor compaction, and consolidated drained triaxial tests—to characterize the soil's strength parameters and deformation properties. Such empirical data serve as critical input for numerical simulations, ensuring realistic modeling of site conditions. The selected landslide site, situated near GREF Phatak below Chandraloke, holds strategic importance as it affects a major transport corridor in the Darjeeling Himalayas, essential for local connectivity and economic activities. The 2021 failure, triggered after heavy rainfall following a prolonged dry spell, exemplifies the sensitivity of Himalayan slopes to sudden hydrological perturbations. This investigation aligns with the growing need for predictive modelling frameworks to forecast landslide occurrences and design mitigation strategies. By quantifying the stability variations under different hydrological states, the research provides valuable insights for disaster risk reduction, infrastructure resilience, and land-use planning in mountainous terrains like Kalimpong. Furthermore, the study contributes to the ongoing discourse on the applicability of different numerical modelling approaches in assessing rainfall-induced slope failures in complex geological environments.

Study Location

One of the touristic hill towns in India's state of West Bengal (WB) is Kalimpong (Fig. 4.1) with Kalimpong's google earth image in Fig. 4.2 also representing the landslide location. It is situated between latitudes 87.47 and 89.47 N and longitude 26.07 and 28.07 E. The region's slope map is shown in Fig. 4.3, with the average elevation being 1247 m. It is a portion of the Darjeeling-Sikkim Himalayan area that has undergone several geo-tectonic movements and is very stiff, folded, complicated, metamorphosed, and disturbed (Gurung & McGowran, 2022; McGowran, 2024). The area experiences a subtropical highland climate with rainy summers and dry winters. The monsoon season, which runs from June through September, is when most of the rain falls (Soja & Starkel, 2007). This area receives 250–300 cm of rainfall on average each year. River Teesta flows in the west, while River Relli flows in the east, encircling Kalimpong, which is located on a curving ridge. While the western face of Kalimpong is mostly steep and rough, the eastern slopes are softer and more stable. This area's yearly temperature fluctuates from a high of 17 °C to a low of 5 °C. Typically reddish in hue, the soil at Kalimpong can occasionally be found with aboundant phyllite and schists (Dikshit & Satyam, 2017). Numerous landslides have severely destabilised the western slope of the Kalimpong Ridge. These places have varied amounts of translation and rockslides and debris from several generations.

The slope configuration, derived from the DEM in Fig. 4.3 indicates steep gradients, particularly along the western flank, which influence runoff and slope instability. Geologically, Kalimpong forms part of a highly deformed Himalayan belt that has undergone multiple episodes of folding, faulting, and metamorphism, resulting in

Fig. 4.1 Location map of the study region Kalimpong in Indian political map (https://asiashope.org/introduction-to-kalimpong)

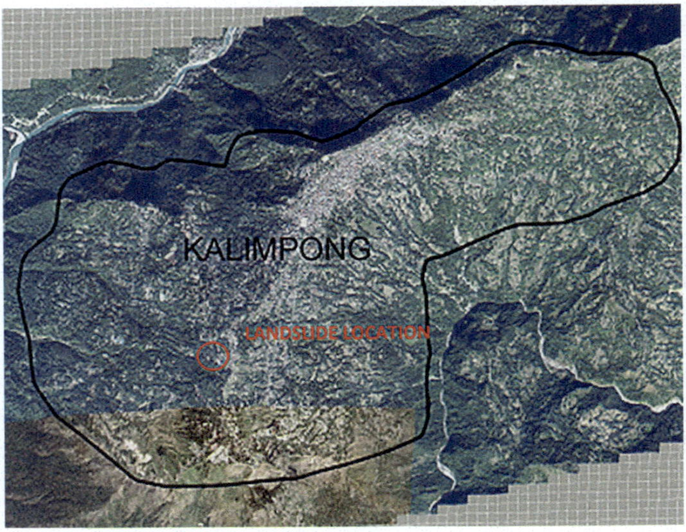

Fig. 4.2 Google image of Kalimpong (www.earth.google.com)

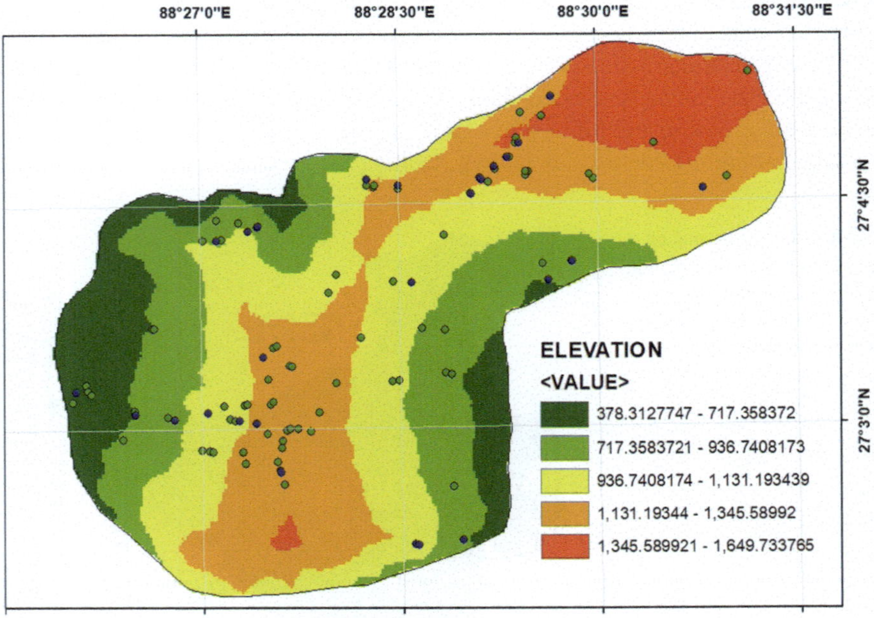

Fig. 4.3 Elevation Contour Map of Kalimpong (in m) (*Source* USGS)

a stiff and complex rock mass structure. The lithology is dominated by phyllites, schists, and quartzites, which are highly weathered and fractured, leading to reduced shear strength and increased susceptibility to failure. The region exhibits pronounced structural discontinuities and jointed rock masses that contribute to its fragile slope stability conditions.

Climatically, Kalimpong experiences a subtropical highland climate with distinct wet and dry seasons. The southwest monsoon (June–September) contributes 250–300 cm of annual rainfall, intensifying infiltration and elevating pore-water pressures within the slope material. Temperature variations between 5 °C and 17 °C influence the weathering processes and hydrological balance (Chatterjee, 2010). The combined effect of intense precipitation and steep slopes frequently initiates rainfall-induced slope failures. Hydrologically, the area is bounded by the Teesta River to the west and the Relli River to the east, with the urban settlement situated on a curvilinear ridge. The western slopes are steep and rugged, while the eastern slopes are comparatively gentle and stable. The dominant soil is reddish-brown residual soil with abundant phyllitic and schistose fragments, indicating moderate to high permeability and low cohesion. The western slopes of the Kalimpong ridge exhibit recurring landslides, including translational and debris slides, influenced by steep gradients, weak lithology, and high monsoonal rainfall. The interplay of tectonic activity, geomorphology, and hydrological factors makes Kalimpong highly prone to rainfall-induced slope instability, warranting detailed geotechnical and numerical analyses for hazard assessment and climate-adaptive slope management. Front view and side view of Landslide near Happy villa is shown in Figs. 4.4 and 4.5 respectively.

Fig. 4.4 Front view of Landslide near Happy Villa, Kalimpong (*Source* Save The Hills)

Fig. 4.5 Side view of Landslide near Happy Villa, Kalimpong (*Source* Save The Hills)

Geohydrological Details

The soil's ability to absorb water results in an increase in the mass of the soil and ultimately the soil's unit area (Bansal & Sarkar, 2023). Due to an increase in pore pressure brought on by increased soil water absorption, the likelihood of landslides increases. Along the sides of streams and rivers, the water flow weathers the rocks, and over time, these processes cause rock and other materials to break, resulting in slides. The monsoon season in Kalimpong brings pockets of rain, which makes it the region's peak landslide season. Numerous minor tributaries called kholas (2nd and 3rd order streams) and jhoras drain the area (mainly 1st and 2nd order stream). The jhoras get their water from a huge number of long-lasting regular springs in the hill's upper section, which are located near the top of the slope (Source: Save The Hills). The whole region is traversed by five subbasins that are all tributaries of the Teesta River (Chatterjee, 2010). Throughout the Teesta basin, first order streams combine to form second and higher order streams and correspondingly drainage map of Kalimpong has been shown in Fig. 4.6.

Methodology

Experimental Work

From the slope, representative soil samples were collected. These soil samples were tested in a lab. The distribution of particle sizes in a soil sample is first and foremost determined by the grain size distribution. A series of standard sieves of various sizes are used in the test, and after the soil has been sieved through them, a particle

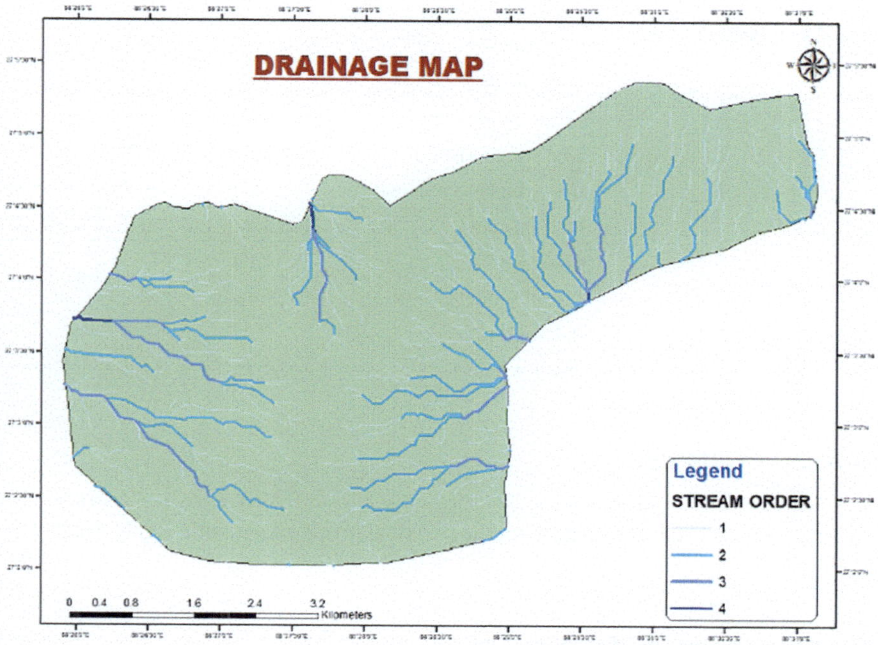

Fig. 4.6 Drainage map of Kalimpong (*Source* USGS)

size distribution curve is plotted after measuring the quantity of soil that has been retained on each sieve. The test offers details on a variety of qualities, including soil permeability, erosion potential, and texture. The maximum soil density and equivalent water content were then determined using the standard proctor test. Three layers of a soil sample are compacted into a cylindrical mould for the test, and each layer is subjected to 25 blows with a standardised hammer. In order to make sure that soil is compacted to the right density during building projects, the test is frequently performed to assess the compaction properties of a soil. Then, in order to ascertain a soil sample's shear strength and stress–strain behaviour, the consolidated drained triaxial test was performed (Bharti et al., 2022). In this test, a cylindrical soil sample is encased with a rubber membrane, and confining pressure is applied while pore water drainage is managed. The sample is then loaded until it breaks axially. Axial and radial strains, pore water pressure, and axial stress are all monitored throughout the test. The test is commonly carried out on saturated soil samples, and it measures the soil's internal friction angle and cohesiveness, two factors that are crucial for geotechnical design and analysis. Since shear strength is one of the crucial factors in slope design, triaxial tests were carried out at varied confining pressures.

Table 4.1 Properties of the soil considered

Properties	Unit	Soil
Soil classification		Clayey soil
Unit weight (Dry)	(kN/m^3)	15.5
Unit weight (Bulk)	(kN/m^3)	18.5
Unit weight (Saturated)	(kN/m^3)	20.29
Friction angle	(degree)	33.2
Cohesion	(kN/m^2)	2
Modulus of elasticity	(MN/m^2)	54
Poisson's ratio		0.23
Dilation angle	(degree)	0

Material Properties

The soil collected is tested to find out the physical properties mentioned in Table 4.1.

Limit Equilibrium Method

A popular technique for examining the slope stability is the limit equilibrium method. It involves dividing a potential failure plane into slices and calculating the forces acting on each slice. The forces which are considered includes the unit weight, the shear strength of the soil along the failure plane, and any external forces acting on the slope, such as water pressure. Following that, the factor of safety is determined, which if less than one, the slope is considered unstable. This method is simple and can provide quick estimates of slope stability, but it has limitations, such as assuming a static condition and not considering soil deformation or complex failure mechanisms (Matthews et al., 2014).

Stability model using Geo Studio (Slope/W) Approach: Slope/W, a part of present limit equilibrium software, enables analysis to manage growing complexity. SLOPE/W employs a broad set of soil models to explain various soil types, complicated stratigraphic and slip surface geometry, and variable pore-water pressure conditions. To analyse the stability of the steep embankment slope, the classic Limit Equilibrium technique is used. It is represented in terms of the moment and force equilibrium factors of the safety equations. The limit equilibrium approach compares the available shear '//strength along the sliding surface with the force required to keep the slope in equilibrium to compute the factor of safety. So as part of this study, Morgenstern Price method was utilized to analyze the slope model and slope failure envelopes for unsaturated and saturated conditions have been shown in Figs. 4.7 and 4.8 respectively.

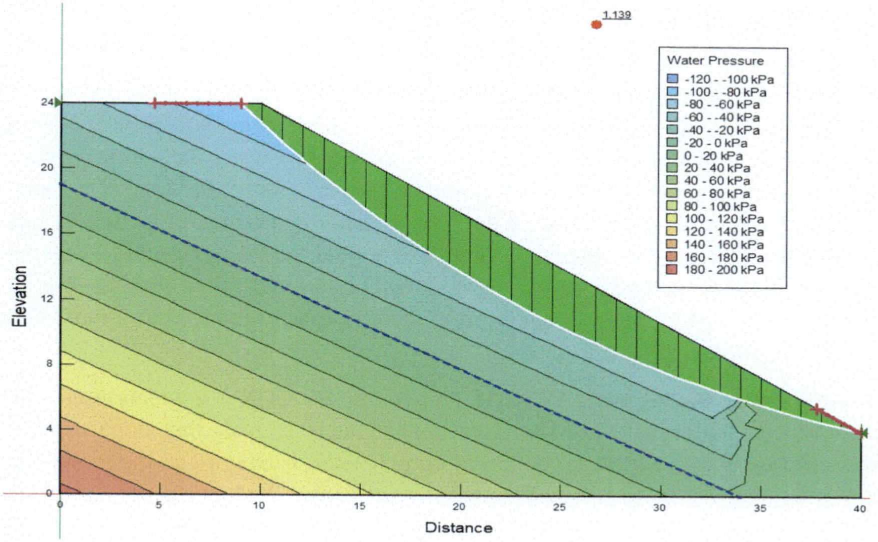

Fig. 4.7 Critical pore water pressure (PWP) variation and FOS in unsaturated conditions in LEM

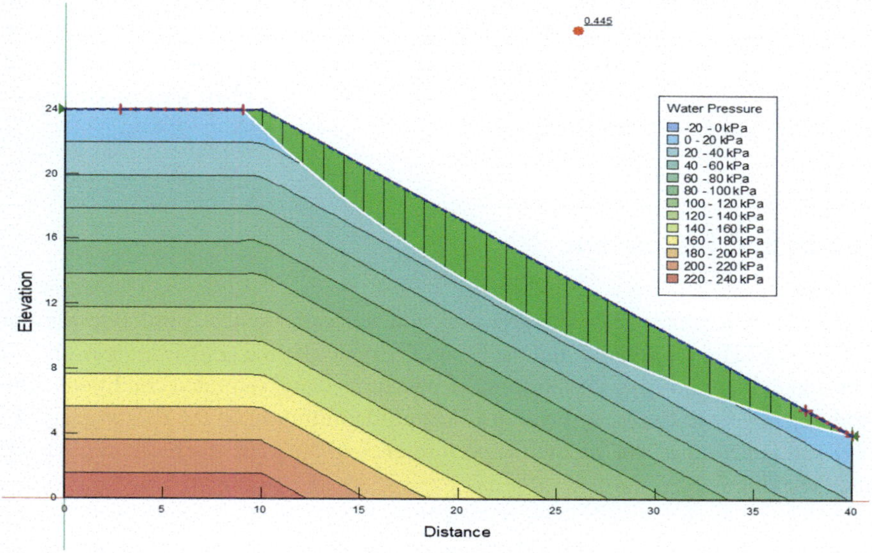

Fig. 4.8 Critical PWP variation and FOS in saturated conditions in LEM

Finite Element Method

FEM is a numerical approach that can be used to analyze the stability of slopes. It involves discretizing the slope into a mesh of small elements and solving for the stress and deformation of each element. The method considers the soil properties, slope geometry, and external loads to simulate the behavior of the slope. The factor of safety is then calculated based on the strength and deformation properties of soil, and the results can be visualized to identify potential failure modes and locations. The finite element method can provide a more accurate and detailed analysis of slope stability compared to the limit equilibrium method, but it requires more input data and computational resources.

Slope Stability model using PLAXIS Approach: PLAXIS is a widely used software package that utilizes FEM for geotechnical analysis, including slope stability. The software allows for the modeling of complex soil behavior and the incorporation of various external loads and boundary conditions. In slope stability analysis, PLAXIS can be used to model different failure modes and mechanisms, such as sliding and soil deformation, and to evaluate the safety factor of the slope under different scenarios. The software also provides visualization tools to help identify potential failure modes and locations. Overall, PLAXIS can provide a more detailed and comprehensive analysis of slope stability than traditional methods, but it requires specialized knowledge and expertise to use effectively so deformation variations and FOS under unsaturated and saturated conditions utilizing FEM have been shown in Figs. 4.9 and 4.10 respectively.

Results and Discussions

A comprehensive numerical analysis was conducted using both the Limit Equilibrium Method (LEM) in SLOPE/W and the Finite Element Method (FEM) in PLAXIS 2D to evaluate the stability of the slope under unsaturated and saturated conditions. The main objective was to assess the effect of rainfall infiltration on the Factor of Safety (FOS) and to compare the predictive performance of both approaches in replicating field behavior.

Under unsaturated conditions, representing the pre-rainfall scenario, the slope showed a FOS of 1.139 using LEM (Morgenstern–Price method), indicating a stable slope with sufficient shear resistance. The FEM analysis, using the Mohr–Coulomb failure criterion, produced a FOS of 1.13, closely matching the LEM result and suggesting only minor differences between the two methods under dry conditions. This stability indicates that, before rainfall, the slope's shear strength and effective stress were adequate to resist gravitational driving forces. After rainfall infiltration and saturation, the FOS decreased significantly due to increased pore-water pressure and reduced matric suction. LEM predicted a FOS of 0.449, while FEM yielded 0.39, both below unity, indicating imminent failure. FEM showed higher sensitivity

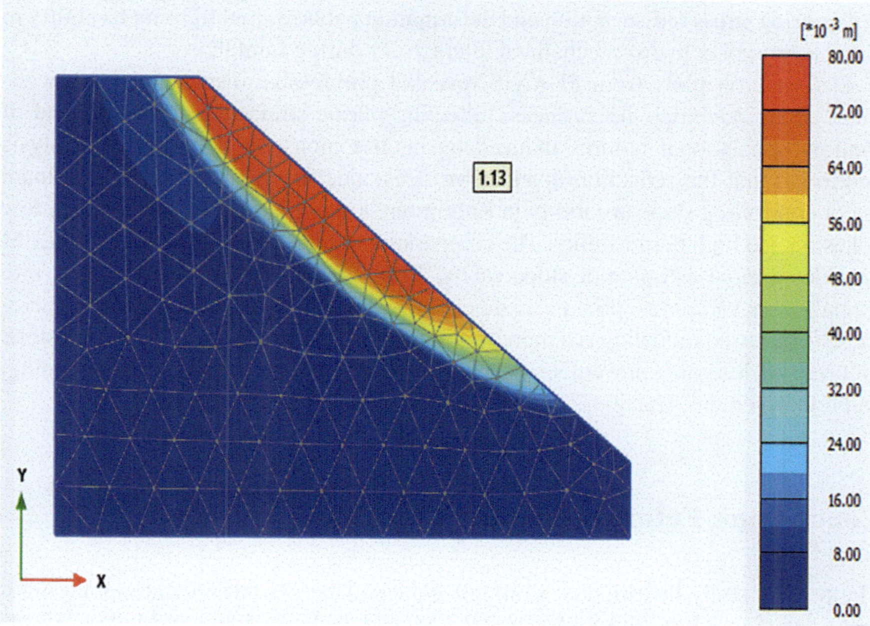

Fig. 4.9 Deformation variation and FOS under unsaturated conditions in FEM

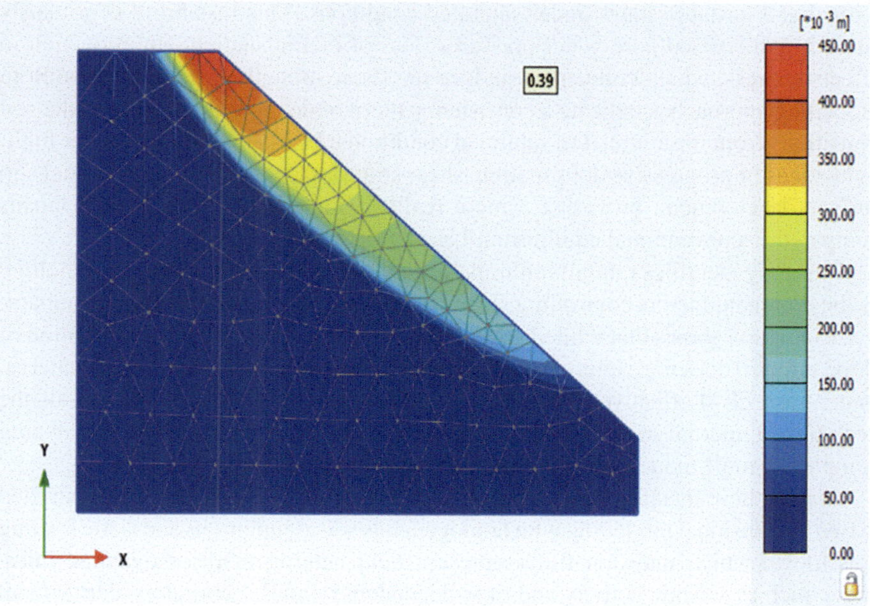

Fig. 4.10 Deformation variation and FOS under saturated conditions in FEM

in capturing stress redistribution and deformation patterns, highlighting its ability to simulate complex hydro-mechanical interactions during rainfall events.

Deformation plots from PLAXIS revealed progressive displacements concentrated along potential slip surfaces, reflecting plastic strain localization typical of shallow translational failures induced by intense monsoon rainfall. The analysis confirmed that the reduction of effective stress due to saturation is the dominant factor controlling slope instability in Kalimpong's clayey soils, characterized by low cohesion and high permeability. The comparative study emphasizes that while LEM provides a rapid estimate of slope safety, FEM offers more detailed insights into spatial stress variations, pore pressure evolution, and displacement patterns. Therefore, FEM-based modeling is a more robust tool for predicting rainfall-induced slope failures and designing preventive measures such as surface drainage, slope regrading, and reinforcement strategies.

Conclusions, Future Scope, and Limitations

The present study investigates a rainfall-induced slope failure in Kalimpong, integrating the Limit Equilibrium Method (LEM) and the Finite Element Method (FEM) to evaluate the impact of rainfall infiltration on slope stability. The analysis demonstrates a clear shift in slope behavior from stable conditions during dry periods to critical and unstable states under saturated conditions. This instability is primarily caused by increased pore-water pressures, loss of matric suction, and reduction in effective stress, which collectively reduce the shear strength of the soil mass. Both modelling approaches indicate a consistent pattern of decreasing stability as the soil transitions from unsaturated to saturated conditions. FEM simulations further highlight zones of progressive deformation, shear strain concentration, and potential slip surface development, providing a more realistic depiction of failure mechanisms compared to conventional equilibrium-based analyses.

The study identifies rainfall infiltration, steep slope geometry, and weak lithology as the dominant factors controlling slope failure in the Kalimpong region. Comparative evaluation shows that while LEM offers a simplified and practical estimation of slope safety, it does not capture stress redistribution and internal deformation patterns. In contrast, FEM effectively models hydro-mechanical interactions and reveals the evolution of internal stresses leading to failure, making it more suitable for analyzing complex rainfall-induced slope failures in hilly terrains.

Despite these insights, the study has certain limitations. The analysis is restricted to two-dimensional modelling with homogeneous soil assumptions and static loading conditions, which may not fully represent field heterogeneity or dynamic influences such as seismic activity and time-dependent rainfall. Laboratory-derived soil parameters may also differ from actual in-situ conditions due to natural variability. Future research should incorporate transient seepage modelling, coupled hydro-mechanical analyses, and three-dimensional simulations to capture real-time slope responses. Integrating remote sensing, GIS-based susceptibility mapping, and rainfall threshold models can strengthen early warning systems. Evaluating mitigation strategies, including surface drainage, soil reinforcement, and bioengineering solutions, will enhance slope resilience and support sustainable hazard management in the Darjeeling–Sikkim Himalayas.

References

Bansal, V., & Sarkar, R. (2023). Numerical analysis and implications of fluctuating water levels on slope stability. *International Conference on Sustainable Infrastructure: Innovation, Opportunities and Challenges*, Springer, 407–419.

Bansal, V., & Sarkar, R. (2024a). Comparative analysis of slope stability for kalimpong region under dynamic loading using limit equilibrium method and machine benchmark learning classifiers. *Iranian Journal of Science Technology, Transactions of Civil Engineering, 48*, 2785–2807.

Bansal, V., & Sarkar, R. (2024b). Prophetical modeling using limit equilibrium method and novel machine learning ensemble for slope stability gauging in kalimpong. *Iranian Journal of Science Technology, Transactions of Civil Engineering, 48*, 411–430.

Bansal, V., Bharti, G., Hurukadli, P., Singh, H., & Shukla, B. K. (2025). Investigation of geotechnical properties of soil using shredded polyethylene terephthalate (PET) wastes. *Lecture Notes in Civil Engineering, 355*, 119–129. https://doi.org/10.1007/978-981-97-8393-9_13

Bharti, G., Shukla, B. K., Srinivasan, V., & Bansal, V. (2022). Experimental study on eccentrically loaded rectangular footing on sand for embedded condition to enhance serviceability of structures. *Materials Today: Proceedings, 61*, 517–522.

Bharti, G., Hurukadli, P., Shukla, B. K., Tripathi, A., Sharma, P. K., Singh, A., Singh, G., Patel, H., & Nigam, K. (2025a). Advanced geotechnical analysis and mitigation of slope stability in Dharchula, Pithoragarh, Uttarakhand, using geostudio software. In *Intelligent Infrastructure and Smart Materials: Sustainable Technologies for a Greener Future* (pp. 233–254). Springer Nature Switzerland. https://doi.org/10.1007/978-3-031-92421-7_14

Bharti, G., Tripathi, A., Hurukadli, P., Bharath, B., Sharma, P. K., & Shukla, B. K. (2025b). Experimental analysis of fly ash and bagasse ash integration as sustainable stabilisers for improving the engineering properties of black cotton soil. In *Intelligent Infrastructure and Smart Materials: Sustainable Technologies for a Greener Future* (pp. 341–358). Springer Nature Switzerland. https://doi.org/10.1007/978-3-031-92421-7_20

Chandrasekaran, S., Sayed Owaise, R., Ashwin, S., Jain, R. M., Prasanth, S., & Venugopalan, R. (2013). Investigation on infrastructural damages by rainfall-induced landslides during November 2009 in Nilgiris, India. *Natural Hazards, 65*, 1535–1557.

Chatterjee, R. (2010). *Landslide hazard zonation mapping of Kalimpong*. VDM Verlag Dr. Muller GmbH & Co. KG.

Cho, S. E., & Lee, S. R. (2002). Evaluation of surficial stability for homogeneous slopes considering rainfall characteristics. *Journal of Geotechnical and Geoenvironmental Engineering, 128*, 756–763.

Collins, B. D., & Znidarcic, D. (2004). Stability analyses of rainfall induced landslides. *Journal of Geotechnical and Geoenvironmental Engineering, 130*, 362–372.

Dar, S. A., Sharma, P. K., & Shukla, B. K. (2023). Enhanced landfill mining waste valorization techniques: A methodological review. *Lecture Notes in Civil Engineering, 281*, 109–125. https://doi.org/10.1007/978-981-19-4731-5_10

Deka, A., Daulagupu, R., Brahma, H., Lodger, H., & Dutta, D. (2020). Effect of antecedent rainfall on slope stability-a case study of Haflong Hill, North East India. *Journal of the Geological Society of India, 96*, 394–398.

Dey, N., & Sengupta, A. (2018). Effect of rainfall on the triggering of the devastating slope failure at Malin, India. *Natural Hazards, 94*, 1391–1413.

Dikshit, A., & Satyam, N. Rainfall thresholds for the prediction of landslides using empirical methods in Kalimpong, Darjeeling, India. *Proceedings of the JTC1 workshop on advances in landslide understanding*, Barcelona, Spain, 2017. 24–26.

Dikshit, A., Satyam, N., & Pradhan, B. (2019). Estimation of rainfall-induced landslides using the TRIGRS model. *Earth Systems and Environment, 3*, 575–584.

Ering, P., & Babu, G. S. (2016). Probabilistic back analysis of rainfall induced landslide-A case study of Malin landslide, India. *Engineering Geology, 208*, 154–164.

Ghosh, S., Carranza, E. J. M., Van Westen, C. J., Jetten, V. G., & Bhattacharya, D. N. (2011). Selecting and weighting spatial predictors for empirical modeling of landslide susceptibility in the Darjeeling Himalayas (India). *Geomorphology, 131*, 35–56.

Gowda, S., Kumar, P. G., Suneetha, K. N., Kavya Sri, G., Gupta, A., & Shukla, B. K. (2025). Optimizing pavement condition index prediction using PSO-LSTM hybrid algorithm: A data-driven approach for road maintenance strategies. In *Intelligent Infrastructure and Smart Materials: Sustainable Technologies for a Greener Future* (pp. 29–53). Springer Nature Switzerland. https://doi.org/10.1007/978-3-031-92421-7_3

Gupta, V., Bhasin, R. K., Kaynia, A. M., Tandon, R. S., & Venkateshwarlu, B. (2016). Landslide hazard in the Nainital township, Kumaun Himalaya, India: The case of September 2014 Balia Nala landslide. *Natural Hazards, 80*, 863–877.

Gurung, L., & Mcgowran, P. (2022). Local perspectives on landslide prevention and management in Kalimpong district, West Bengal, India. *Disaster Prevention Management: An International Journal, 31*, 193–201.

Huang, Y. H. (2014). Slope stability analysis by the limit equilibrium method: Fundamentals and methods. American Society of Civil Engineers.

Kanungo, D., Arora, M., Sarkar, S., & Gupta, R. (2009). A fuzzy set based approach for integration of thematic maps for landslide susceptibility zonation. *Georisk, 3*, 30–43.

Kanungo, D., Sarkar, S., & Sharma, S. (2011). Combining neural network with fuzzy, certainty factor and likelihood ratio concepts for spatial prediction of landslides. *Natural Hazards, 59*, 1491–1512.

Kaynia, E. E. H. A. M., A Comparative Study of Empirical Models for Landslide Prediction using Historical Case Histories.

Khanduri, S., Sajwan, K., Rawat, A., Dhyani, C., & Kapoor, S. (2018). Disaster in Rudraprayag District of Uttarakhand Himalaya: A special emphasis on geomorphic changes and slope instability. *Journal of Geography Natural Disasters, 8*, 1–9.

Kristo, C., Rahardjo, H., & Satyanaga, A. (2017). Effect of variations in rainfall intensity on slope stability in Singapore. *International Soil and Water Conservation Research, 5*, 258–264.

Matthews, C., Farook, Z., & Helm, P. (2014). Slope stability analysis–limit equilibrium or the finite element method. *Ground Engineering, 48*, 22–28.

Mcgowran, P. (2024). Geopolitical assemblages and disasters-in-the-making in Kalimpong. *India. Political Geography, 108*, Article 103008.

Moni, M. & Sazzad, M. 2015. Stability analysis of slopes with surcharge by LEM and FEM. *International Journal of Advances in Structural and Geotechnical Engineering, 4*, 216–225.

Mukherjee, A., & Mitra, A. (2001). Geotechnical study of mass movements along the Kalimpong approach road in the Eastern Himalayas. *Indian Journal of Geology, 73*, 271–280.

Pradhan, S. P., Vishal, V., & Singh, T. (2018). Finite element modelling of landslide prone slopes around Rudraprayag and Agastyamuni in Uttarakhand Himalayan terrain. *Natural Hazards, 94*, 181–200.

Rahardjo, H., Li, X., Toll, D., & Leong, E. (2001). The effect of antecedent rainfall on slope stability. *Geotechnical and Geological Engineering, 19*, 371–399.

Rahardjo, H., Ong, T., Rezaur, R., & Leong, E. C. (2007). Factors controlling instability of homogeneous soil slopes under rainfall. *Journal of Geotechnical and Geoenvironmental Engineering, 133*, 1532–1543.

Raj, M., & Sengupta, A. (2014). Rain-triggered slope failure of the railway embankment at Malda, India. *Acta Geotechnica, 9*, 789–798.

Sarkar, S., Roy, A. K., & Martha, T. R. (2013). Soil depth estimation through soil-landscape modelling using regression kriging in a Himalayan terrain. *International Journal of Geographical Information Science, 27*, 2436–2454.

Sarma, C. P., Dey, A., & Krishna, A. M. (2019). *Investigation of rainfall-induced landslides at the Hillslopes of Guwahati Region*. Springer.

Senthilkumar, V., Chandrasekaran, S., & Maji, V. (2018). Rainfall-induced landslides: Case study of the Marappalam landslide, Nilgiris District, Tamil Nadu. *India. International Journal of Geomechanics, 18*, 05018006.

Soja, R., & Starkel, L. (2007). Extreme rainfalls in Eastern Himalaya and southern slope of Meghalaya Plateau and their geomorphologic impacts. *Geomorphology, 84*, 170–180.

Umrao, R. K., Singh, R., Sharma, L., & Singh, T. (2017). Soil slope instability along a strategic road corridor in Meghalaya, North-Eastern India. *Arabian Journal of Geosciences, 10*, 260.

Chapter 5
Climate Change and Air Pollution: Mechanisms, Health Impacts, and Integrated Policy Solutions

Udit Mittal, Vaishnavi Bansal, and Amit Roy

Abstract Climate change and air pollution represent converging threats to global health, with interlinked sources and effects. Both are among the most pressing global environmental challenges, exerting profound effects on human health, ecosystems, and economies, and are tightly interlinked through shared emission sources, feedback mechanisms, and atmospheric processes. This review synthesizes current scientific understanding of the mechanisms by which climate change influences air pollution patterns and the consequent health impacts. Increasing temperatures and altered weather patterns enhance the formation and concentration of harmful air pollutants, such as ground-level ozone and fine particulate matter (PM2.5). Specifically, the review explores the mechanisms linking meteorological parameters with air pollutant dispersion, chemical transformation, and the formation of these secondary pollutants. These pollutants are significant contributors to respiratory and cardiovascular morbidity and mortality worldwide. Quantitative projections indicate that the increase in ozone and PM2.5 concentrations could lead to higher rates of mortality and morbidity, particularly under high greenhouse gas emission scenarios. Feedback loops, such as black carbon's warming effect, further complicate these interactions. The review emphasizes integrated climate and air quality policies targeting short-lived climate pollutants (SLCPs) for dual health and climate benefits, while identifying research gaps for future management strategies.

Keywords Climate change · Air pollution · Health impacts · Particulate matter · Air quality · Policy integration

U. Mittal (✉) · A. Roy
Department of Electrical Engineering, JSS Academy of Technical Education, Noida, India
e-mail: uditmittal.eed@jssaten.ac.in

A. Roy
e-mail: amitroy.eed@jssaten.ac.in

V. Bansal
Department of Civil Engineering, JSS Academy of Technical Education, Noida, India
e-mail: vaishnavibansal@jssaten.ac.in

R. Sarkar et al. (eds.), *Climate Change, Environmental Hazards and Community-Based Resilience*, Disaster Risk Reduction, https://doi.org/10.1007/978-981-95-9485-6_5

Introduction

Climate change and air pollution are primarily driven by the combustion of fossil fuels and by the change in the usage of land (Shenoy et al, 2023; Verma et al., 2025). Such changes are interlinked global challenges with profound impacts on human health, ecosystems, and economies. Anthropogenic emanations of greenhouse gases and air pollutants, such as nitrogen oxides (NOx), volatile organic compounds (VOCs), and fine particulate matter (PM2.5), create synergistic environmental and health risks (Fiore et al., 2015). Climate-driven changes in temperature, precipitation, and wildfire frequency exacerbate the formation and persistence of harmful pollutants, including ground-level ozone and PM2.5, which are linked to respiratory and cardiovascular morbidity and mortality worldwide (Orru et al., 2017). These pollutants, formed through complex atmospheric processes, contribute to nearly seven million untimely deaths annually due to air pollution. The bidirectional relationship between climate and air quality—where climate influences pollutant dynamics and pollutants like black carbon and ozone contribute to climate forcing—underscores the need for integrated study and management (Rawat et al., 2024).

Understanding the intricate interactions between climate variables and air pollution is essential to anticipate future health risks, inform regulatory standards, and design effective mitigation policies. This review integrates insights from atmospheric science, epidemiology, and policy analysis to synthesize the evidence on climate change influences on air quality and related health outcomes.

Climate change and air pollution are intrinsically linked global challenges that share common causes—primarily the combustion of fossil fuels and industrial processes. Both contribute significantly to premature mortality worldwide, with air pollution alone responsible for nearly seven million deaths annually. The interplay between rising global temperatures, changing meteorological conditions, and atmospheric chemistry alters pollutant formation, dispersion, and persistence. Consequently, human exposure to harmful pollutants such as ozone and PM2.5 is expected to increase in several regions (Orru et al., 2017; Kinney, 2008).

The link between air quality and climate is bidirectional. While climate influences pollutant dynamics through temperature, humidity, and wind patterns, air pollutants such as black carbon and ozone themselves contribute to climate forcing. This interconnectedness underscores the necessity of studying these two environmental crises together rather than in isolation.

Mechanisms and Climate Processes Linking Climate and Air Quality

Climate change influences air pollution through a complex interplay of meteorological variables and atmospheric processes that govern the formation, transformation, transport, and removal of pollutants such as ground-level ozone and fine particulate matter (PM2.5). These processes, driven by changes in temperature, solar radiation, precipitation, wind patterns, and atmospheric stability, amplify pollutant concentrations and exacerbate health risks (Fiore et al., 2015). This section synthesizes the key mechanisms and climate-driven processes affecting air quality, highlighting their implications for pollutant dynamics.

Temperature and Solar Radiation

Elevated temperatures and stronger solar radiation, intensified by climate change, accelerate photochemical reactions that form tropospheric ozone through interactions between nitrogen oxides (NO_x) and volatile organic compounds (VOCs) in the presence of sunlight. These conditions are particularly pronounced in urban areas during warmer months, leading to higher ozone concentrations (Fiore et al., 2015). Similarly, increased temperatures enhance the formation of secondary aerosols, such as sulfates and organic aerosols, contributing to elevated PM2.5 levels, although high heat may volatilize some organic aerosol components, reducing their particulate form (Jacob & Winner, 2009; Singh et al., 2025). Rising methane emissions, a potent greenhouse gas and ozone precursor, further elevate background ozone levels globally (Fiore et al., 2015).

Precipitation and Wind Patterns

Precipitation acts as a cleansing mechanism via wet deposition, removing soluble gases and particles like PM2.5 from the atmosphere. Climate-driven shifts in rainfall patterns alter deposition rates, affecting pollutant lifetimes and concentrations (Jacob & Winner, 2009). Wind patterns influence pollutant dispersion, with stronger winds diluting pollutants and stagnant conditions causing localized accumulation, particularly in urban environments (Fiore et al., 2015). Changes in atmospheric circulation due to climate change can thus exacerbate or mitigate air quality issues in specific regions.

Atmospheric Stability and Mixing

Climate change alters vertical atmospheric mixing and boundary layer dynamics, impacting pollutant dispersion near the surface. Stable atmospheric conditions and low mixing heights trap pollutants, leading to poor air quality episodes, while enhanced mixing can dilute concentrations (Fiore et al., 2015). Climate models suggest that increased frequency of stable conditions may worsen pollutant exposure in some regions, particularly during heatwaves (Bloomer et al., 2009).

Natural Emissions

Climate-driven events, such as droughts and wildfires, increase natural emissions of PM2.5 and its precursors, including dust and organic aerosols from biomass burning. Higher temperatures also elevate biogenic VOC emissions from vegetation, contributing to secondary pollutant formation and deteriorating air quality. For example, wildfire frequency and intensity are projected to rise, significantly increasing PM2.5 levels in affected regions (Fiore et al., 2015).

Key Pollutants and Sources

The ozone, which is formed at the ground under with photochemical reactions involves NO_x and VOCs. This phenomenon is at its peak during hot, sunny conditions, making urban and downwind rural areas particularly vulnerable (Kinney, 2008). PM2.5 arises from primary emissions (e.g., combustion, dust) and secondary formation from precursor gases like sulphur dioxide (SO_2) and NOx, with its small size enabling deep lung penetration and health impacts (Orru et al., 2017). Other pollutants, such as nitrogen dioxide (NO_2) and carbon monoxide (CO), also result from combustion and contribute to respiratory and cardiovascular effects, while climate change extends pollen seasons, exacerbating allergic diseases (D'Amato et al., 2014). Table 5.1 summarizes how key meteorological variables influence ozone and PM2.5 concentrations:

Table 5.1 Summary of meteorological variables

Meteorological variable	Effect on ozone	Effect on PM2.5	Citations
Temperature	Increases photochemical reaction rates, elevating ozone levels	Enhances secondary aerosol formation; may volatilize organic components	Fiore et al. (2015)
Solar Radiation	Drives photochemical reactions, increasing ozone	Promotes secondary PM2.5 formation	Fiore et al. (2015)
Precipitation	Reduces ozone via wet deposition	Removes PM2.5 via wet deposition	Jacob and Winner (2009)
Wind Patterns	Disperses or concentrates ozone	Disperses or concentrates PM2.5	Fiore et al. (2015)
Atmospheric Stability	Traps ozone under stable conditions	Traps PM2.5, worsening air quality	Fiore et al. (2015), Bloomer et al. (2009)

Feedback Loops Between Air Pollution and Climate

Air pollutants exhibit a bidirectional relationship with climate: they both respond to climate conditions and actively influence the climate system themselves through various feedback loops. Understanding these interactions is essential to grasping the complexity of the climate–air pollution nexus and the implications for mitigation policies.

Black Carbon

The incomplete combustion of fossil fuels, biofuels, and biomass produces particulate matter, namely, Black carbon (BC). It strongly absorbs solar and terrestrial radiation, leading to atmospheric warming. BC's ability to absorb sunlight makes it a significant short-lived climate pollutant contributing to both atmospheric heating and surface dimming, which can affect regional climate patterns (Bond et al., 2013). Its deposition on snow and ice hastens melting by decreasing surface albedo, thereby amplifying warming in sensitive Arctic and mountainous regions.

Sulphate Aerosols

Sulphate aerosols, formed predominantly from sulphur dioxide (SO_2) emissions, which are generated by power plants driven by coal and by various industrial sources, scatter sunlight and exert a net cooling effect on Earth's climate. This negative radiative forcing offsets some greenhouse gas warming, contributing to complex trade-offs in environmental and health impacts. For example, while reducing SO_2 emissions improves air quality and public health, it diminishes the cooling influence of sulphates, which may result in short-term regional warming and altered precipitation patterns, often referred to as the 'climate penalty' (Myhre et al., 2013).

Tropospheric Ozone

Tropospheric ozone serves a dual role as both a harmful pollutant and a potent greenhouse gas. It contributes to radiative forcing by absorbing terrestrial infrared radiation, accounting for a significant fraction of short-lived climate forcing agents. Increases in ozone concentrations enhance warming, which can further exacerbate ozone formation, creating a positive feedback loop. The interplay between pollutants

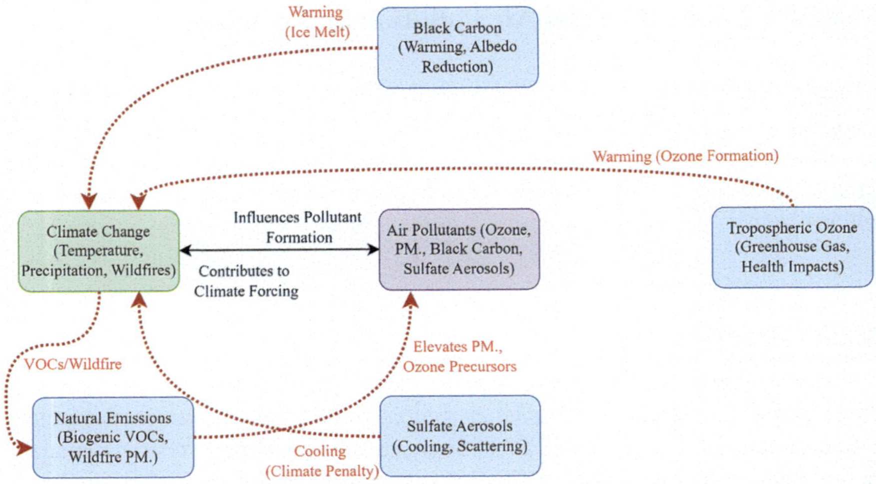

Fig. 5.1 Block Diagram of Feedback Loops Between Climate Change and Air Pollutants

and climate is characterized by multiple feedback mechanisms. For example, mitigating pollutants like black carbon and ozone precursors simultaneously benefits air quality and climate, known as a "*win–win*" scenario (Melamed et al., 2016).

However, addressing sulphate emissions requires balancing immediate health gains against potential warming in sensitive regions (Bond et al., 2013). These interactions complicate policy decisions as emission reductions have differential impacts on climate forcing agents with varying atmospheric lifetimes and radiative properties. Furthermore, changing climate conditions influence natural emissions (e.g., biogenic VOCs) and atmospheric chemistry, altering pollutant formation processes and feedback strengths (Fiore et al., 2015). These complex feedback loops, including climate-driven increases in VOCs and wildfires, are illustrated in the block diagram (Fig. 5.1), which highlights the bidirectional interactions between climate change and air pollutants. Arrows indicate bidirectional influences, with dashed lines representing key feedback mechanisms.

Health Impacts of Climate-Driven Air Pollution

Climate change significantly exacerbates the public health burden posed by air pollution by influencing pollutant concentrations, exposure patterns, and population susceptibility (Fiore et al., 2015). Ambient air pollution, which comprises harmful components like fine particulate matter (PM2.5), ground-level ozone (O_3), nitrogen dioxide (NO_2), carbon monoxide (CO), and other toxic compounds, is recognized as a leading environmental risk factor for the global disease burden (Cohen et al., 2017).

Respiratory and Cardiovascular Effects

Respiratory disorders such as bronchitis, asthma flare-ups, chronic obstructive pulmonary disease (COPD), and heightened susceptibility to respiratory infections are closely linked to PM2.5 exposure. (Fiore et al., 2015; Orru et al., 2017). PM2.5's small aerodynamic diameter allows penetration deep into the pulmonary alveoli, triggering systemic inflammation and oxidative stress that elevate risks for cardiovascular diseases such as ischemic heart disease and stroke (Cohen et al., 2017).

Ozone, a potent respiratory irritant formed via photochemical reactions intensified by climate change, contributes to airway inflammation, reduced lung function, and respiratory mortality (Fiore et al., 2015; Kinney, 2008). Epidemiological studies estimate that thousands of ozone-related premature deaths occur annually in regions such as the US and Europe, with climate-driven increases in ambient ozone concentrations projected to further elevate this toll (Fann et al., 2015; Orru et al., 2017).

Neurological and Developmental Health

Emerging evidence links long-term air pollution exposure to adverse neurological outcomes, including cognitive decline, neurodevelopmental disorders, and increased risk of neurodegenerative diseases such as Alzheimer's and Parkinson's (Wilson et al., 2017). PM2.5 exposure is highly related to respiratory diseases such as exacerbations of asthma, chronic obstructive pulmonary disease (COPD), bronchitis, and increased vulnerability to respiratory infections.

Mortality and Morbidity Burden

Globally, ambient air pollution is estimated to cause more than 4 million premature deaths each year., primarily through cardiovascular and respiratory ailments, but also involving multi-organ impacts (Cohen et al., 2017). Climate change-driven increases in pollutant concentrations and extreme events such as wildfires and heatwaves may exacerbate this burden (Liu et al., 2016).

Vulnerable Populations

Certain demographics, which are vulnerable to pollution, include small age groups, elderly individuals and persons with pre-existing health conditions, exhibit heightened vulnerability to pollution health effects. Socioeconomically disadvantaged populations additionally face disproportionate exposure due to environmental injustice and limited healthcare access (Melamed et al., 2016).

Synergistic and Indirect Effects

Air pollution also indirectly affects health by damaging crops and ecosystems, contributing to food insecurity and malnutrition, particularly in vulnerable regions (Tai et al., 2014). Psychosocial impacts of pollution and climate change add further health stressors.

Epidemiological Approaches and Health Impact Assessment

Health Impact Assessment (HIA) is a critical tool for systematically quantifying the potential health consequences of projects, policies, or programs, particularly those related to environmental exposures such as air pollution under changing climate conditions. HIAs vary in methodological complexity, ranging from rapid desktop assessments utilizing existing data to comprehensive evaluations that incorporate new data collection, advanced modelling, and extensive stakeholder engagement. The principal types of HIA are often summarized (e.g., in Table 5.2) based on their duration, data requirements, and typical applications in the context of assessing health effects from climate-driven air pollution (Cohen et al., 2017; Fann et al., 2015; Fiore et al., 2015).

Policy Integration and Co-benefits

Addressing the intertwined challenges of climate change and air pollution demands an integrated policy frameworks that leverage synergies to maximize public health benefits while mitigating environmental impacts. Policy integration refers to coordinating actions across sectors, such as energy, transportation, urban planning, and health, to simultaneously diminish greenhouse gas emissions and air pollutant levels.

Climate and Air Quality Policy Linkages

Emission sources for greenhouse gases and air pollutants often overlap, enabling policies targeting fossil fuel combustion reductions to yield both climate and health improvements. Renewable energy deployment, energy efficiency programs, and

Table 5.2 Types of Health Impact Assessment (HIA) for climate-air pollution studies

HIA type	Duration	Description	Typical data sources	Key applications	References
Rapid (Desktop)	Weeks to months	Uses existing published information or secondary datasets with limited or no new data collection. Suitable for quick screening of health impacts	Published epidemiological studies, health statistics, and existing pollutant concentration data	Preliminary screenings, project scoping, and initial policy inputs	Fiore et al. (2015)
Intermediate	Several months	Combines document review with some data collection, stakeholder engagement, scenario analysis, and basic modeling	Literature, targeted surveys, focus groups, and basic chemical transport models	Urban planning, medium-scale project assessments, policy options evaluation	Fann et al. (2015), Melamed et al. (2016)
Comprehensive	6–12 months or more	In-depth, quantitative assessments using primary data collection, advanced atmospheric and health modeling, and comprehensive uncertainty analysis	Cohort studies, population health databases, and high-resolution model outputs	Large infrastructure projects, national policy development, and regulatory impact assessments	Cohen et al. (2017), Turner et al. (2016)

transportation electrification reduce both carbon dioxide emissions and co-emitted pollutants like NOx and particulate matter (Melamed et al., 2016; Rao et al., 2013).

Targeting Short-Lived Climate Pollutants (SLCPs)

Mitigation strategies focused on short-lived climate pollutants such as methane and black carbon can yield immediate and substantial benefits for both climate and public health. Integrated approaches include transitioning to cleaner energy sources, promoting sustainable transport systems with reduced vehicular emissions, and implementing joint regulatory frameworks that address greenhouse gases and air quality management in tandem. These co-benefits strengthen the case for coordinated policymaking across environmental and health sectors, facilitating more effective climate action with enhanced societal returns (Shindell et al., 2012; UNEP, 2011).

Health Co-Benefits of Climate Mitigation

Several studies document significant health co-benefits from climate change mitigation policies. For instance, cleaner energy transitions and stricter vehicle emissions standards have been associated with reductions in respiratory and cardiovascular diseases due to lower ambient pollutant concentrations (Fann et al., 2015; Garcia-Menendez et al., 2015). Such co-benefits often provide strong economic incentives, sometimes outweighing direct climate mitigation savings.

Challenges and Trade-Offs

While the alignment of air quality and climate policies produces synergistic effects, trade-offs can occur. For example, reductions in sulphate aerosols improve air quality but reduce their cooling effect on the climate, potentially increasing regional warming temporarily (Bond et al., 2013). Policymakers must navigate such complexities, ensuring balanced strategies that minimize unintended consequences.

Policy Frameworks and Examples

The popular international frameworks, such as the Paris Agreement and the Sustainable Development Goals (SDGs), emphasize co-benefits in climate and health, encouraging multisector approaches. Regional examples include the European Union's air quality directives integrated with climate action plans and US initiatives linking greenhouse gas reduction to air quality improvements (Maione et al., 2016; Melamed et al., 2016).

Recommendations for Improved Integration

Effective integration requires cross-sector collaboration, robust data sharing, and health impact assessment incorporation in climate policymaking. Incentivizing clean technologies, expanding public transportation, and enhancing urban green spaces provide multiple benefits by improving air quality, reducing emissions, and mitigating heat island effects (Erickson & Jennings, 2017).

Research Gaps and Future Directions

Even though milestone advancements in understanding the interplay between climate change and air pollution, several critical gaps remain that must be addressed to inform effective policy and mitigation strategies. First, there is a limited understanding of the synergistic effects of multiple climate and pollution stressors, such as the combined impacts of heatwaves, wildfires, and elevated pollutant levels on human health. Furthermore, the long-term chemical evolution of PM2.5 under changing climate conditions is insufficiently characterized, hindering accurate projections of future air quality and health impacts.

Emerging pollutants such as allergenic pollen and wildfire smoke demand greater attention, given their increasing prevalence and complex interactions with traditional pollutants like ozone and particulate matter. Assessing these non-traditional pollutants requires expanding monitoring networks and epidemiological research to quantify their health impacts comprehensively. Additionally, future models should incorporate behavioural responses and socioeconomic adaptation strategies, including community resilience and technology adoption, to produce more realistic health impact projections.

A key priority for upcoming research is enhancing model resolution, especially through regional downscaling of climate and air quality models. Higher-resolution models can generate granular, actionable data for policymakers at the local level, facilitating targeted interventions. Simultaneously, there is a pressing need to integrate epidemiological data with climate projections to accurately assess the chronic health impacts of long-term exposure to climate-sensitive pollutants.

Investigating the co-benefits of mitigation efforts—such as emission reductions that simultaneously improve air quality and slow climate change—can provide compelling economic and health incentives for policy adoption. Finally, research must embed socioeconomic trends, projections, and societal vulnerability factors into modeling frameworks; a comprehensive understanding of societal capacity for adaptation and mitigation will support more equitable and effective policymaking (Smith et al., 2023; Global Commission on Adaptation, 2024).

Future research should focus on multi-stressor interactions, long-term chemical dynamics, non-traditional pollutants, model resolution improvement, and socio-economic integration. Addressing these gaps will underpin more nuanced, context-specific policies crucial for safeguarding health in a changing climate.

Conclusion

The interconnected nature of climate change and air pollution requires coordinated scientific, policy, and societal responses to address the health challenges emerging from environmental aspects. Rising global temperatures amplify the formation and persistence of harmful pollutants like ground-level ozone and PM2.5, contributing to millions of premature deaths annually and compounding health risks, especially among vulnerable populations, such as children, senior citizens, and socioeconomically disadvantaged communities. This review demonstrates that integrated mitigation policies targeting both greenhouse gases and air pollutants offer significant cobenefits for climate and public health, illustrating that environmental protection and health goals can be achieved simultaneously through coordinated action. Successful examples include renewable energy transitions, sustainable transportation systems, and regulatory frameworks addressing short-lived climate pollutants, all of which provide immediate health dividends alongside long-term climate benefits.

Despite substantial progress in understanding these linkages, critical research gaps persist in quantifying multi-stressor interactions, characterizing long-term pollutant evolution under changing climate conditions, and assessing emerging threats such as wildfire smoke and allergenic pollens. Addressing these knowledge gaps through enhanced interdisciplinary research, improved regional modeling resolution, and comprehensive health impact assessments will strengthen evidence-based policymaking.

Given projected climate scenarios and accelerating urbanization trends globally, translating this scientific evidence into actionable policy requires sustained collaboration between researchers, policymakers, and communities. Ultimately, holistic, evidence-based strategies that recognize the dual nature of these challenges are urgently needed to protect public health, achieve climate goals, and secure environmental sustainability for future generations. Only through such integrated approaches can we effectively confront the converging crises of climate alteration and air pollution in an era of unprecedented universal environmental change.

References

Bloomer, B. J., Stehr, J. W., Salathé, E. P., Jr., & Piety, C. A. (2009). Changes in ozone air pollution over the western United States due to effects of ENSO on meteorology. *Atmospheric Environment, 43*(23), 3764–3772.

Bond, T.C., Doherty, S.J., & Fahey, D.W. et al. (2013). Bounding the role of black carbon in the climate system: a scientific assessment. Journal of *Geophysical Research*: *Atmospheres, 118*(11), 5380–5552.

Cohen, A. J., Brauer, M., Burnett, R., et al. (2017). Estimates and 25-year trends of the global burden of disease attributable to ambient air pollution: An analysis of data from the Global Burden of Diseases Study 2015. *Lancet, 389*(10082), 1907–1918.

Erickson, P., & Jennings, M. (2017). Quantifying the co-benefits of climate policies. Energy Policy. 106:301–309. Pettersen K, Fleck R. 2014. Urban green infrastructure for air quality and climate adaptation. *Urban Forestry Urban Greening, 13*(4):661–670.

Fann, N., Lamson, A.D., & Anenberg, S.C. et al. (2015). Estimating the national public health burden associated with exposure to ambient PM2.5 and ozone. *Environ Health Perspect, 123*(10), 1139–1149.

Fiore, A. M., Naik, V., & Leibensperger, E. M. (2015). Global air quality and climate: A coupled perspective. *Chemical Society Reviews, 44*(12), 122–145.

Garcia-Menendez, F., Hopper, T., Tebaldi, C., et al. (2015). Impact of climate change mitigation on air quality and human health. *Environmental Science and Technology, 49*(16), 9803–9810.

Global Commission on Adaptation. (2024). Adaptation report.

Kinney, P. L. (2008). Climate change, air quality, and human health. *American Journal of Preventive Medicine, 35*(5), 459–467.

Liu, Y., Goodrick, S. L., & Heilman, W. E. (2016). Wildfire smoke and climate change: Exploring the health and environmental impacts. *Environment International, 92–93*, 114–123.

Maione, G., Russo, F., Menna, F., et al. (2016). Co-benefits of the EU air quality and climate policies in the Mediterranean urban context. *Environmental Science & Policy, 64*, 1–11.

Melamed, M. L., Noordzij, M., Criqui, M. H., et al. (2016). Air pollution as a risk factor in cardiovascular disease: Current knowledge and future directions. *Environmental Science & Policy, 62*, 59–66.

Myhre, G., Shindell, D., & Bréon, F.M. et al. (2013). Anthropogenic and natural radiative forcing. In *Climate Change 2013: The Physical Science Basis*. IPCC. Cambridge University Press.

Orru, H., Ebi, K. L., & Forsberg, B. (2017). The effects of climate change on respiratory and allergic disease. *International Journal of Environmental Research and Public Health, 14*(5), 523.

Rao, S., Klimont, Z., Smith, S. J., et al. (2013). Integrated assessments of air quality and climate policy co-benefits. *Environmental Science & Policy, 27*, 61–73.

Rawat, S., Shukla, B. K., Sihag, P., Srivastav, Y., & Singh, H. K. (2024, February). Environmental effects and sustainable hydropower development: a review of deployment research, environmental valuation and IoT sensors integration. In *AIP Conference Proceedings* (Vol. 3050, No. 1, p. 040008). AIP Publishing LLC.

Shenoy, A., Shukla, B. K., Kumar, A., Verma, A., & Bajpai, A. (2023, September). Interplay between criteria pollutants and Covid-19 pandemic. In *AIP Conference Proceedings* (Vol. 2800, No. 1, p. 020101). AIP Publishing LLC.

Shindell, D., Kuylenstierna, J. C. I., Vignati, E., et al. (2012). Simultaneously mitigating near-term climate change and improving human health and food security. *Science, 335*(6065), 183–189.

Singh, H. K., Malik, A., Karanwal, A., Dhama, A., Pandey, A. K., Verma, S., & Shukla, B. K. (2025). Strategic planning and operational analysis of PV-wind hybrid systems in emerging Asian markets: A detailed assessment of India and China. *Lecture Notes in Networks and Systems, 1236*, 117–138. https://doi.org/10.1007/978-981-96-1012-9_8

Smith, K. R., Woodward, A., Campbell-Lendrum, D., et al. (2023). Health, equity, and climate change vulnerability: Socio-economic dimensions. *Nature Clinical Practice Endocrinology & Metabolism, 13*(1), 1–9.

Tai, A.P.K., Mickley, L.J., & Jacob, D.J. (2014). Air pollution impacts on crop yields under increasing ozone exposure. Nat Clim Chang. 4:112–116. Patz JA, Campbell-Lendrum D, Holloway T, et al. 2014. Climate change and food security risks. *Lancet, 374*(9698), 1633–1634.

Turner, M. C., Jerrett, M., Pope, C. A., 3rd., et al. (2016). Long-term ozone exposure and mortality in a large prospective study. *Environmental Health Perspectives, 124*(10), 1484–1489.

UNEP. (2011). Near-term climate protection and clean air benefits. United Nations Environment Programme.

Verma, R., Parashar, B., Kulshrestha, P., & Shukla, B. K. (2025). Machine learning perspectives on solar panel efficiency: the impact of pollutants and environmental factors. In intelligent infrastructure and smart materials: sustainable technologies for a greener future (pp. 135–150). Springer Nature Switzerland. https://doi.org/10.1007/978-3-031-92421-7_8

Wilson, J., Lawson, S., Linnenluecke, M. K., et al. (2017). The effects of climate change on neurological health. *Environmental Research, 159*, 202–209.

Part II
Environmental Hazards

Chapter 6
Flood Susceptibility Assessment in Tehri Garhwal District of Uttarakhand: An Application of RS-GIS and Machine Learning Algorithms

Subhendu Jana, Nirmal Biswas, Anik Saha, Sunil Saha, and Sanjit Kumar Shil Sharma

Abstract This study used a comparative multi-model approach to investigate flood susceptibility in Uttarakhand's Tehri Garhwal district. Three different approaches were used: a conventional statistical model Logistic Regression (LR); Random Forest (RF), a machine learning methodology and a hybrid ensemble approach Bagging. In order to map geographical vulnerability, 14 flood-conditioning factors were included into these models. Using the Jenks natural breaks approach for ideal calibration, susceptibility was divided into five levels: Very High, High, Moderate, Low, and Very Low. The spatial flood susceptibility modeling of the region placed 595 mouza (particularly Devprayag and Jakhanidhar) within hazardous high-risk zones, despite the fact that RF and LR data suggest 55% of the district remains relatively safe. The Bagging approach offers a more cautious viewpoint, flagging nearly 17% of the land as highly susceptible to flood. These environmental hazards are driven by a "cascading" effect: extreme storms and melting ice create debris flows, while landslides further congest river channels to exacerbate flooding. Model validation via Receiver Operating Characteristic (ROC) curves demonstrated that the RF model achieved the highest predictive accuracy with an Area Under the Curve (AUC) of 94.48%. The Bagging and Logistic Regression models followed with AUC values of 91.16% and 85.88%, respectively, confirming the superior performance of ensemble machine learning in complex Himalayan terrains. These susceptibility maps serve as essential tools for local authorities to enhance community preparedness against increasing extreme hydrological events.

S. Jana · N. Biswas · S. Saha (✉)
Department of Geography, University of Gour Banga, Malda, West Bengal, India
e-mail: sunilgeo.88@gmail.com

A. Saha
Department of Geography, Harishchandrapur College, Malda, West Bengal, India

S. K. S. Sharma
Department of Geography, Mathabhanga College, Mathabhanga, West Bengal, India

R. Sarkar et al. (eds.), *Climate Change, Environmental Hazards and Community-Based Resilience*, Disaster Risk Reduction, https://doi.org/10.1007/978-981-95-9485-6_6

Keywords Flood susceptibility mapping · Tehri Garhwal district · GIS and remote sensing · Random Forest · Bagging and logistic regression · Himalayan flood hazard assessment

Introduction

Flooding represents a critical hydrological imbalance where discharge outpaces a channel's physical limits, inundating the active floodplain (Islam et al., 2026). According to Haghizadeh et al. (2017) these incidents usually stem from meteorological extremes or drainage obstructions, posing a high risk of developing into devastating regional crises. However, when the intensity and magnitude of floods remain within certain limits, they can provide ecological benefits by recharging groundwater, enhancing soil fertility, enriching riverine nutrients, and improving the productivity of floodplain areas (Kraus et al., 2019; Zhang et al., 2017). Floods have an extreme amount of capacity for negative impacts on individuals, farming, urban areas, mode of transportation, dental bridges ecological systems, archaeological sites, and additional assets (Huang, 2008; Hassan et al., 2023). The Riverside civilization has increasingly confronted a consistent risk of flooding due to urbanization, growing forestry loss, and artificial involvement in the last century (Hassan et al., 2023; Islam et al., 2026). The Floods are among the most current crucial challenges arising in the world from climate change and natural disasters (Rahmati et al., 2016b; Pal et al., 2024). The Himalayan areas' limited drainage basins and steep vertical slopes greatly increase the infrastructure's and residents' susceptibility to unexpected flooding (Thapliyal et al., 2024). Although the geography of Uttarakhand is characterised by mass waste and torrential floods, these hydrological phenomena frequently serve as a catastrophic catalyst, leading to severe financial instability and structural destruction (Changmai et al., 2026). Moreover, the increasing frequency of these overflows exacerbates already-existing environmental risks, particularly those resulting from localised atmospheric bursts and cryospheric changes. Hazardous flooding threatens 1/3 of the global geographical area and 82% of the citizens of 90 nations. Floods affect about 140 million people worldwide every year, more than any other natural disaster (Rahmati & Pourghasemi, 2017). Flooding threatens and destroys over 20,000 people every year around the world (Paul et al., 2019; Sarhadi et al., 2012). In Asia, flooding is one of the most hazardous risks since it is responsible for 90% of anthropogenic losses and destruction (Paul et al., 2019; Thapliyal et al., 2024). There were 1840 major flood disasters between 1900–2022; this represented 29% of all the important natural disasters. The flood catastrophes had 6,884,491 fatalities and 1,109,311,917,000 USD gross monetary damages (Jonkman et al., 2024). India belongs to the category among the country's most susceptible to flooding. The nation is inherently vulnerable to water-related disasters due to its large territory and location in the center of a monsoon-heavy climatic zone. India is at the vanguard of the world's climate sensitivity due to this interaction of weather patterns and topography (Paul et al., 2019). The geographical area of India stands at 40.0 million hectares, and

1/8 of its surface is vulnerable to flooding (Thapliyal et al., 2024). Floods jeopardies approximately thirty million individuals in the country, and greater than 1500 people pass away from them every year (Wu et al., 2020; Pal et al., 2024; Changmai et al., 2026). In Uttarakhand, flash flooding is the most pervasive and frequent of all meteorological natural hazards (Changmai et al., 2026). Uttarakhand holds the leading position in India in terms of the risk from life-threatening flash floods (Islam et al., 2026; Saha et al., 2024). Statistics from the USDMA (2003) reveal the severe impact of Uttarakhand's floods, which neutralize 75 lakh hectares of land and result in Rs. 1,805 crores in property and crop losses during an average season. The mortality rate is equally severe, averaging 1,600 victims annually. However, the 1977 flood stands out as a statistical anomaly in terms of severity, accounting for 11,316 of the lives lost in the region's recorded history (Gupta et al., 2014). According to the presented study by the Council on Energy, Environment, and Water (CEEW), catastrophic flooding and related meteorological events throughout the state are a problem in more than 85% of Uttarakhand's districts (Gupta et al., 2014). The lack of thorough information collected at hydrometric stations is a major obstacle for developing countries when evaluating flood hazards (Khosravi et al., 2016). Despite these difficulties, mapping flood susceptibility is still crucial for sustainable development and regional planning since it identifies possible hazards in watershed regions (Feng & Wang, 2011; Saha et al., 2024). Due to the intrinsic complexity of this mapping process, several complicated factors must be included (Kuriqi et al., 2020). This susceptibility is highlighted by historical data: Gwar Village, Lwarkha, Kirti Nagar, Thayud, Devprayag, Byaasi, and Pinjal were among the areas badly affected by the 2021 floods (Kansal & Singh, 2022; Chevuturi et al., 2016). Additionally, the devastation caused by these occurrences was seen in 2014 when a bridge across the Henwal River in Nagini was destroyed due to flooding. It is awkward to determine the risk of flooding assignments in developing nations because of the lack of data assembled at hydrometric stations (Khosravi et al., 2016). The map of flood susceptibility has implications for strategy, sustainable development, and floods that are possible in the vicinity of a watershed (Feng & Wang, 2011). A number of distinct and complicated factors are required by the invention of mapping flood susceptibility (Kuriqi et al., 2020). Since the accessibility of the conditioning factor of the flood is a necessity for the models for flood susceptibility assessment, numerous flood-predisposing factors are taken under consideration for modeling (Koskei et al., 2022). Advanced GISs are extensively employed in complex planning techniques, decision-making, environmental management, modeling, and holistic management (Hong et al., 2018; Yufeng & Fengxiang, 2009). In the case of susceptibility mapping, a number of statistical approaches with univariate or multivariate models are currently developed (Poussin et al., 2014; Paul et al., 2019). The evolution of flood susceptibility mapping has led to a triad of modeling strategies (Paul et al., 2019). This includes the application of hydrological software (e.g., SWAT and LISFLOOD-FP), as well as various statistical measures ranging from logistic regression to fuzzy logic (Nandi et al., 2016; Zou et al., 2013). More recently, the field has been transformed by machine learning techniques such as Random Forest and Support Vector Machines (Tehrany et al., 2015; Lee et al., 2017). However, the pursuit of more reliable data has shifted the focus toward hybrid systems. By merging different algorithmic strengths, these hybrid models outperform traditional methods (Pham et al., 2016), fulfilling the modern requirements for high-precision flood risk visualization (Kuriqi et al., 2020). To predict the Flood susceptibility assessment many researchers, suggest to utilized the Bagging model (Yariyan et al., 2020) for interpretation. In this context, the current research is going to identify flood and flooding hazard-susceptible areas on a spatial scale in Tehari Garhwal district, Uttarakhand in India using a Logistic Regression statistical model (Nandi et al., 2016) and machine learning models like Random Forest (Lee et al., 2017), the Bagging model (Yariyan et al., 2020) as the hybrid machine learning model. After constructing those models, their accuracy has been evaluated by implementing the ROC curve.

The research main object is to prepare a flood inventory for the Tehri Garhwal district of Uttarakhand, analyze the key elements that cause floods in the area, and identify the zones most susceptible to flood hazards.

Study Area

The Tehri Garhwal region is a rugged highland region of Uttarakhand that serves as a spiritual sanctuary as well as an important area for scientific flood susceptibility research. Its impressive territory covers 3,642 square kilometers and reaches altitudes ranging from 243 meters to more than 7,000 meters. Tehri Garhwal, a well-known Himalayan district in Uttarakhand that borders Uttarkashi, Chamoli, Pauri, Rudraprayag, and Dehradun, is bounded to the west and east by the rivers Yamuna and Alaknanda. This 'heartland of the hills' is defined by the sacred rivers that flow through it, especially the Bhagirathi, which bisects the district and draws thousands of pilgrims to the headwaters of the Ganges and Yamuna. Life in this district is primarily rural in spirit but urban in residence, with almost 89% of the population living in urban areas such as Chamba, New Tehri and Narendra Nagar. The district, which stretches from the lower foothills to the high-altitude Himalayan peaks, is carved out by the Bhagirathi River. A history of "forts" and spiritual purification is reflected in the term itself. The region's steep topography and vast timbers, which are vital resources for the district's ecological balance and economic income, characterize it both historically and physically. Regional topography has a significant impact on how climate change develops, frequently helping one region while destabilising another. Because people living in mountainous or foothill regions encounter specific environmental challenges, their survival strategies differ greatly. To analyse these processes, this study combines existing material (including government handbooks and research papers) with new research. These research work, flood susceptibilities are primarily designed to raise awareness among the local community on how climatic fluctuations affect their everyday economic and social life (Fig. 6.1).

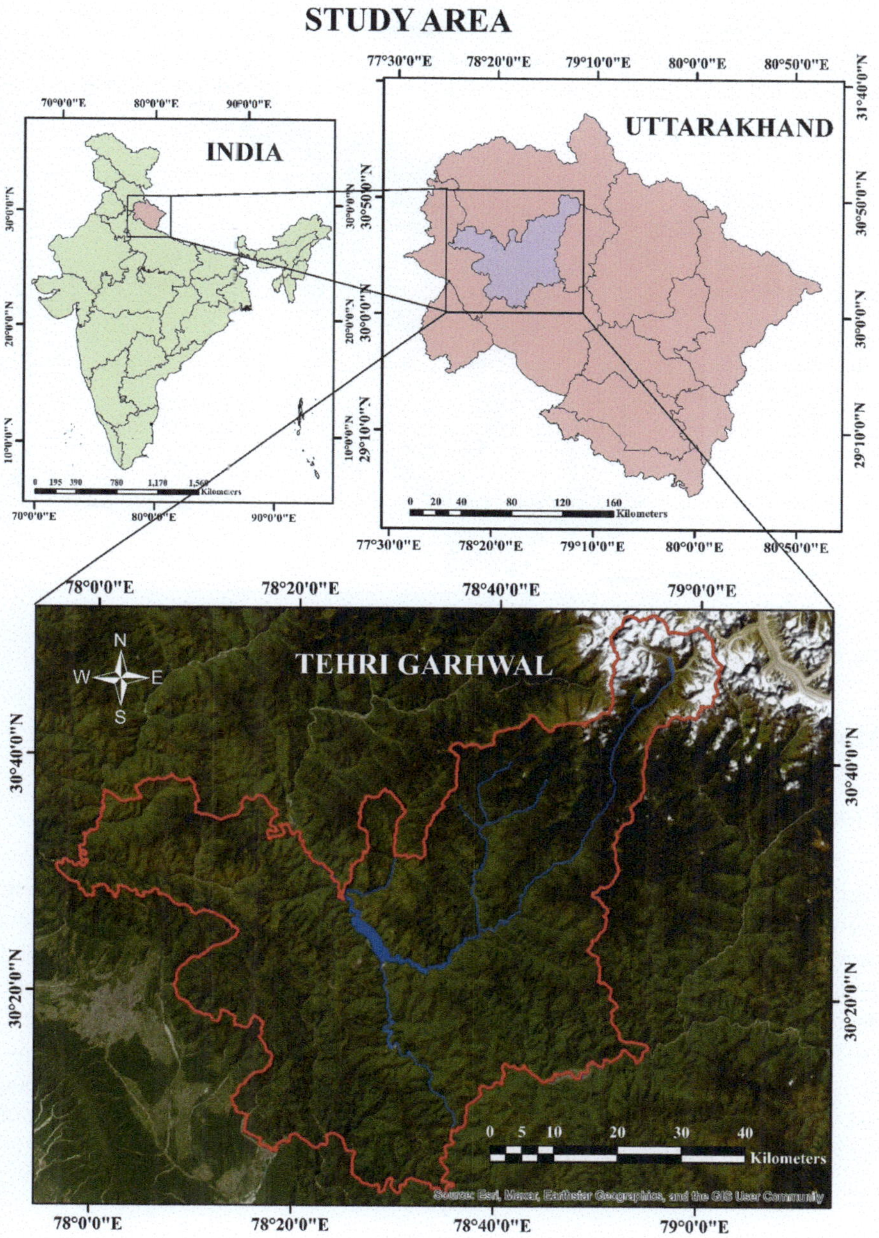

Fig. 6.1 Study area

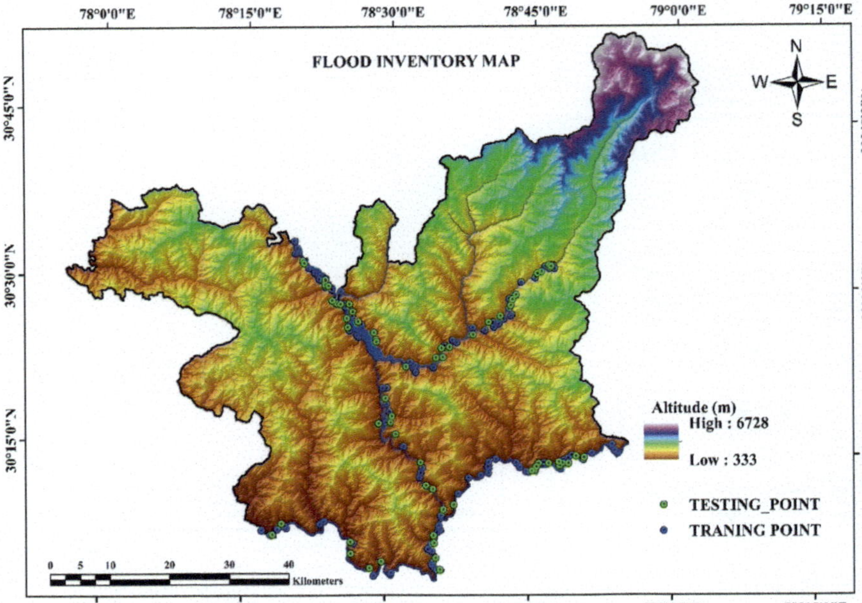

Fig. 6.2 Flood inventory map

Materials and Methods

Flood Inventory Map

Past flood events are very much important for the forecasting of Future floods that will occur and the area is likely to flood (Paul et al., 2019). FIM or floods inventory maps show the relationship of floods and their causes and are used to forecast future flood events (Rahmati et al., 2016a, 2016b). In this present study, the FIM Alternatively, the Flood Inventory Map was made using GAR Atlas Flood Map (wesr.unepgrid.ch, 2018). In all 88 flood damage sites were identified Global Positioning Systems' assistance or GPS. 70% of samples thereafter were used to produce this map of flood risk and the rest 30% of the sample locations for assessing the ability to predict of the applied suitable models (Saha et al., 2024; Paul et al., 2019) (Fig. 6.2).

Flood Conditioning Factors

Flood occurrences are usually controlled by a number of interrelated factors and intricate threshold systems. Our team discovered and examined a number of key drivers in order to precisely map vulnerability. From physical topography (altitude,

Table 6.1 The sources of the specific data utilized and the flood conditioning factors (FLCFs)

Factor for Flood Susceptibility	Article references	Data used	Sources
Altitude	Paul et al. (2019)	SRTM DEM	USGS Earth Explorer (https://earthexplorer.usgs.gov/)
Slope	Rahmati and Pourghasemi (2017)		
Aspect			
Geology	Jonkman et al. (2024)	Reference geological map	Geological Survey of India
Geomorphology	Jonkman et al. (2024)	Reference geomorphological map	Geological Survey of India (https://bhukosh.gsi.gov.in/Bhukosh/Public)
LULC	Paul et al. (2019)	LANDSAT 8 OLI IMAGE (30*30)	USGS Earth Explorer (https://earthexplorer.usgs.gov/)
Soil texture	Rahmati and Pourghasemi (2017)	Reference district soil map	National Bureau of Soil Survey and Land Use Planning
SPI	Paul et al. (2019)	SRTM DEM	USGS Earth Explorer (https://earthexplorer.usgs.gov/)
TWI	Paul et al. (2019)		
TRI	Rahmati and Pourghasemi (2017)		
NDVI	Sarhadi et al. (2012)	LANDSAT 8 OLI IMAGE (30*30)	
Distance from river	Paul et al. (2019)	Open Street Map	https://extract.bbbike.org/
TPI	Rahmati and Pourghasemi (2017)	SRTM DEM	USGS Earth Explorer (https://earthexplorer.usgs.gov/)
Rainfall intensity	Thapliyal et al. (2024)	Nasa Power Data Access Viewer	https://power.larc.nasa.gov/data-access-viewer/

slope, and TRI) to hydrological and physiological indicators (rainfall intensity, TWI, and NDVI), we incorporated twenty distinct environmental and topographical layers. These criteria, which are listed in Table 6.1, were chosen after a thorough analysis of scholarly research and the accessibility of local data.

Geomorphology

The Zone of the Central Himalayas, located The Main Central Thrust is north of (MCT), which is exposed in the northeastern part of the district, and the Zone of the Lower Himalayas, which is exposed in the rest of the region south of the Main Central Thrust, are two major physiographic divisions. The district's physiography is marked by lofty summits of mountains, narrow valleys, and gorges. In accessible terrain is common because of the severely rocky topography and thick cover of forests. The range of altitude is 369–6672 m. Major ridges in the area often follow an NNW–SSE trend, Concurrent with the country rocks' strike. But there are also ridges that trend in the NW–SE, N–S, and E–W directions that are primarily influenced by structural factors. Geomorphology has the capability to assist in providing a deterministic, supplementary way to determining flood risk or hazard for land use planning purposes. However, the greatest potential for utilizing geomorphology is most likely through the combination of these methodologies with traditional engineering and hydrological procedures. A method for examining how catchment hydrology, floodplains, and river courses react to long-term changes in the land and climate use is provided by geomorphology. In this investigation, the geomorphology map was driven from Bhukosh—Geological Survey of India.

Altitude

Altitude is a significant geographic criterion that decides the vulnerability related to floods (Ghanbarpour et al., 2013). Altitude controls the frequency of flooding. Generally, areas with lower elevations and slopes increase the runoff height and flood the area. Low-lying depressions are prone to flooding. For this study, we obtained elevation maps using 30* spatial resolution 30 m from the SRTM DEM. In study region, the altitude is ranged between 333 and 6628 m (Fig. 6.3a).

Slope

Slope is one of the critical criteria affecting flood propagation speed (Vie-gas 2004). Low-elevation flat areas are prone to flooding. These kinds of places have been found in the middle and northern portion of the district range of elevation in this area is 333–6728 m. The valley where the Bhagirathi, Bhilangana and Alaknanda rivers run has the research area's lowest elevation. Steep slopes tend to reduce water intrusion into the soil. This water quickly flows into the river as a land stream and hence flood occurred. In this work, the slope map was prepared using the SRTM DEM used to Surface Analysis in ArcGIS software (Fig. 6.3b).

Aspect

Aspect is the basic direction of tilt. Refers to a point and is measured in degrees clockwise from north. Its value ranges from 0° to 360° (Magliulo et al. 2008). The map of aspect is created from SRTM DEMs and used to ascertain several hydrological events like roots, vegetation, weathering, and evapotranspiration growth and meteorological occurrences like precipitation, insolation, and local. The area's morphological

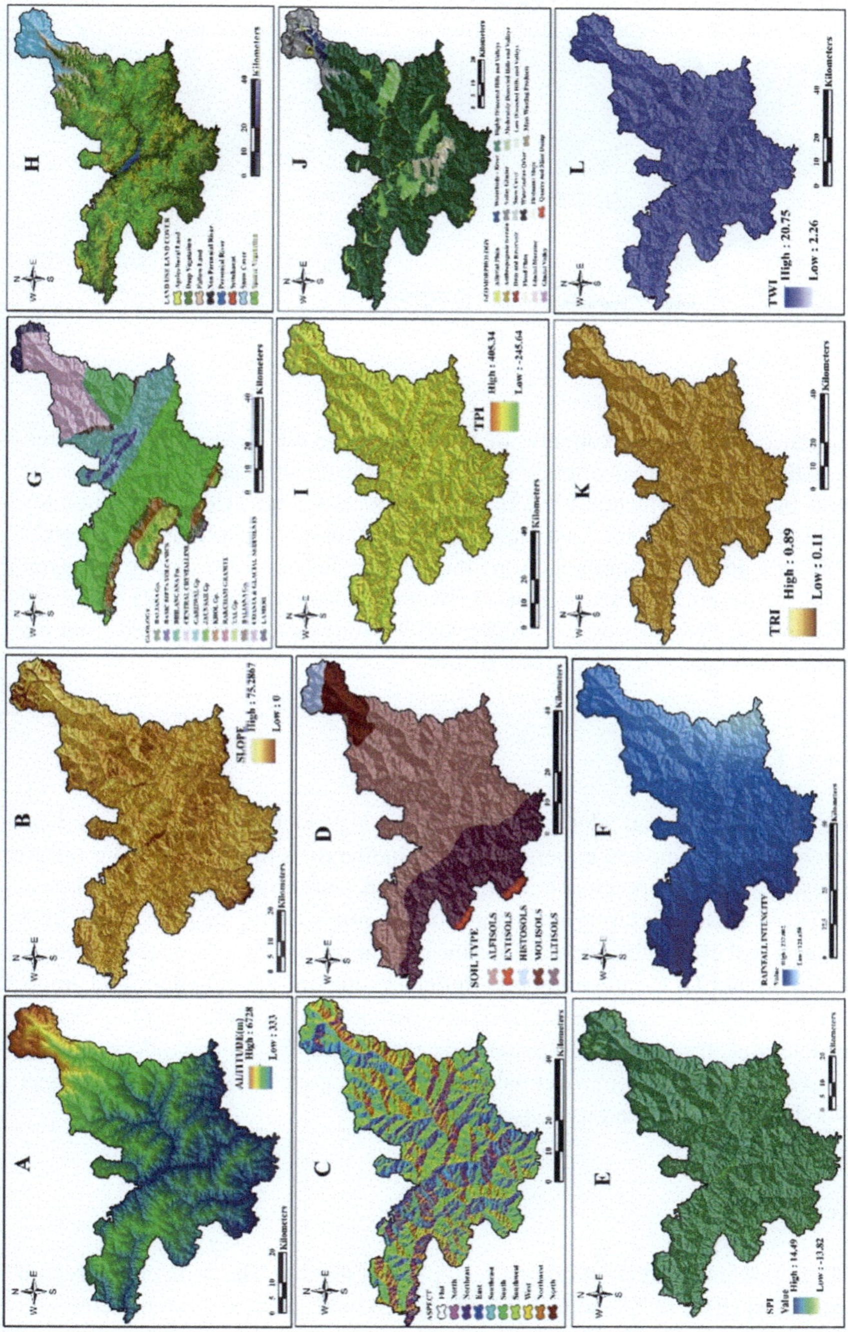

Fig. 6.3 Flood conditioning factors

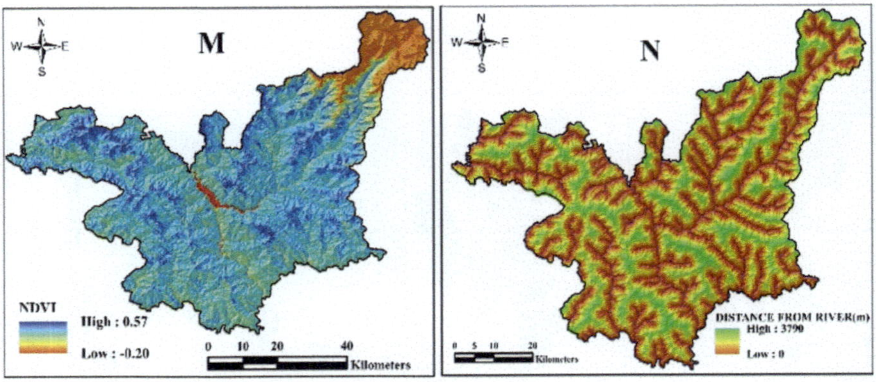

Fig. 6.3 (continued)

structure also impact the likelihood of flooding. Since the chosen northern hemisphere is the research area, whereas the southern-facing side receives highest solar radiation, but the north-facing side is remaining coolest (Cerdà et al., 1995; Courtney Mustaphi & Pisaric, 2013; Lloyd Hayes, 1941). In addition, the east side receives more UV rays and direct sunlight during the day than the west side (Beaty & Taylor, 2001). There are nine classes: 1 (Flat), 2 (North), 3 (Northeast), 4 (East), 5 (Southeast), 6 (South), 7 (Southwest), 8 (West), and 9 (Northwest) were created from the aspect map using equal intervals (Fig. 6.3c).

Land-Use Land-Cover

Among the most crucial elements in determining flood susceptibility is land cover, which either impacts either directly or indirectly a number of hydrological processes, including evapotranspiration, infiltration, and surface runoff. The purpose of LANDSAT 8 OLI was to produce the study region's LULC map for April 2023 using supervised image classification and the maximum likelihood approach imagery (Fig. 6.3h). Eight LULC categories were defined in this study e.g., Fallow land, Snow cover, Sparse vegetation, perennial river, non-perennial river, Built-up, Deep Forest, and Agricultural cropland. LULC patterns determine runoff volume produced during rainfall events and thus affect water balance within a region. Hence, LULC can affect both flood susceptibility and its consequences.

Soil Texture

Another important flood conditioning factor is soil texture. The soils of any region were produced either by the both chemical and physical characteristics minerals and rocks are known as paedogenetic processes by soil transport (rivers and sediments) (Jaiswal et al., 2002). In this research the soil map of the region was created using maps from the National Bureau of Land Use Planning and Soil Survey Uttarakhand, India, and Seven The following types were mapped for soil texture. Fine loamy, coarse loamy, loamy shallow, loamy skeletal, fine, and loamy. Some soil structures like clay and silt have low infiltration rates and create a rapid runoff of heavy rainfall. Thereupon, runoff can be faster during heavy rains (Saha et al., 2024). The Soil texture in the Tehari Gharwal district varies from 2.26 to 20.75 km.

TWI and SPI

Beven and Kirkby stated that TWI in 1979 is an a crucial factor is the frequency of floods. Researchers expressed that TWI presents the volume of water accumulating in a cell of catchment area (Beven & Kirkby, 1979). SPI and TWI values range from 2.26 to 20.75 and −13.81 to 14.48 in the study area respectively (Fig. 6.3l). The greater the TWI value, the greater the likelihood that water will accumulate as it shows the wet conditions in the region. SPI has a profound impact on river systems, demonstrating the erosive rushing water's power (Tehrany et al., 2015). The equation presents calculations for TWI:

$$TWI = \ln \frac{As}{\tan(slope)} \tag{6.1}$$

where 'As' is slope in degrees and the precise catchment area in meters. "SPI" was computed through the use of Eq. (6.2)

$$\text{SPI} = \text{As} \times \tan(\text{slope}) \tag{6.2}$$

TRI (Topographic Roughness Index)

TRI contributes significantly to making the region prone to flooding (Kalantari et al., 2017; Weiss, 2001). SRTM DEM was put into action to create TRI using Eq. (6.2).

$$\text{TRI} = \sqrt{Crb\left(maximum^2 - minimum^2\right)} \tag{6.3}$$

where The lowest and maximum are the lowest and highest elevations in the neighbouring pixels, and Crb represents the min and max values. In Tehri Gharwal district the TRI differs from 0.1 to 0.8. It displays the difference between the heights of adjacent pixels of the DEM grid (Fig. 6.3k). TRI was selected as an eventful parameter since it depicts the topographic features as well as the both concavity and convexity of the terrain (Kalantari et al., 2017; Riley et al., 1999).

NDVI

NDVI is one of the essential parameters for flood probability because of its capacity to regulate surface runoff and infiltration (García-Ruiz et al., 2008; Shahabi et al., 2020). While the NDVI value in this research varied from −0.2 to 0.6, the index value's range is −1 to +1. According to Khosravi et al. (2016), floods are negatively correlated with NDVI because negative values indicate water and positive ones imply vegetation. Low NDVI and high flood probability are indicated by higher NDVI and flooding likelihood, respectively. The NDVI calculated using the February 3rd LANDSAT 8 OLI images, 2023 symbolizes the research area's vegetation cover (Fig. 6.3m). The equation for computing NDVI is Eq. (6.4).

$$\mathrm{NDVI} = \frac{NIR - RED}{NIR + RED} \tag{6.4}$$

where, NIR = near infrared band and RED = the red colour spectrum's band.

Distance from River

Flooding is common along the rivers. Therefore, distance to rivers is also regarded as one of the geomorphological conditioning factors (Ghanbarpour et al., 2013). A distance map to the river map was then created, as rivers undermining levees and saturating some of the material in the water table of the river affect levee stability. Distance to A river is portrayed by the adjacency of the streams and drainages within a region. In the present study, I used Euclidean distance in software ArcGIS to create a distance map from vector maps of rivers (Hossein Mojaddadi et al., 2017) (Fig. 6.3n). Distance from river ranges 0–3.7 km in the study area respectively.

TPI (Topographic Positioning Index)

TPI is a very simple, repeatable, and powerful method for terrain classification into slope and surface parts (Pande 2010). It compares The cell's height against The mean height of a cell with a given adjacent size. The calculated values of TPI factor define and identify different relief terrains such as valleys, slopes, and undulating ridges. Here, SRTM DEM was used to create TPI using Eqs. (6.5) and (6.6).

$$\mathrm{TPI} = \mathrm{M}_0 - \mathrm{M} \tag{6.5}$$

$$\mathrm{M} = \frac{1}{ar}\sum_i ErZ_i \tag{6.6}$$

where small r represents the tiny peaks and valleys, an is the total number surrounding the points, M_0 is The central point's height, and M is the average height within a specified radius (R) used for determine (Weiss, 2001) The TPI in Tehari Gharwal district varies from −245.64 to 405.34 km (Fig. 6.3i).

Rainfall Intensity

Intensity of precipitation is one of the contributory flood factors vulnerability. Precipitation intensity depends on topographical features and prevailing weather conditions because of the extensive time and space fluctuations of precipitation events falling within this region. To calculate the rainfall intensity for the basin, rainfall data for the last 5 years were gathered from NASA Power's Data Viewer Access. Utilizing the Modified Fournier Index, it was determined (MFI) (Saha et al., 2024) (Eq. 6.7).

$$\text{MFI} = \frac{\sum_{i=1}^{12} qi}{Q} \tag{6.7}$$

where Q represents the average yearly precipitation and qi is the i-month mean precipitation. In in this field of research, the values of RI varied from 121 to 237 mm respectively (Fig. 6.3).

Factors that contribute to flooding: (A) Elevation, (B) Slope, (C) Aspect, (D) Soil type, (E) Stream power index, (F) Rainfall intensity, (G) Geology, (H) Land use/land cover, (I) Topographical positioning index, (J) Geomorphology, (K) Topographical ruggedness index, and (L)Topographical wetness index, (M) Normalized difference vegetation index, and (N) Distance from the river.

Methods

Random Forest Model

Breiman (1996) formulated the machine-learning algorithm known as a random forest for non-parametric multivariate classification. RF methods have been used remote sensing picture categorization (Ham et al., 2005; Pal, 2024) and psychological research (Strobl et al., 2009). As it has been shown in the literature, RF model has been advocated as a powerful tool regarding regression and classification problems (Pourghasemi et al., 2017; Youssef et al., 2015). The RF model designed as an ensemble of one bootstrap consolidation (Micheletti et al., 2014). This represents the bootstrap or modified bagging combining technique. In this algorithm, random binary trees are used that incorporate bootstrapping techniques to use a subset of the observations. The algorithm samples a random sample of the training information derived from the first dataset and uses these data to build the model; the remaining data are known as "out-of-bag" (OOB) data (Breiman, 1996). Integrative training information derived from the first dataset and huge amount 'forest' by the quantity of trees, and they include generated using a bootstrapped sample and the classification regression tree framework Given a subset of variables chosen at random selected at each node. Decisions regarding model building and class membership are formed by considering the results of all decision trees being chosen at the end (Strobl et al., 2009). Regarding running the RF model in the present research, it was done using the 'Random Forest' package of 'R Studio' along with ArcGIS 10.3.1.

Bagging

The Bagging method was first introduced by Breiman (1996). This machine learning benchmark method makes use of bootstrap sampling method (Breiman, 1996). Model used by several researchers for the evaluation of vulnerability (e.g., flood, landslide, Forest fire, etc.) because there is sufficient prediction in this model capability (Hong et al., 2018). In this research, this approach was chosen because it uses each generated subset to build the decision tree before putting them all together to form final model (Hong et al., 2018). Minimizing the inconsistent behaviour of alignment error improves the alignment accuracy (Wu et al., 2020). According to Breiman (1996) and Yariyan et al. (2020), a bagging classifier is regarded as a 3-step bagging method. The relocation strategy was adopted to generate the bootstrap sample of the substantial training samples. Integrating them into the genuine feature process for the bagging classification stage can support the accuracy of each array of subsets for these machine learning algorithms (Breiman, 1996; Yariyan et al., 2020).

Logistic Regression (LR)

In logistic regression, a problem is studied in which the solution has been predicted depending on a binary response of the dependent factor, especially if a landslide occurrence that one or a group had anticipated occurred in this study of causal factors, with true or false and 0 or 1 as the possible outcomes (Akgun, 2012). The correlative performance Relating a dependent variable (Y) to a number (n) of independent variables (V1, V2, Vn) is utilised to forecast a logit modification of probability depending on landslide existence or absence (). Although the susceptibility is not explicitly defined by the LR model, the probability values that are produced from it may be used to infer it. The Eq. (6.8) was used in the model's fitting.

$$Y = Logit(p) = ln\left\{\frac{p}{1-p}\right\} = C0 + C1V1 + C2V2 + \cdots + CnVn, \quad (6.8)$$

where p is the dependent variable (the chance of landslide event Y); p/(1 − p) is the likelihood ratio; C0 is intercepted; and the coefficients used to determine the independent variables V1, V2, and Vn's contributions are C1, C2,..., and Cn.

Validation Technique

Receiver Operating Characteristics Curve

ROC curves and precision probabilistic models have been used by several researchers (Yariyan et al., 2020). The prediction power of the model with diagnostic tests is shown graphically here (Singha et al., 2020). ROC shows the specificity-location of non-landslides and the sensitivity-location of landslides, forest fires, and flood occurrences., non-forest fires, non-flood events on Y and X axes, respectively. Prediction rate curve and success rate curve should be calculated by applying datasets for training and validation, respectively. AUC computed the Model's predictions for landslides, forest fires, flood events, and non-events. Value AUC varies from 0.5 to 1. Value will be close to 0.5, indicating poor model performance (Singha et al., 2020). Further, the higher accuracy values enable the model to predict the landslide, forest fires and flood vulnerability better (Chapi et al., 2017). Sensitivity, specificity, and AUC were computed. using Eqs. (6.9–6.12).

$$\text{Sensitivity} = \text{AR}/(\text{AR} + \text{EB}) \tag{6.9}$$

$$\text{Specificity} = \text{AB}/(\text{PF} + \text{ED}) \tag{6.10}$$

$$\text{Precision} = \text{AR}/(\text{AR} + \text{PF}) \tag{6.11}$$

$$\text{AUC} = \left(\sum \text{AR} + \sum \text{ED}\right)/(\text{L} + \text{Q}) \tag{6.12}$$

where, AB and AR are true negatives and true positives. PF and EB represent false positives and false negatives, L is the number of flood or landslide or forest fire event pixels and Q is the total number of non-flood or non-landslides, or non-forest fires event pixels.

Results and Discussion

In the first phase of the investigation, FSMs were generated with the single-model-based assessment approach. Thus, using the random forest as a machine learning method, Bagging as a hybrid machine learning method, and logistic regression models as a statistical model alone, three different FSMs for the Tehri Garhwal district were generated. All the models were implemented in RStudio using the R programming language and SPSS version 17, where the susceptibility in each model was divided into five groups: Very High (VH), High (H), Moderate (M), Low (L), and Very Low (VL), classified based on the classification technique for Jenks's natural

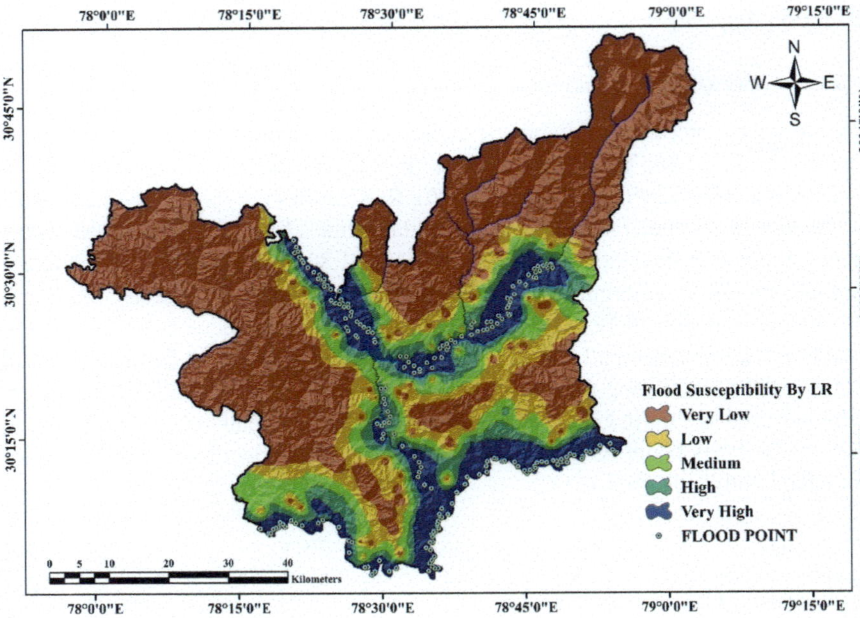

Fig. 6.4 Flood susceptibility map by RF

breaks. Four different strategies, namely the geometric interval, natural breaks, equal-interval, and quantile methods, were analyzed for the categorization of FSMs into the Very High, High, Moderate, Low, and Very Low (VH–VL) categories. Following the findings of Rahmati and Pourghasemi 2017 and Pal et al. (2024), the Jenks natural breaks method stands out as the superior calibration tool for flood distribution. Its effectiveness stems from a unique ability to maximize differences between categories while maintaining high internal consistency and eliminating subjective bias. The values of the FSMs produced through all models vary between 0.0 and 1.0.

The prime causes of flash floods in mountainous areas include heavy rains, melting glaciers, sudden landslides, and avalanches. The amount of rainfall, its location and distribution, landscape, form of vegetation and density, soil type, and soil water content all affect where and how soon flash flooding may occur. The flash flood mainly occurs in most of regions within the downstream part of the Alaknanda River and Bhagirathi River in the Tehri Garhwal district due to short, high-intensity rainstorms, and glacier breaks. This cascading effect collaborates with the slaughterous flooding in the terrain instability of the district. During cloudburst or heavy rainfall, the occurrence of landslide and boulder displacement makes the path of the river narrow and amplifies the river's flow pattern. Again, during the rainy season, the water level of Tehri dam rose menacingly, and the release of waters from its massive reservoir to avoid breaches threatened to flood nearly villages downstream along the Bhagirathi River (Table 6.2).

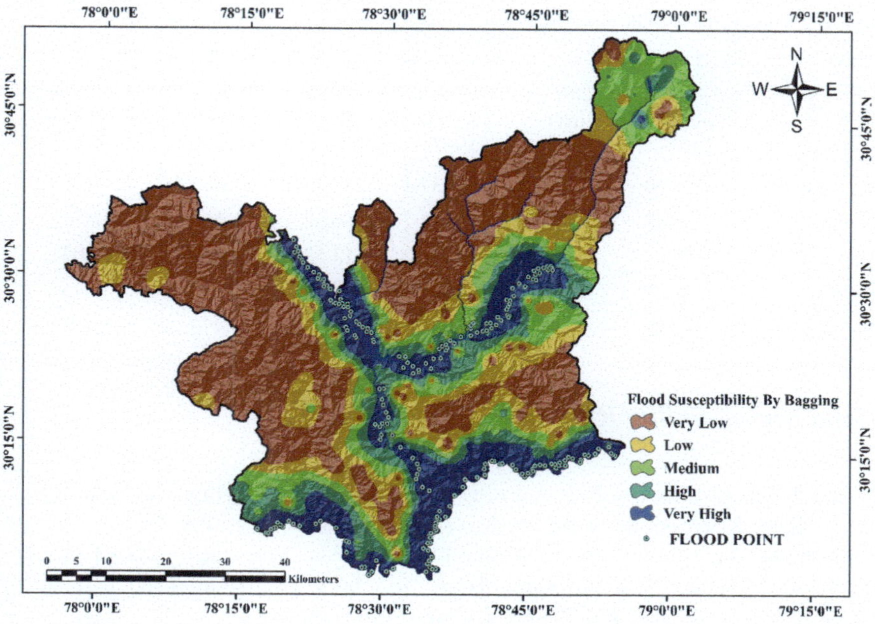

Fig. 6.5 Flood susceptibility map by bagging

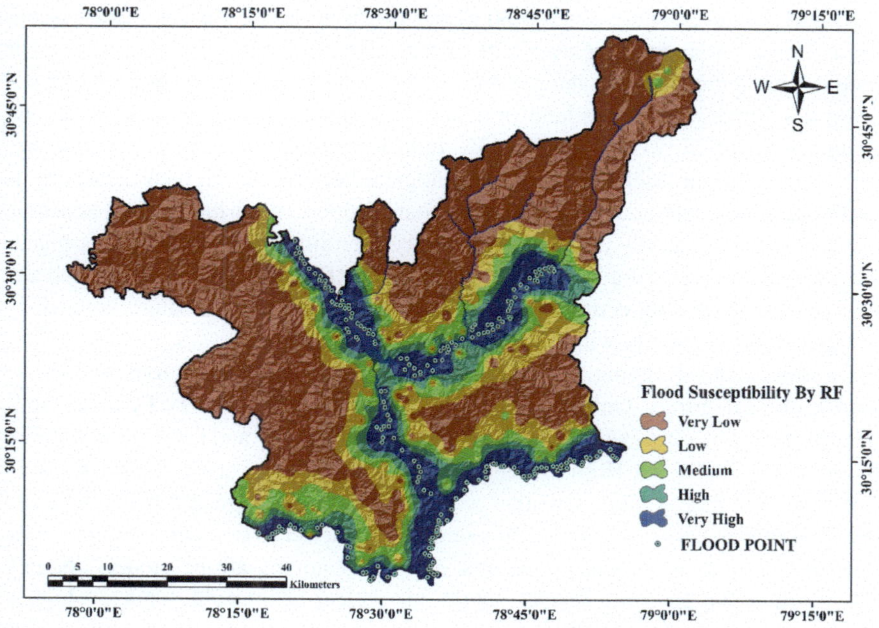

Fig. 6.6 Flood susceptibility by LR

Table 6.2 Flood susceptibility results

Flood susceptibility zone	RF		Bagging		LR	
	Effected area (Sq. Km)	Effected area (%)	Effected area (Sq. Km)	Effected area (%)	Effected area (Sq. Km)	Effected area (%)
Very low	2109.60	55.40	1723.64	45.27	2104.28	55.26
Low	555.88	14.60	610.97	16.05	514.18	13.50
Medium	359.07	9.43	514.70	13.52	381.67	10.02
High	301.00	7.90	401.03	10.53	333.61	8.76
Very high	482.22	12.66	557.44	14.64	473.91	12.45

Flood Susceptibility Results

Spatial Three distinct machine learning-based models—Bagging, Random Forest, and Logistic Regression—were used to study vulnerability to flooding in order to comprehend the spatial pattern of flood-prone areas. Five categories of susceptibility—Very Low, Low, Medium, High, and Very High—are used to classify output (Fig. 6.4, 6.5 and 6.6). The predictive skill is shown in the class distribution area and sensitivity of each model toward the flood-conditioning variables under study.

The RF model categorizes 2109.60 sq. km (55.40%) under the Very Low susceptibility class, indicating that over half of the landscape shows low propensity for flooding due to favorable physiographic characteristics such as gentle slopes, better infiltration zones, and stable landforms. Low susceptibility stands at 14.60%, and Medium susceptibility is as high as 9.43%, indicating transitional susceptibility zones with moderate hydrological or geomorphological limitations. High susceptibility is 7.90%, and Very High susceptibility stands at 12.66%, capturing pockets sensitive to flooding that generally correspond with steep terrain, fragile lithology, or places lying adjacent to major drainage channels. Overall, RF hence displays a balanced and reliable distribution, reflecting its capability to identify nonlinear interrelationships and spatial heterogeneity within the dataset.

Eventually, the Bagging model yields relatively higher proportions within the flood-prone categories. Only 45.27% is categorized as having Very Low susceptibility in the region class, which is substantially lower than RF and LR. Low (16.05%), Medium (13.52%), and High (10.53%) classes characterize a marked increase to show that Bagging is more sensitive to variation in landscape and hence tends to identify broader zones as potentially vulnerable. The Very High susceptibility zone represents the largest portion of the area of 14.64% among all the models, thereby showing that Bagging emphasizes subtle but critical geomorphic or hydrological weaknesses. Thus, this model gives quite a risk-oriented output, useful for regions where mitigation and preparedness strategies need to be worked out on a priority basis. The Logistic Regression model provides a susceptibility pattern very close to Random Forest. The Very Low susceptibility category covers about 55.26% of the region, which further shows that LR outlines similar low-risk topography conditions. Classes with Low (13.50%), Medium (10.02%), and High (8.76%) susceptibility depict a rather uniform increase of gradient. The Very High susceptibility class is 12.45%, somewhat lower than the value obtained by RF, but much lower compared with Bagging, confirming that in defining the most critical zones, LR gave a more conservative classification. This characterizes LR, where the whole model depends on linear relationships and gives smoother and much more stable susceptibility patterns. A comparative evaluation of the three models reveals evident methodological differences. RF and LR both classify a much larger proportion of the landscape as low-susceptible, thereby depicting stable topographic and hydrological settings over much of the region. In contrast, Bagging allocates more area to higher susceptibility classes, perhaps indicating that this algorithm detects more geomorphic and hydrological features that could potentially contribute to flood risk. The Very High category of susceptibility is the most widespread in Bagging (14.64%), followed by RF (12.66%) and then LR (12.45%), which reflects its increased sensitivity for identifying more areas at high risk. This variation emphasizes the importance of ensemble modeling for flood-related research since different algorithms capture various characteristics of landscape behavior and environmental constraints.

In all, the three models demonstrate the spatial heterogeneity of flood susceptibility in the study area. Both Random Forest and Logistic Regression emphasize large areas with low risk, but the model from the Bagging ensemble outlines a wider distribution of vulnerable terrain. These combined insights strengthen the reliability of the susceptibility assessment and provide a robust foundation for flood hazard management, zoning regulations, early-warning planning, and sustainable land-use decision-making in the region.

Model Validation

The analysis revealed that model accuracy was checked by ROC curve, through the use Among the datasets used for training and validation for three hazards, i.e., landslide, forest fire, and flood. The AUC was generated utilizing the datasets for training and validation as well as is represented in Fig. 6.7. The highest AUC value shows that he models shows reliable predictive strength. The success curve indicates If the accuracy of the RF model is superior to that of the other models, applied in flood hazard model validation. The value of RF (AUC) is 0.944 (94.48%) followed by Bagging (AUC) 0.911 (91.16%), and LR (AUC) 0.858 (85.88%) respectively which shows high-performance level of this models.

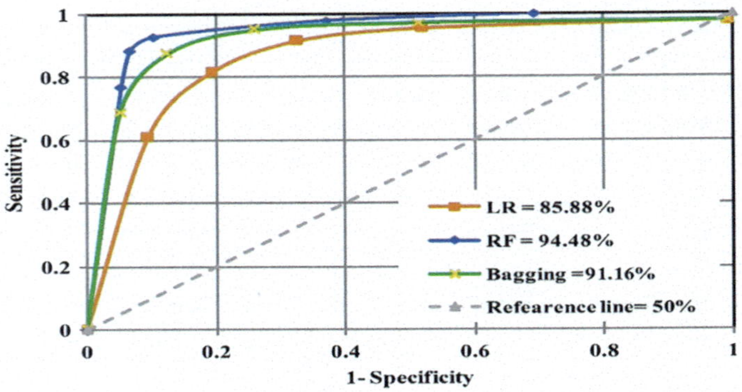

Fig. 6.7 ROC curve (success rate curve utilizing training dataset) validation of flood susceptibility maps

Conclusion

In this study, Random Forest and Logistic Regression, and Bagging models present a spatially explicit and in Descriptive assessment on flood susceptibility and its nature in Uttarakhand's Tehri Garhwal district. The work illustrates and highlights the regions of varying flood risk in the district through mapping flood inventory data with an extensive set of fourteen flood-conditioning factors. Accordingly, although more than half of the district falls within a very low susceptibility category, a substantial portion falls under the zone of high to very dangerous areas in the research region-including nearly 595 villages that show considerable exposure to flood hazards in case of extreme rainfall events.

Thus, among the applied models, Random Forest had The maximum predictive Reliability an AUC of 94.48%, and Bagging followed by 91.16% and Logistic Regression with 85.88%. This clearly indicates that ensemble machine-learning techniques are more efficient than others in representing complex environmental relationships and spatial variability in the Himalayan terrain. Though methodologically different, the models consistently identified the critical flood-prone pockets, especially in the Devprayag and Jakhanidhar blocks where steep slopes combine with fragile lithology and high-intensity precipitation to enhance flood sensitivity.

The overall study provided a valid scientific Guiding structure for understanding the distribution related to flood risk in the district. These susceptibility maps are expected to function as an effective instrument for disaster management, local government, and planners, agencies by providing focused attention to mitigation efforts, proper land-use regulation, early warning systems, and community preparedness. This type of assessment is urgently needed to guide sustainable development and protect settlement areas in the face of recent increases in extreme hydrological events in the Tehri Garhwal region.

References

Akgun, A. (2012). A comparison of landslide susceptibility maps produced by logistic regression, multi-criteria decision, and likelihood ratio methods: A case study at İzmir, Turkey. *Landslides, 9*(1), 93–106.

Ay, N., & Amari, S. I. (2015). A novel approach to canonical divergences within information geometry. *Entropy, 17*(12), 8111–8129.

Beaty, R. M., & Taylor, A. H. (2001). Spatial and temporal variation of fire regimes in a mixed conifer forest landscape, Southern Cascades, California, USA. *Journal of Biogeography, 28*(8), 955–966.

Beven, K. J., & Kirkby, M. J. (1979). A physically based, variable contributing area model of basin hydrology/Un modèle à base physique de zone d'appel variable de l'hydrologie du bassin versant. *Hydrological Sciences Journal, 24*(1), 43–69.

Breiman, L. (1996). Bagging predictors. *Machine Learning, 24*(2), 123–140.

Census of India. (2011). https://censusindia.gov.in

Cerdà, A., Imeson, A. C., & Calvo, A. (1995). Fire and aspect induced differences on the erodibility and hydrology of soils at La Costera, Valencia, southeast Spain. *CATENA, 24*(4), 289–304.

Changmai, J., Patidar, A. K., & Kothyari, G. C. et al. (2026). Geohazard dynamics in Kumaon and Garhwal regions of Uttarakhand Himalayas: A critical review of trends and socio-environmental impacts. *Spatial Information Research, 34*, 11. https://doi.org/10.1007/s41324-025-00662-9

Chapi, K., Singh, V. P., Shirzadi, A., Shahabi, H., Bui, D. T., Pham, B. T., & Khosravi, K. (2017). A novel hybrid artificial intelligence approach for flood susceptibility assessment. *Environmental Modelling & Software, 95*, 229–245.

Chevuturi, A., & Dimri, A. P. (2016). Investigation of Uttarakhand (India) disaster-2013 using weather research and forecasting model. *Natural Hazards, 82*(3), 1703–1726.

Courtney Mustaphi, C. J., & Pisaric, M. F. (2013). Varying influence of climate and aspect as controls of montane forest fire regimes during the late Holocene, south-eastern British Columbia, Canada. *Journal of Biogeography, 40*(10), 1983–1996.

Feng, C. C., & Wang, Y. C. (2011). GIScience research challenges for emergency management in Southeast Asia. *Natural Hazards, 59*(1), 597–616.

García-Ruiz, J. M., Regüés, D., Alvera, B., Lana-Renault, N., Serrano-Muela, P., Nadal-Romero, E., Navas, A., Latron, J., Martí-Bono, C., &Arnáez, J. (2008). Flood generation and sediment transport in experimental catchments affected by land use changes in the central Pyrenees. *Journal of Hydrology, 356*(1–2), 245–260.

Ghanbarpour, M. R., Salimi, S., & Hipel, K. W. (2013). A comparative evaluation of flood mitigation alternatives using GIS-based river hydraulics modelling and multicriteria decision analysis. *Journal of Flood Risk Management, 6*(4), 319–331.

Gupta, A. K., Nair, S. S., Wajih, S. A., Chopde, S., Gupta, G., & Aggrawal, G. (2014). *Mainstreaming climate change adaptation and disaster risk reduction into district level development plans.* NIDM New Delhi (India), GEAG Gorakhpur (UP, India) and ISET, Colorado (US), 114 p.

Haghizadeh, A., Siahkamari, S., Haghiabi, A. H., & Rahmati, O. (2017). Forecasting flood-prone areas using Shannon's entropy model. *Journal of Earth System Science, 126*(3), 39.

Ham, J., Chen, Y., Crawford, M. M., & Ghosh, J. (2005). Investigation of the random forest framework for classification of hyperspectral data. *IEEE Transactions on Geoscience and Remote Sensing, 43*(3), 492–501.

Hassan, Z. A., Shahrin, F., Ogunna, C., & Foden, G. (2023). Preserving heritage, historical artefacts and culture for social resilience and tourism amid flood disasters: A case study of Morten Village of Sungai Melaka, Malaysia. In *Handbook of flood risk management and community action* (pp. 36–50). Routledge.

Hong, H., Panahi, M., Shirzadi, A., Ma, T., Liu, J., Zhu, A. X., Chen, W., Kougias, I., & Kazakis, N. (2018). Flood susceptibility assessment in Hengfeng area coupling adaptive neuro-fuzzy inference system with genetic algorithm and differential evolution. *Science of the total Environment, 621*, 1124–1141.

Horritt, M. S., & Bates, P. D. (2002). Evaluation of 1D and 2D numerical models for predicting river flood inundation. *Journal of Hydrology, 268*(1–4), 87–99.

Hossein Mojaddadi, et al. (2017). https://www.researchgate.net/publication/340172084

Huang, X., Tan, H., Zhou, J., Yang, T., Benjamin, A., Wen, S. W., Li, S., Liu, A., Li, X., Fen, S., & Li, X. (2008). Flood hazard in Hunan province of China: An economic loss analysis. *Natural Hazards, 47*(1), 65–73. https://doi.org/10.1007/s11069-007-9197-z

Islam, M. S., Akter, F., Hossain, M. T., & Hossain, M. R. (2026). Decadal morphological change analysis of the Padma river and local perceptions at Char-Janajat using NDWI and remote sensing. *Discover Geoscience, 4*(1), 83. https://doi.org/10.1007/s44288-026-00461-1

Jaiswal, R. K., Mukherjee, S., Raju, K. D., & Saxena, R. (2002). Forest fire risk zone mapping from satellite imagery and GIS. *International Journal of Applied Earth Observation and Geoinformation, 4*(1), 1–10.

Jonkman, S. N., Curran, A., & Bouwer, L. M. (2024). Floods have become less deadly: An analysis of global flood fatalities 1975–2022. *Nature Hazards, 120*, 6327–6342. https://doi.org/10.1007/s11069-024-06444-0

Kalantari, Z., Ferreira, C. S. S., Walsh, R. P. D., Ferreira, A. J. D., & Destouni, G. (2017). Urbanization development under climate change: Hydrological responses in a peri-urban Mediterranean catchment. *Land Degradation & Development, 28*(7), 2207–2221.

Kazakis, N., Kougias, I., & Patsialis, T. (2015). Assessment of flood hazard areas at a regional scale using an index-based approach and Analytical Hierarchy Process: Application in Rhodope-Evros region, Greece. *Science of the Total Environment, 538*, 555–563.

Khosravi, K., Nohani, E., Maroufinia, E., & Pourghasemi, H. R. (2016). A GIS-based flood susceptibility assessment and its mapping in Iran: A comparison between frequency ratio and weights-of-evidence bivariate statistical models with multi-criteria decision-making technique. *Natural Hazards, 83*(2), 947–987.

Kansal, M. L., & Singh, S. (2022). Flood management issues in hilly regions of Uttarakhand (India) under changing climatic conditions. *Water, 14*(12), 1879.

Kia, M. B., Pirasteh, S., Pradhan, B., Mahmud, A. R., Sulaiman, W. N. A., & Moradi, A. (2012). An artificial neural network model for flood simulation using GIS: Johor River Basin, Malaysia. *Environmental Earth Sciences, 67*(1), 251–264.

Koskei, A., Watanabe, T., Makanji, D. L., & Eshiamwata, G. W. (2022). Flood vulnerability and risk mapping in Tana River River County, Kenya. Using multi-criteria decision analysis and GIS. *Journal of the Kenya National Commission for UNESCO*.

Kowalzig, J. (2008). *Climate, Poverty, and Justice: What the Poznan UN climate conference needs to deliver for a fair and effective global deal*. Oxfam.

Kraus, C. N., Bonnet, M. P., de Souza Nogueira, I., Morais Pereira Souza Lobo, M. T., da Motta Marques, D., Garnier, J., & Cardoso Galli Vieira, L. (2019). Unraveling flooding dynamics and nutrients' controls upon phytoplankton functional dynamics in Amazonian floodplain lakes. *Water, 11*(1), 154.

Kuriqi, A., Koçileri, G., & Ardiçlioğlu, M. (2020). Potential of Meyer-Peter and Müller approach for estimation of bed-load sediment transport under different hydraulic regimes. *Modeling Earth Systems and Environment, 6*(1), 129–137.

Lee, S., Kim, J. C., Jung, H. S., Lee, M. J., & Lee, S. (2017). Spatial prediction of flood susceptibility using random-forest and boosted-tree models in Seoul metropolitan city, Korea. *Geomatics, Natural Hazards and Risk, 8*(2), 1185–1203.

Liu, K., Li, Z., Yao, C., Chen, J., Zhang, K., & Saifullah, M. (2016). Coupling the k-nearest neighbor procedure with the Kalman filter for real-time updating of the hydraulic model in flood forecasting. *International Journal of Sediment Research, 31*(2), 149–158.

Lloyd Hayes, G. (1941). Influence of altitude and aspect on daily variations in factors of forest-fire danger.

Magliulo, P., Di Lisio, A., Russo, F., & Zelano, A. (2008). Geomorphology and landslide susceptibility assessment using GIS and bivariate statistics: A case study in southern Italy. *Natural Hazards, 47*(3), 411–435.

Micheletti, N., Foresti, L., Robert, S., Leuenberger, M., Pedrazzini, A., Jaboyedoff, M., & Kanevski, M. (2014). Machine learning feature selection methods for landslide susceptibility mapping. *Mathematical Geosciences, 46*(1), 33–57.

Nandi, A., Mandal, A., Wilson, M., & Smith, D. (2016). Flood hazard mapping in Jamaica using principal component analysis and logistic regression. *Environmental Earth Sciences, 75*(6), 465.

Negese, A., Worku, D., Shitaye, A., & Getnet, H. (2022). Potential flood-prone area identification and mapping using GIS-based multi-criteria decision-making and analytical hierarchy process in Dega Damot district, northwestern Ethiopia. *Applied Water Science, 12*(12), 255. https://doi.org/10.1007/s13201-022-01772-7

Pal, S., Saha, A., Gogoi, P., & Saha, S. (2024). An ensemble of J48 decision tree with adaboost and bagging for flood susceptibility mapping in the sundarbans of West Bengal, India. In *Geomorphic risk reduction using geospatial methods and tools* (pp. 117–133). Springer.

Pande, R. K. (2010). Flash flood disasters in Uttarakhand. *Disaster Prevention and Management: An International Journal, 19*(5), 565–570. https://doi.org/10.1108/09653561011091896

Paul, G. C., Saha, S., & Hembram, T. K. (2019). Application of the GIS-based probabilistic models for mapping the flood susceptibility in Bansloi Sub-basin of Ganga-Bhagirathi River and their comparison. *Remote Sensing in Earth Systems Sciences, 2*, 120–146. https://doi.org/10.1007/s41976-019-00018-6

Pham, B. T., Tien Bui, D., Prakash, I., & Dholakia, M. B. (2016). Rotation forest fuzzy rule-based classifier ensemble for spatial prediction of landslides using GIS. *Natural Hazards, 83*(1), 97–127.

Pourghasemi, H. R., Yousefi, S., Kornejady, A., & Cerdà, A. (2017). Performance assessment of individual and ensemble data-mining techniques for gully erosion modeling. *Science of the Total Environment, 609*, 764–775.

Poussin, J. K., Botzen, W. W., & Aerts, J. C. (2014). Factors of influence on flood damage mitigation behaviour by households. *Environmental Science & Policy, 40*, 69–77.

Pulvirenti, L., Pierdicca, N., Chini, M., & Guerriero, L. (2011). An algorithm for operational flood mapping from Synthetic Aperture Radar (SAR) data using fuzzy logic. *Natural Hazards and Earth System Sciences, 11*(2), 529–540.

Rahmati, O., & Pourghasemi, H. R. (2017). Identification of critical flood prone areas in data-scarce and ungauged regions: A comparison of three data mining models. *Water Resources Management, 31*(5), 1473–1487.

Rahmati, O., Pourghasemi, H. R., & Zeinivand, H. (2016a). Flood susceptibility mapping using frequency ratio and weights-of-evidence models in the Golastan Province, Iran. *Geocarto International, 31*(1), 42–70.

Rahmati, O., Zeinivand, H., & Besharat, M. (2016b). Flood hazard zoning in Yasooj region, Iran, using GIS and multi-criteria decision analysis. *Geomatics, Natural Hazards and Risk, 7*(3), 1000–1017.

Riley, S. J., DeGloria, S. D., & Elliot, R. (1999). Index that quantifies topographic heterogeneity. *Intermountain Journal of sciences, 5*(1–4), 23–27.

Saha, S., Saha, A., Agarwal, A., Kumar, A., & Sarkar, R. (2024). Spatial flash flood modeling in the Beas River Basin of Himachal Pradesh, India. Using GIS-based machine learning algorithms. In: Sarkar, R., Saha, S., Adhikari, B. R., & Shaw, R. (Eds.), *Geomorphic risk reduction using geospatial methods and tools. Disaster risk reduction.* Springer. https://doi.org/10.1007/978-981-99-7707-9_8

Sarhadi, A., Soltani, S., & Modarres, R. (2012). Probabilistic flood inundation mapping of ungauged rivers: Linking GIS techniques and frequency analysis. *Journal of Hydrology, 458*, 68–86.

Shahabi, H., Shirzadi, A., Ghaderi, K., Omidvar, E., Al-Ansari, N., Clague, J. J., & Ahmad, A. (2020). Flood detection and susceptibility mapping using sentinel-1 remote sensing data and a machine learning approach: Hybrid intelligence of bagging ensemble based on k-nearest neighbor classifier. *Remote Sensing, 12*(2), 266.

Singha, P., Das, P., Talukdar, S., & Pal, S. (2020). Modeling livelihood vulnerability in erosion and flooding induced river island in Ganges riparian corridor, India. *Ecological Indicators, 119*, 106825.

Smith, K. (1997). Environmental hazards: Assessing risk and reducing disaster. *Environment International, 23*(3), 422–422.

Stefanidis, S., & Stathis, D. (2013). Assessment of flood hazard based on natural and anthropogenic factors using analytic hierarchy process (AHP). *Natural Hazards, 68*(2), 569–585.

Strobl, C., Malley, J., & Tutz, G. (2009). An introduction to recursive partitioning: Rationale, application, and characteristics of classification and regression trees, bagging, and random forests. *Psychological Methods, 14*(4), 323.

Tehrany, M. S., Pradhan, B., & Jebur, M. N. (2015). Flood susceptibility analysis and its verification using a novel ensemble support vector machine and frequency ratio method. *Stochastic Environmental Research and Risk Assessment, 29*(4), 1149–1165.

Thapliyal, A., Kimothi, S., Dumka, U. C., & Das, I. C. (2024). Climate change-derived environmental and physical factors influencing the socioeconomic development in the Himalayan region. *Environmental Research, 241*, 117552. https://doi.org/10.1016/j.envres.2023.117552

USDMA. (2003). https://cdnbbsr.s3waas.gov.in/s34a06d868d044c50af0cf9bc82d2fc19f/uploads/2025/08/20250805137616996.pdf

Weiss, A. (2001, July). Topographic position and landforms analysis. In *Poster Presentation, ESRI User Conference, San Diego, CA* (Vol. 200).

Wesr.unepgrid.ch. (2018). https://wesr-climate.unepgrid.ch/climate/home

Wu, H. L., Cheng, W. C., Shen, S. L., Lin, M. Y., & Arulrajah, A. (2020). Variation of hydro-environment during past four decades with underground sponge city planning to control flash floods in Wuhan, China: An overview. *Underground Space, 5*(2), 184–198.

Yariyan, P., Janizadeh, S., Van Phong, T., Nguyen, H. D., Costache, R., Van Le, H., Pham, B. T., Pradhan,B., & Tiefenbacher, J. P. (2020). Improvement of best first decision trees using bagging and dagging ensembles for flood probability mapping. *Water Resources Management, 34*(9), 3037–3053.

Youssef, A. M., Pradhan, B., Jebur, M. N., & El-Harbi, H. M. (2015). Landslide susceptibility mapping using ensemble bivariate and multivariate statistical models in Fayfa area, Saudi Arabia. *Environmental Earth Sciences, 73*(7), 3745–3761.

Yufeng, S., & Fengxiang, J. (2009, July). Landslide stability analysis based on generalized information entropy. In *2009 International Conference on Environmental Science and Information Application Technology* (Vol. 2, pp. 83–85). IEEE.

Zhang, Q., Gu, X., Shi, P., & Singh, V. P. (2017). Impact of tropical cyclones on flood risk in southeastern China: Spatial patterns, causes and implications. *Global and Planetary Change, 150*, 81–93.

Zou, Q., Zhou, J., Zhou, C., Song, L., & Guo, J. (2013). Comprehensive flood risk assessment based on set pair analysis-variable fuzzy sets model and fuzzy AHP. *Stochastic Environmental Research and Risk Assessment, 27*(2), 525–546.

Chapter 7
GIS-Based Landslide Susceptibility Zonation of Kohima District Using the Frequency Ratio Model

Gaurav Bharti and Vaishnavi Bansal

Abstract This study develops a district-scale landslide susceptibility map for Kohima, Nagaland, using the Frequency Ratio (FR) model within a GIS environment. A landslide inventory compiled from published sources, satellite interpretation, and limited field checks was randomly split into training (70%) and validation (30%) sets. Ten causative factors were prepared from DEM and thematic datasets: slope, aspect, elevation, land-use/land-cover (LULC), geomorphology, rainfall, and drainage density. FR values were computed for each factor class, and a pixel-wise Landslide Susceptibility Index (LSI) was obtained by summing the class FRs. The LSI raster was classified into five zones—Very Low, Low, Moderate, High, and Very High using natural breaks. Validation against the independent set shows an Area Under the ROC Curve (AUC) of 0.61, indicating good predictive skill. Spatially, susceptibility concentrates along major highway corridors, steep dissected hillslopes, weathered flyash formations, and rapidly urbanizing fringes around Kohima town. Approximately ~21% of the district falls within the High–Very High classes, highlighting priority tracts for slope stabilization, drainage management, and development control. The FR approach provides an interpretable baseline for hazard screening; future work can integrate time-varying rainfall thresholds and ensemble learning to refine predictions and quantify uncertainty for operational planning.

Keywords Kohima district · Frequency ratio · GIS · Landslide susceptibility · DEM · ROC–AUC · Road-cut hazards

Introduction

Landslide susceptibility mapping (LSM) provides a spatially explicit estimate of zones most prone to slope failures, accounting for terrain, lithology, hydrology, and

G. Bharti · V. Bansal (✉)
Department of Civil Engineering, JSS Academy of Technical Education, Noida, India
e-mail: vaishnavibansal@jssaten.ac.in

G. Bharti
e-mail: gauravbharti000@gmail.com

R. Sarkar et al. (eds.), *Climate Change, Environmental Hazards and Community-Based Resilience*, Disaster Risk Reduction, https://doi.org/10.1007/978-981-95-9485-6_7

anthropogenic influences rather than precise temporal triggers (Nagavi et al., 2024; Shukla et al., 2024). It forms the cornerstone of landslide risk reduction frameworks by supporting land-use zoning, infrastructure corridor planning, and the prioritization of mitigation efforts where resources are limited (Shukla et al., 2025a, 2025d). Conceptually, LSM establishes empirical or analytical relationships between historical landslide occurrences and predisposing factors such as slope angle, aspect, curvature, elevation, lithology, soils, land use/land cover (LULC), and proximity to drainage and roads, thereby classifying terrain into categories of very low to very high susceptibility. Over several decades, research has shown that mapping unit choice, factor representation, and modeling technique collectively determine map quality and interpretability. Hence, systematic evaluation and continuous refinement of LSM methodologies are essential to produce decision-ready maps suitable for field application in the Himalayan context (Shukla et al., 2025c, 2025e). LSM methodologies have evolved through heuristic, statistical, and machine learning (ML) paradigms, each offering distinct strengths and limitations (Bansal & Sarkar, 2024a, 2024b).

Knowledge-driven approaches, such as Analytical Hierarchy Process (AHP) and fuzzy logic, depend on expert judgment to assign factor weights and are widely applied in data-scarce regions (Myronidis et al., 2016; Sarkar & Kanungo, 2004; Sema et al., 2017). Statistical and data-driven frameworks, notably the Frequency Ratio (FR) and logistic regression, infer relationships directly from landslide inventories and remain popular due to their simplicity, transparency, and probabilistic interpretation (Avinash & Ashamanjari, 2010; Fayez et al., 2018; Gautam, 2022; Pal & Chowdhuri, 2019; Pham et al., 2015).The FR method has shown consistently robust results across diverse Himalayan terrains, including the Aglar watershed and Kinnaur District, where slope, drainage proximity, and lithology emerged as dominant controlling factors (Keshri et al., 2023; Ramesh & Anbazhagan, 2015). In the Eastern Himalaya, LULC transformation, road construction, and deforestation have been documented as key accelerants of susceptibility (Ritse et al., 2020). Geoinformatics-based zonation studies in Kohima and adjoining areas confirm that steep slopes and road corridors cluster the highest risk zones (Bansal & Sarkar, 2023; Bilwa et al., 2015; Chang et al., 2021). Recent decades have seen a rapid expansion of machine learning and deep learning methods in LSM. Support Vector Machines (SVM) demonstrate high predictive capability in complex multidimensional spaces (Huang & Zhao, 2018; Lee et al., 2017), while Artificial Neural Networks (ANNs) capture nonlinear interactions effectively (Lee et al., 2004; Moayedi et al., 2019). More advanced techniques such as Convolutional Neural Networks (CNNs) can directly learn spatial hierarchies from terrain images, offering improved generalization for large datasets (Azarafza et al., 2021; Wang et al., 2019). Comparative analyses reveal that while these ML approaches improve predictive accuracy, they require dense, high-quality inventories and rigorous cross-validation to prevent overfitting (Ado et al., 2022; Kavzoglu et al., 2019; Lee, 2019; Liu et al., 2023). In the context of Uttarakhand and adjoining Himalayan regions, FR-based and ML-based approaches both show promise. Studies employing receiver operating characteristic (ROC) curves, success/prediction rate analysis, and k-fold cross-validation have underscored the importance of independent verification datasets to ensure reliability (Broeckx et al., 2018; Petschko et al., 2014; Wilde et al., 2018). The increasing availability of high-resolution optical, radar, and InSAR data is also enhancing future prospects for dynamic LSM and climate-linked hazard forecasting (Azarafza et al., 2021; Liu et al., 2023; Wang et al., 2019).

Study Area

Kohima District as shown in Fig. 7.1, located in the south-central part of Nagaland in northeast India, is characterized by rugged, closely spaced hill ranges with steep valley sides and narrow intermontane valleys. The relief, dense drainage, and active geomorphic processes make slopes inherently susceptible to failure, particularly along road corridors and expanding urban fringes around Kohima city. Regional geology is dominated by highly weathered flysch sequences—interbedded shale, siltstone, and sandstone—together with colluvial and residual soils that weaken rapidly under saturation, jointing, and excavation (Kumar et al., 2008; Sema et al., 2017). Field reports in and around Kohima consistently link cut-slope instability to the combination of lithological weakness, intense monsoon rainfall, and anthropogenic modification such as slope undercutting for roads and building foundations (Bilwa et al., 2015; Khatsu & van Westen, 2002).

Climatically, the district experiences a humid monsoon regime with the bulk of annual precipitation concentrated between June and September, when short-duration, high-intensity storms trigger widespread shallow slides and debris flows. Vegetation pattern and land-use/land-cover (LULC) dynamics—shifting cultivation, conversion

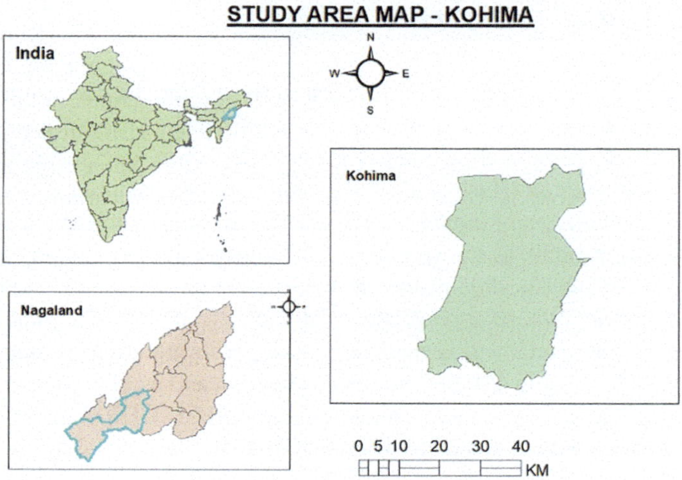

Fig. 7.1 Study area map—Kohima

to built-up areas, and degradation of forest cover—modulate near-surface hydrology and root reinforcement, thereby influencing spatial landslide susceptibility (Ritse et al., 2020). The transport network, including national and state highways that follow ridge lines and steep valley flanks, introduces additional hazard via repeated cut-and-fill operations, inadequate surface drainage, and toe erosion along stream crossings (Kumar et al., 2008; Sema et al., 2017). Against this physiographic, geological, and anthropogenic backdrop, Kohima offers a representative setting for district-scale landslide susceptibility assessment using GIS-based statistical approaches (Bharti et al., 2022; Broeckx et al., 2018; Lee, 2019; Shukla et al., 2025c).

Materials and Methods

Landslide Inventory Map

A consolidated landslide inventory was compiled from published reports, bhukosh website and maps for the Kohima region (e.g., Bilwa et al., 2015; Chang et al., 2021; Khatsu & van Westen, 2002; Kumar et al., 2008; Notoka et al., 2025; Sema et al., 2017), complemented by visual interpretation of high-resolution satellite imagery and limited field checks. In line with best practice, the inventory was randomly partitioned into training ($\approx$70%) and validation ($\approx$30%) subsets, as model performance critically depends on the quality and completeness of this input dataset (Gaidzik & Ramírez-Herrera, 2021).

Landslide Causative Factors

Slope angle, aspect, curvature and elevation derived from Digital evelation map (DEM) and geomorpholical map have been derived from Bhukosh website, land use/land cover (LULC) from satellite imagery given LULC's well-documented influence on stability in the region (Ritse et al., 2020). Rainfall data has been collected from Metrological Department of India for 2011–2020. The main hydrological factors are drainage density, rainfall, as shown in Fig. 7.2. Drainage density is a very important controlling factor for landslide events. It means for a unit area, how many stream lengths are present. Where appropriate, continuous variables were classified and categorical variables retained as mapped classes, with attention to mapping-unit effects (Erener & Düzgün, 2012). The slope, elevation and aspect are topographical factors. These factor shave been generated using ASTER GDEM of 30 m spatial resolution with the help of spatial tools in ArcGIS10.4.

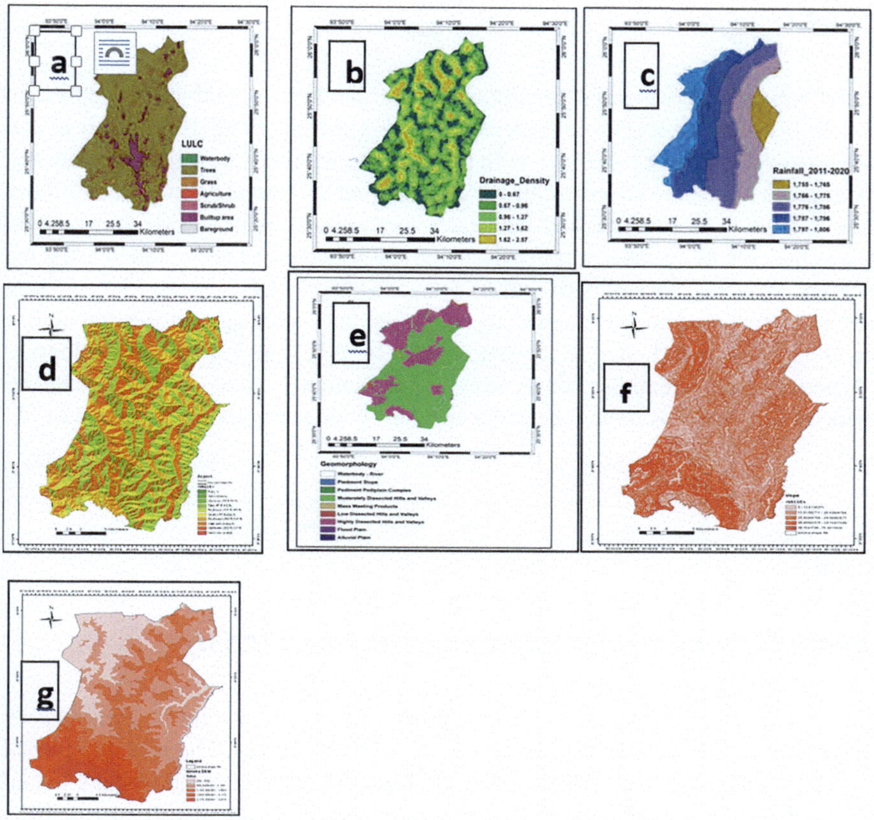

Fig. 7.2 Causative factor maps: **a** LULC, **b** drainage density, **c** rainfall, **d** slope, **e** geomorphology, **f** slope, **g** DEM

Frequency Ratio (FR) Model

Susceptibility was modeled using the FR method, a robust bivariate statistical approach widely used in landslide studies for its conceptual simplicity and interpretability (Avinash & Ashamanjari, 2010; Fayez et al., 2018; Gautam, 2022; Keshri et al., 2023; Pal & Chowdhuri, 2019; Ram Mohan et al., 2011; Ramesh & Anbazhagan, 2015; Shano et al., 2021; Sonker & Tripathi, 2022). For each factor class i, FR is computed as the ratio between (i) the percentage of landslide pixels in class i and (ii) the percentage of all pixels in class i. Values > 1 indicate positive association with landslide occurrence, whereas values < 1 indicate negative association.

Model Validation

Model validation is essential for evaluating the predictive reliability of statistical models. In this study, we employ receiver operating characteristic (ROC) analysis and the area under the curve (AUC) for both success-rate and prediction-rate assessments. The success-rate curve is derived by cross-tabulating the training landslide inventory (n = 243) with the reclassified landslide susceptibility map (LSM) and ranking classes in descending susceptibility. The prediction-rate curve is computed using the independent testing inventory (n = 106). For each curve, AUC is obtained by plotting the cumulative percentage of correctly identified landslides against the cumulative percentage of mapped area. AUC values range from 0.5 (no better than random) to 1.0 (perfect discrimination); hence, values approaching 1.0 indicate superior model performance, whereas values near 0.5 indicate poor predictive skill (Chen et al., 2014).

Results

Landslide Susceptibility Map Using FR Model (LSM)

The pixel-wise LSI field was classified to produce the final LSM for Kohima District using standard GIS procedures (He & Beighley, 2008). A large share of inventoried landslides falls within the High and Very High zones, confirming the model's spatial discrimination as shown in Fig. 7.3. In Geomorphology, Moderately Dissected Hills and Valleys concentrate the bulk of failures (83.33% of landslides on 66.82% of area; FR = 1.25) as shown in Table 7.1, marking them as the dominant susceptible backdrop. Highly Dissected Hills and Valleys cover a large share (30.96%) but host proportionally fewer failures (13.75%; FR = 0.44), implying comparatively lower per-unit-area propensity—potentially due to lithology/structure or mapping/denominator effects. Waterbody–River occupies just 0.98% of area yet accounts for 2.08% of failures (FR = 2.13), a strong local hotspot consistent with toe erosion and saturation along channels. The remaining classes—Low Dissected Hills and Valleys (1.04% area; 0.83% failures; FR = 0.80), Piedmont Slope (0.01% area; FR = 0), Mass Wasting Products (0.02%; FR = 0), Pediment–Pediplain Complex (0.08%; FR = 0), Alluvial Plain (0.08%; FR = 0), and Flood Plain (0.01%; FR = 0) show either below-average association or no inventoried failures—reflecting intrinsic stability and/or very limited areal exposure. In Aspect, Flats show no recorded failures (FR = 0 on 0.59% area). Among oriented slopes, S (112.5–157.5) is most failure-prone (12.01% area; 15.83% failures; FR = 1.32), with SE (157.5–202.5) and SW (202.5–247.5) moderately elevated (9.94% and 10.79% area; 10.83% and 11.67% failures; FR = 1.09 and 1.08). E (67.5–112.5) and W (247.5–292.5) are near proportional (13.83% and 12.59% area; 13.75% and 12.50% failures; FR $\approx$ 0.99 each). The N (0–22.5), NE, and NW quadrants are comparatively stable (6.75%, 14.18%, 13.05% area vs. 5.42%,

12.50%, 11.67% failures; FR = 0.80, 0.88, 0.89), and the wrap-around N (337.5–360) is slightly below average (6.27% area; 5.83% failures; FR = 0.93). Overall, south-exposed slopes (S, SE, SW) show a consistent positive association, while northern exposures trend negative to neutral. Land Use/Land Cover (LULC). Built-up Area is the most prominent anthropogenic hotspot—4.19% of the landscape concentrates 34.58% of failures (FR = 8.25), implicating cut-and-fill, surcharge, drainage concentration, and surface sealing. Grass also shows elevated susceptibility (0.18% area; 0.42% failures; FR = 2.32), consistent with shallow, weakly rooted covers favoring shallow slips. Scrub/Shrub is modestly above average (2.94% area; 3.75% failures; FR = 1.28). In contrast, Trees appear protective (92.55% area; 61.25% failures; FR = 0.66), reflecting root reinforcement and better hydrological regulation across the dominant forested matrix. Classes with no inventoried failures—Waterbody (0.02% area; FR = 0), Agriculture land (0.11%; FR = 0), and Bareground (0.01%; FR = 0)—likely reflect true absence in the mapped period and/or minimal areal footprint.

$$\begin{aligned} \text{LSM} &= \text{FR}_{\text{DEM}} + \text{FR}_{\text{slope}} + \text{FR}_{\text{aspect}} + \text{FR}_{\text{rainfall}} + \text{FR}_{\text{drainage density}} \\ &\quad + \text{FR}_{\text{geomorphology}} + \text{FR}_{\text{LUL}} \end{aligned}$$

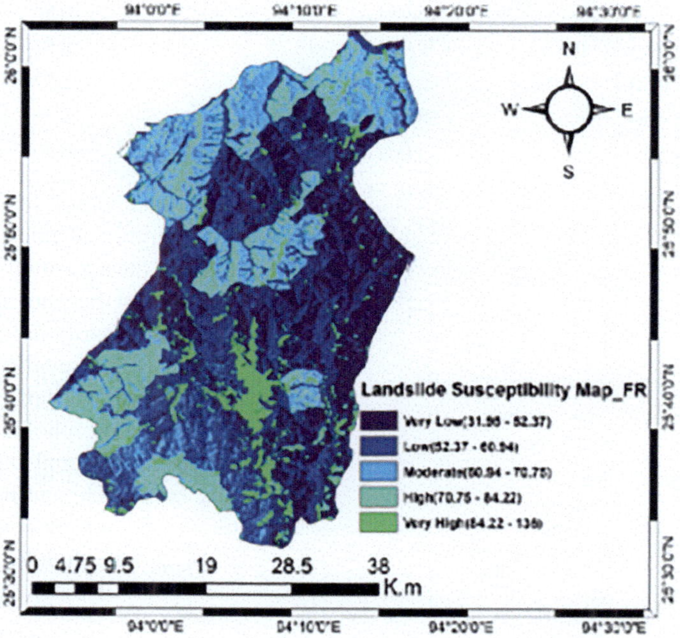

Fig. 7.3 LSM of Kohima district, Nagaland, India

Table 7.1 Frequency ratio classes of causative factor

Factor	Class	Landslide occurrence	% of Landslide occurrence	Class area	% of class area	Frequency ratio
Elevation	200–802	16,501.81	7.92	253,336	16.8	0.47
Elevation	802–1155	53,848.02	25.83	550,473	36.5	0.71
Elevation	1155–1603	130,277.46	62.5	431,636	28.62	2.18
Elevation	1603–2175	6948.13	3.33	166,446	11.04	0.3
Elevation	2175–3015	868.52	0.42	106,188	7.04	0.06
Slope	0–13.01	22,581.43	10.83	252,982	16.84	0.64
Slope	13.01–20.40	63,401.7	30.42	465,553	30.99	0.98
Slope	20.40–28.09	73,823.89	35.42	428,004	28.49	1.24
Slope	28.09–38.15	33,872.14	16.25	265,934	17.7	0.92
Slope	38.15–75.42	14,764.78	7.08	89,776	5.98	1.19
Rainfall	1,755–1,765	9553.68	4.6	110,125	7.34	0.63
Rainfall	1,766–1,775	46,031.37	22.18	335,622	22.38	0.99
Rainfall	1,776–1,786	79,903.51	38.49	416,109	27.75	1.39
Rainfall	1,787–1,796	57,322.08	27.62	389,624	25.99	1.06
Rainfall	1,797–1,806	15,633.29	7.53	254,628	16.98	0.44
Drainage density	0–0.67	39,083.24	18.75	271,669	18.01	1.04
Drainage density	0.67–0.96	60,796.15	29.17	430,838	28.57	1.02
Drainage density	0.96–1.27	68,612.79	32.92	395,692	26.24	1.25
Drainage density	1.27–1.62	33,003.62	15.83	290,849	19.29	0.82
Drainage density	1.62–2.57	6948.13	3.33	119,031	7.89	0.42
Geomorphology	Moderately dissected hills and valleys	173,703.3	83.33	1,000,000	66.82	1.25
Geomorphology	Low dissected hills and valleys	1737.03	0.83	15,701	1.04	0.8
Geomorphology	Piedmont slope	0	0	174	0.01	0
Geomorphology	Mass wasting products	0	0	363	0.02	0
Geomorphology	Pediment Pediplain complex	0	0	1151	0.08	0
Geomorphology	Alluvial plain	0	0	1198	0.08	0
Geomorphology	Flood plain	0	0	133	0.01	0
Geomorphology	Waterbody-river	4342.58	2.08	14,736	0.98	2.13
Geomorphology	Highly dissected hills and valleys	28,661.04	13.75	466,875	30.96	0.44
Aspect	Flat (−1)	0	0	8959	0.59	0

(continued)

Table 7.1 (continued)

Factor	Class	Landslide occurrence	% of Landslide occurrence	Class area	% of class area	Frequency ratio
Aspect	N (0–22.5)	11,290.713	5.42	101,778	6.75	0.8
Aspect	NE (22.5–67.5)	26,055.492	12.5	213,836	14.18	0.88
Aspect	E (67.5–112.5)	28,661.041	13.75	208,628	13.83	0.99
Aspect	S (112.5–157.5)	33,003.623	15.83	181,099	12.01	1.32
Aspect	SE (157.5–202.5)	22,581.426	10.83	149,856	9.94	1.09
Aspect	SW (202.5–247.5)	24,318.459	11.67	162,666	10.79	1.08
Aspect	W (247.5–292.5)	26,055.492	12.5	189,899	12.59	0.99
Aspect	NW (292.5–337.5)	24,318.459	11.67	196,853	13.05	0.89
Aspect	N (337.5–360)	12,159.229	5.83	94,505	6.27	0.93
LULC	Waterbody	0	0	344	0.02	0
LULC	Trees	127,671.91	61.25	1,000,000	92.55	0.66
LULC	Grass	868.51639	0.42	2703	0.18	2.32
LULC	Agriculture land	0	0	1621	0.11	0
LULC	Scrub/shrub	7816.6475	3.75	44,335	2.94	1.28
LULC	Builtup area	72,086.86	34.58	63,184	4.19	8.25
LULC	Bareground	0	0	179	0.01	0

Model Performance and Spatial Patterns

The FR model was validated using ROC analysis based on both success-rate and prediction-rate curves. For the success-rate curve, the training landslide inventory ($n = 243$) was cross-tabulated with the reclassified LSM and classes ranked in descending susceptibility to obtain the cumulative percentage of correctly captured landslides versus cumulative mapped area. For the prediction-rate curve, an independent testing set ($n = 106$) was used to evaluate generalization. In both cases, AUC was computed by plotting the cumulative percentage of landslide occurrence against the cumulative percentage of area. The reported AUC for the FR model's prediction-rate curve is 0.61 (Fig. 7.4), indicating performance better than random discrimination (0.5) but only modest predictive skill (Table 7.2).

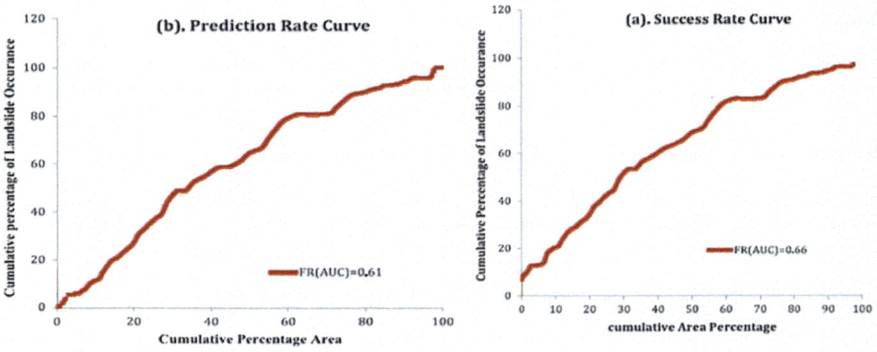

Fig. 7.4 Prediction and success rate curves

Table 7.2 % landslide and % area for each LSM class for FR model

Class	Range	% Landslide	% Area
Very low	32–52	8.74	19.54
Low	52–61	8.76	16.35
Moderate	61–71	18.33	26.74
High	71–84	27.08	31.12
Very high	84–136	37.08	6.245

Discussion and Conclusion

In this study, the Frequency Ratio (FR) model—a bivariate statistical approach—was applied across the entire Kohima district to produce a landslide susceptibility map (LSM). The FR analysis integrated eleven thematic layers (elevation, aspect, slope, rainfall, drainage density, geomorphology and LULC) prepared in ArcGIS from ASTER DEM–derived datasets. The resulting FR-based LSM delineates 6.24% of the total area as landslide-prone. Model validation using ROC/AUC shows a success performance of 66%**,** while independent testing indicates a prediction rate of 61% as shown in Fig. 7.4.

These figures confirm FR as a useful baseline method for regional susceptibility mapping, with potential accuracy gains expected if additional conditioning factors (e.g., lineament density, road network) are incorporated.

References

Ado, M., Amitab, K., Maji, A. K., Jasińska, E., Gono, R., Leonowicz, Z., & Jasiński, M. (2022). Landslide susceptibility mapping using machine learning: A literature survey. *Remote Sensing, 14*(13), 3029. (Note: Old [4] and [5] were identical; only one entry is needed).

Avinash, K. G., & Ashamanjari, K. G. (2010). A GIS and frequency ratio-based landslide susceptibility mapping: Aghnashini river catchment, Uttara Kannada, India. *International Journal of Geomatics and Geosciences, 1*(3), 343–354.

Azarafza, M., Azarafza, M., Akgün, H., Atkinson, P. M., & Derakhshani, R. (2021). Deep learning-based landslide susceptibility mapping. *Scientific Reports, 11*(1), 24112.

Bansal, V., & Sarkar, R. (2023). Numerical analysis and implications of fluctuating water levels on slope stability. In *International Conference on Sustainable Infrastructure: Innovation, Opportunities and Challenges* (pp. 407–419) Springer.

Bansal, V., & Sarkar, R. (2024a). Comparative analysis of slope stability for Kalimpong region under dynamic loading using limit equilibrium method and machine benchmark learning classifiers. *Iranian Journal of Science Technology, Transactions of Civil Engineering, 48*, 2785–2807.

Bansal, V., & Sarkar, R. (2024b). Prophetical modeling using limit equilibrium method and novel machine learning ensemble for slope stability gauging in Kalimpong. *Iranian Journal of Science Technology, Transactions of Civil Engineering, 48*, 411–430.

Bharti, G., Shukla, B. K., Srinivasan, V., & Bansal, V. (2022). Experimental study on eccentrically loaded rectangular footing on sand for embedded condition to enhance serviceability of structures. *Materials Today: Proceedings, 61*, 517–522.

Bilwa, L. M., Yanthan, M. A., Madesh, P., & Hutti, B. (2015). Identification of potential landslide hazard zonation mapping using geoinformatics for Kohima region, Nagaland, India. *International Journal of Innovative Science, Engineering & Technology, 2*(11), 1–10.

Broeckx, J., Vanmaercke, M., Duchateau, R., & Poesen, J. (2018). A data-based landslide susceptibility map of Africa. *Earth-Science Reviews, 185*, 102–121.

Chang, C. N., Ezung, M., Apon, M., Supongtemjen, Walling, T., & Thong, G. T. (2021). Assessment of landslides along NH 29 in the Kevüza area, Kohima, Nagaland. *Indian Geotechnical Journal, 51*(4), 841–860.

Chen, W., Li, W., Hou, E., Zhao, Z., Deng, N., Bai, H., & Wang, D. (2014). Landslide susceptibility mapping based on GIS and information value model for the Chencang District of Baoji, China. *Arab Journal of Geoscience, 7*(11), 4499–4511.

Erener, A. R. Z. U., & Düzgün, H. S. B. (2012). Landslide susceptibility assessment: What are the effects of mapping unit and mapping method. *Environmental Earth Sciences, 66*, 859–877.

Fayez, L., Pazhman, D., Pham, B. T., Dholakia, M. B., Solanki, H. A., Khalid, M., & Prakash, I. (2018). Application of frequency ratio model for the development of landslide susceptibility mapping at part of Uttarakhand State, India. *International Journal of Applied Engineering Research, 13*(9), 6846–6854.

Gaidzik, K., & Ramírez-Herrera, M. T. (2021). The importance of input data on landslide susceptibility mapping. *Scientific Reports, 11*(1), 19334.

Gautam, N. (2022). Landslide susceptibility mapping of Kinnaur District in Himachal Pradesh, India using probabilistic frequency ratio model. *Journal of the Geological Society of India, 98*(11), 1595–1604.

He, Y., & Beighley, R. E. (2008). GIS-based regional landslide susceptibility mapping: A case study in southern California. *Earth Surface Processes and Landforms: The Journal of the British Geomorphological Research Group, 33*(3), 380–393.

Huang, Y., & Zhao, L. (2018). Review on landslide susceptibility mapping using support vector machines. *CATENA, 165*, 520–529.

Kavzoglu, T., Colkesen, I., & Sahin, E. K. (2019). Machine learning techniques in landslide susceptibility mapping: A survey and a case study. *Landslides: Theory, Practice and Modelling*, 283–301.

Keshri, D., Sarkar, K., & Chattoraj, S. L. (2023). Landslide susceptibility mapping in parts of Aglar watershed, Lesser Himalaya based on frequency ratio method in GIS environment. *Journal of Earth System Science, 133*(1), 1.

Khatsu, P., & van Westen, C. J. (2002). Urban multi-hazard risk analysis using GIS and Remote Sensing: A case study from Kohima Town, Nagaland, India. Paper presented at the Regional

Workshop on Best Practices in Disaster Mitigation, Lessons learned from the Asian Urban Disaster Mitigation Program and other initiatives, Bali, Indonesia.

Kumar, K., Prasad, P. S., Goyal, N., & Mathur, S. (2008). A study on debris and earth flow on National Highway-39, near Kohima, Nagaland. In *Proceedings of the 12th International Conference of International Association for Computer Methods and Advances in Geomechanics (IACMAG)* (pp. 4453–4460).

Lee, S. (2019). Current and future status of GIS-based landslide susceptibility mapping: A literature review. *Korean Journal of Remote Sensing, 35*(1), 179–193.

Lee, S., Hong, S. M., & Jung, H. S. (2017). A support vector machine for landslide susceptibility mapping in Gangwon Province, Korea. *Sustainability, 9*(1), 48.

Lee, S., Ryu, J. H., Won, J. S., & Park, H. J. (2004). Determination and application of the weights for landslide susceptibility mapping using an artificial neural network. *Engineering Geology, 71*(3–4), 289–302.

Liu, S., Wang, L., Zhang, W., He, Y., & Pijush, S. (2023). A comprehensive review of machine learning-based methods in landslide susceptibility mapping. *Geological Journal, 58*(6), 2283–2301.

Moayedi, H., Mehrabi, M., Mosallanezhad, M., Rashid, A. S. A., & Pradhan, B. (2019). Modification of landslide susceptibility mapping using optimized PSO-ANN technique. *Engineering with Computers, 35*, 967–984.

Myronidis, D., Papageorgiou, C., & Theophanous, S. (2016). Landslide susceptibility mapping based on landslide history and analytic hierarchy process (AHP). *Natural Hazards, 81*, 245–263.

Nagavi, J. C., Shukla, B. K., Bhati, A., Rai, A., & Verma, S. (2024). Harnessing geospatial technology for sustainable development: A multifaceted analysis of current practices and future prospects. In *Sustainable development and geospatial technology: Volume 1: Foundations and innovations* (pp. 147–170). Springer Nature Switzerland. https://doi.org/10.1007/978-3-031-65683-5_8

Notoka, K., Chang, C. N., Ezung, M., Thong, G. T., Kirha, V., Dzuvichu, K., Miachieo, K., & Walling, T. (2025). Appraisal of a complex landslide at Kohima Town, Nagaland, India. *Journal of Geosciences Research, 10*(1), 14–21.

Pal, S. C., & Chowdhuri, I. (2019). GIS-based spatial prediction of landslide susceptibility using frequency ratio model of Lachung River basin, North Sikkim, India. *SN Applied Sciences, 1*, 1–25.

Petschko, H., Brenning, A., Bell, R., Goetz, J., & Glade, T. (2014). Assessing the quality of landslide susceptibility maps-case study Lower Austria. *Natural Hazards and Earth System Sciences, 14*(1), 95–118.

Pham, B. T., Tien Bui, D., Indra, P., & Dholakia, M. (2015). Landslide susceptibility assessment at a part of Uttarakhand Himalaya, India using GIS-based statistical approach of frequency ratio method. *Int J Eng Res Technol, 4*(11), 338–344.

Ram Mohan, V., Jeyaseelan, A., Raj, T. N., Narmatha, T., & Jayaprakash, M. (2011). Landslide susceptibility mapping using frequency ratio method and GIS in south eastern part of Nilgiri District, Tamilnadu, India. *International Journal of Geomatics and Geosciences, 1*(4), 951–961.

Ramesh, V., & Anbazhagan, S. (2015). Landslide susceptibility mapping along Kolli hills Ghat road section (India) using frequency ratio, relative effect and fuzzy logic models. *Environmental Earth Sciences, 73*, 8009–8021.

Ritse, V., Basumatary, H., Kulnu, A. S., Dutta, G., Phukan, M. M., & Hazarika, N. (2020). Monitoring land use land cover changes in the Eastern Himalayan landscape of Nagaland, Northeast India. *Environmental Monitoring and Assessment, 192*, Article 711.

Sarkar, S., & Kanungo, D. P. (2004). An integrated approach for landslide susceptibility mapping using remote sensing and GIS. *Photogrammetric Engineering & Remote Sensing, 70*(5), 617–625.

Sema, H. V., Guru, B., & Veerappan, R. (2017). Fuzzy gamma operator model for preparing landslide susceptibility zonation mapping in parts of Kohima Town, Nagaland, India. *Modeling Earth Systems and Environment, 3*(2), 573–589.

Shano, L., Raghuvanshi, T. K., & Meten, M. (2021). Landslide susceptibility mapping using frequency ratio model: The case of Gamo highland, South Ethiopia. *Arabian Journal of Geosciences, 14*, 1–18.

Shukla, B. K., Gupta, L., Sharma, P. K., Tyagi, K., Yadav, H., Singh, S., & Yadav, Y. (2025a). Advancements in remote sensing for water quality assessment: A comprehensive exploration. In *Geo-data revolution: Advances in spatial analysis and natural hazard mapping* (pp. 59–82). Springer Nature Switzerland. https://doi.org/10.1007/978-3-031-94078-1_4

Shukla, B. K., Gupta, L., Singh, H. K., Malik, A., Karanwal, A., & Dhama, A. (2025b). Integrating multi-criteria decision making and advanced technologies in water quality index development. In *Geo-data revolution: Advances in spatial analysis and natural hazard mapping* (pp. 329–351). Springer Nature Switzerland. https://doi.org/10.1007/978-3-031-94078-1_18

Shukla, B. K., Gupta, L., Parashar, B., Sharma, P. K., Sihag, P., & Shukla, A. K. (2025c). Integrative assessment of surface water contamination using GIS, WQI, and machine learning in urban-industrial confluence zones surrounding the national capital territory of the Republic of India. *Water, 17*(7), 1076. https://doi.org/10.3390/w17071076

Shukla, B. K., Sharma, P. K., Singh, C., Kumari, N., Chaudhary, A., & Garg, A. (2025d). Integrated urban water management: Optimizing rainwater harvesting and stormwater infrastructure for sustainable urban development. In *Geo-data revolution: Advances in spatial analysis and natural hazard mapping* (pp. 103–118). Springer Nature Switzerland. https://doi.org/10.1007/978-3-031-94078-1_6

Shukla, B. K., Yadav, H., Singh, S., Verma, S., Shukla, A. K., & Sharma, C. (2025e). Sustainable engineering of recycled aggregate concrete: Structural performance and environmental benefits under circular economy frameworks. *Construction Materials, 5*(3), 67.

Shukla, B. K., Tripathi, A., Nagavi, J. C., Prasad, S., Sahoo, M., Chaudary, S., Ayush & Verma, S. (2024). Integrated water resources management (IWRM) in the geospatial epoch: An ontological dive into sustainable hydrological governance. In *Sustainable development and geospatial technology: Volume 2: Applications and future directions* (pp. 1–22). Springer Nature Switzerland. https://doi.org/10.1007/978-3-031-65703-0_1

Sonker, I., & Tripathi, J. N. (2022). Remote sensing and GIS-based landslide susceptibility mapping using frequency ratio method in Sikkim Himalaya. *Quaternary Science Advances, 8*, 100067. (Note: Old [34] and [47] were identical).

Wang, Y., Fang, Z., & Hong, H. (2019). Comparison of convolutional neural networks for landslide susceptibility mapping in Yanshan County, China. *Science of the Total Environment, 666*, 975–993.

Wilde, M., Günther, A., Reichenbach, P., Malet, J. P., & Hervás, J. (2018). Pan-European landslide susceptibility mapping: ELSUS Version 2. *Journal of Maps, 14*(2), 97–104.

Chapter 8
Spatio-temporal Pattern of Drought in Maharashtra, India

Sipra Chowdhury and Priyanka Gogoi

Abstract Drought is an incident that occurs in almost all climatic regions. It is not just an absence of rainfall but a continued period of little rainfall. Maharashtra, despite being the financial capital, is not well prepared to handle the undesirable effects of drought thus proper drought vulnerability studies are required there. For the present study, from 1981 to 2022 mainly rainfall data was downloaded from NASA Power Access and then SPI (Standardized Precipitation Index) was analysed. Using the SPI values parameters like EDF (Frequency of Extreme Drought), SDF (Frequency of Severe Drought), DM (Magnitude of Drought), MDI (Mean Intensity of Drought), RPED (Extreme Drought Return Period) and RPSD (Severe Drought Return Period) for the time scale of 3, 6, 12, 24 months each were evaluated. Using these parameters spatial layers were prepared using kriging function in Arc GIS software. These spatial maps give a rough sketch of the drought scenario in Maharashtra. Thus proper mitigation and management policies are required to be adapted in Maharashtra to diminish the unpleasant effects of drought.

Keywords Drought · SPI · Kriging · Maharashtra

Introduction

Various reports of international bodies like the Intergovernmental Panel on Climate Change's 5th report states that because of extreme weather events and global warming, droughts are becoming more frequent and extreme all over the world (Bai et al., 2020). One of the most commonly going on disasters globally is drought, with wide regional variation in its characteristics and impacts. Drought like conditions are caused by little rainfall due delays in the rainy season and irregular supply of rain. Drought impacts society on different levels like human health, agriculture, biodiversity, food security, economics and water resources and is the leading of all costly natural disasters (Hayes et al., 1999). The most often used description

S. Chowdhury · P. Gogoi (✉)
Department of Geography, University of Gour Banga, Malda, West Bengal, India
e-mail: priyankag19161@gmail.com

R. Sarkar et al. (eds.), *Climate Change, Environmental Hazards and Community-Based Resilience*, Disaster Risk Reduction, https://doi.org/10.1007/978-981-95-9485-6_8

of drought is "the naturally occurring phenomenon that exists when precipitation has been significantly below normal recorded levels, causing serious hydrological imbalances that adversely affect land resource production systems". Drought is a slow-occurring hazard occurring naturally that evidences a phase of notably below-normal precipitation, causing water insufficiency and extensive interference of agricultural, environmental, and socio-economic structures (Wilhite & Glantz, 1985). Droughts are primarily classified into hydrological, agricultural, meteorological, and socio-economic droughts (Wilhite & Glantz, 1985). Hydrological, meteorological, socio-economic and agricultural droughts are the main types of droughts (Wilhite & Glantz, 1985). Drought has a cascading effect, as below-normal rain directs to meteorological drought, and its continuance over broad periods influences groundwater, soil wetness, and streamflows, resulting into hydrological, agricultural and socioeconomic droughts. Different drought indicators have different reaction times like scarcity of soil moisture, streamflows and rainfall response differently (Umran, 1999; Zhu et al., 2016). Drought assessments spatio-temporally consigns to the scientific approach of evaluating drought patterns by considering both their spatial distribution and temporal variability. According to Zargar et al. (2011), it entails the methodical assessment of drought features, including DF, MDI, DM, and RPD. It is predicted that will become further recurrent and extreme, leading to food insecurity and risky water resources (Banerjee and Bera, 2021). Dissimilar climatic zones display changing drought conditions (Lan et al., 2022). Drought is caused by a number of reasons, including the region's characteristics and water constraint, in addition to periods with low precipitation (Querner & van Lanen, 2001; Saadat et al., 2011). Widespread modeling studies have predicted drought features, comprising intensity, severity and duration (Dikshit et al., 2022). Researchers have used various tools to detect drought conditions, including the Standardized Precipitation Evapotranspiration Index (Vicente-Serrano et al., 2010), Soil Moisture Index (Herrera-Estrada et al., 2017), Moisture Adequacy Index (Kumar et al., 2019), Normalized Difference Vegetation Index (Huete et al., 2002), and Standardized Precipitation Index (McKee et al., 1993). Drought vulnerability has been measured and predicted using the Palmer Drought Severity Index (Palmer, 1965), pattern recognition methods, and the ARIMA model (Wu et al., 2021). SPI is now globally accepted and is also considered as the chief meteorological marker used by most of the organizations to measure meteorological droughts at the World Meteorological Organization conference in 2009. In the present study spatial layers using temporal data for the whole of Maharashtra were prepared. These drought frequency, magnitude, intensity and return period maps help us to evaluate the drought scenario in the whole state. Maharashtra being the financial capital of India is severely drought hit which when fixed can help in boosting the economic setting thus proper mitigation and management policies should be adapted.

Study Area

Over 5 crore people live in India, where droughts occur most frequently in about 16% of the total country's geographical span (Dutta et al., 2015). The majority of these people rely on farming and related facilities for their livelihood (Udmale et al., 2014). For the present study Maharashtra state has been picked (Fig. 8.1). Being the 3rd biggest state in India it wraps an area of 307,713 km^2 and has a 720 km coastline in the west. Its neighbouring states are Chhattisgarh and Telangana in the east, Gujarat and Madhya Pradesh in the north, Karnataka and Goa in the south. Maharashtra is located between 15° 44′ to 22° 06′ North Latitude and 72° 56′ to 80° 54′ East Longitude. Maharashtra has a tropical monsoon climate, which varies from region to region due to its vast topography, from the Western Ghats to the Deccan plateau. Maharashtra, located in western India, has a diverse climate, while some regions receive adequate rainfall, significant parts particularly in its Vidarbha and Marathwada region is drought prone. Maharashtra experiences most of its showers through the south-west monsoon season, which runs for the months June–September. However, delays, early withdrawal, or unequal rainfall can cause meteorological droughts. The Western Ghats receive high rainfall and eastern regions' annual rainfall is less than 500 mm. The Western Ghats Mountains block the humid monsoon winds, resulting in semi-arid conditions in coastal districts like Marathwada, Vidarbha and parts of Solapur.

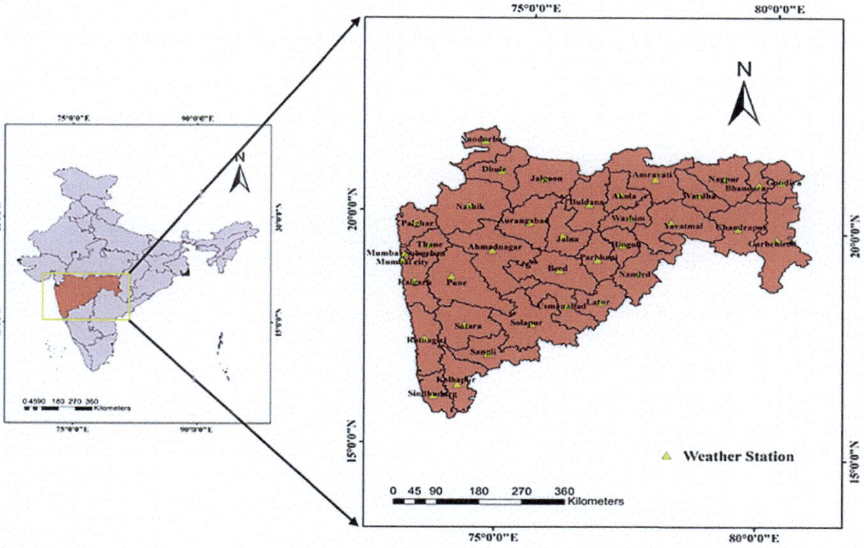

Fig. 8.1 Location map of the study area

Materials and Methods

Materials

A total of twenty-five parameters including annual rainfall, extreme drought frequency of 3, 6, 12, 24 months, severe drought frequency of 3, 6, 12, 24 months, drought magnitude of 3, 6, 12, 24 months, mean drought intensity of 3, 6, 12, 24 months, return period of extreme drought and return period of severe drought for 3, 6, 12, 24 months were used for the present assessment of spatio-temporal pattern of drought (Table 8.1).

Table 8.1 Variables selected for the spatio-temporal assessment of drought

Sl. No.	Variables	Sources
1	Annual rainfall/monthly rainfall data	NASA power access
2	EDF of 3 month	Calculated from SPI values
3	EDF of 6 month	Calculated from SPI values
4	EDF of 12 month	Calculated from SPI values
5	EDF of 24 month	Calculated from SPI values
6	SDF of 3 month	Calculated from SPI values
7	SDF of 6 month	Calculated from SPI values
8	SDF of 12 month	Calculated from SPI values
9	SDF of 24 month	Calculated from SPI values
10	DM of 3 month	Calculated from SPI values
11	DM of 6 month	Calculated from SPI values
12	DM of 12 month	Calculated from SPI values
13	DM of 24 month	Calculated from SPI values
14	MDI of 3 month	Calculated from SPI values
15	MDI of 6 month	Calculated from SPI values
16	MDI of 12 month	Calculated from SPI values
17	MDI of 24 month	Calculated from SPI values
18	RPED of 3 month	Calculated from SPI values
19	RPED of 6 month	Calculated from SPI values
20	RPED of 12 month	Calculated from SPI values
21	RPED of 24 month	Calculated from SPI values
22	RPSD of 3 month	Calculated from SPI values
23	RPSD of 6 month	Calculated from SPI values
24	RPSD of 12 month	Calculated from SPI values
25	RPSD of 24 month	Calculated from SPI values

Methods

The 25 parameters used and formula used for their calculation are shown in (Table 8.2).

Standardised Precipitation Index (SPI)

McKee was credited for inventing SPI in 1993 by for managing the consequences of drought incidents. The index of SPI was permitted by the WMO. Standardised Precipitation Index is used to measure drought in which negative value represent drought and positive value represents wet conditions (Table 8.3). Here for assessing meteorological drought for different districts of Maharashtra SPI has been used. Here four different time scale, viz. 3, 6, 12 and 24 month has been used for each district. In this study the 1981–2022 annual rainfall data were examined to calculate SPI value. Using SPI values of different districts of Maharashtra state, DF, DM, MDI and RPED and RPSD were calculated and their logical consideration and directionality of influence is given in (Table 8.4). This data was used for drought vulnerability mapping using Kriging function in Arc GIS software. The formula for SPI is as follows (McKee et al., 1993)

$$\mathrm{SPI} = \frac{Pi - Pm}{SD} \tag{8.1}$$

where, average precipitation is represented by Pm, precipitation is represented by Pi, and SD means standard deviation.

Table 8.2 Methods used to assess Spatio-temporal assessment of drought

Indicators	Methodology	Remarks
EDF and SDF for 3, 6, 12, 24 months	$N_{i,100} = \frac{Ni}{i.\,n} * 100$	Where, $N_{i,100}$ equals to the number of droughts for a timescale i in 100 years, N_i equals to the number of months with droughts for a timescale i in the n year set and i equals to time scale of 3, 6, 12, 24 months
DM for 3, 6, 12, 24 months	$M_D = \sum_{i=1}^{m} SPI_{ij}$	Where, SPI_{ij} equals to SPI values of a drought event at j time scale and m equals to number of months
MDI for 3, 6, 12, 24 months	$MI = \frac{MD}{Mean}$	Where, MD equals to drought magnitude
RPED and RPSD for 3, 6, 12, 24 months	$RP = \frac{n}{Er}$	Where, n equals to number of droughts, and Er equals to rank of drought

Table 8.3 Wet or dry drought classes according to SPI values

Class	Drought category
2.0+	Extremely wet
1.5 to 1.99	Very wet
1.0 to 1.49	Moderately wet
0.99 to 0	Mildly drought
0 to −0.99	Mild dry
−1.0 to −1.49	Moderate dry
−1.5 to −1.99	Severely dry
−2 and less	Extremely dry

Source World Meteorological Organization, 2012

Table 8.4 Spatio-temporal estimation of drought vulnerability parameters and their directionality of influence towards drought vulnerability

Variables	Logical consideration	Directionality of influence
DF (Drought Frequency)	More drought frequency means more vulnerable	Direct
DM (Drought Magnitude)	More drought frequency means more vulnerable	Direct
MDI (Mean Drought Intensity)	More mean drought intensity means more vulnerable	Direct
RPD (Return Period of Drought)	More return period of drought means less vulnerable	Inverse

Result and Discussion

In this present study, DF, DM, MDI, RPD were selected for mapping spatio-temporal assessment of drought. The aforesaid parameters interpolate 36 districts covering the entire Maharashtra. Kriging technique was applied to interpolate the datasets (Nasrollahi et al., 2018). RPED and RPSD are the parameters which are negatively or indirectly related to drought vulnerability whereas parameters like DF of extreme and severe drought, MDI, and DM are directly or positively related to drought vulnerability.

Drought Frequency

Among different criteria layers used for assessing Spatio-temporal assessment of drought, 3 months EDF is high in Chandrapur, Yavatmal, Gadchiroli (Fig. 8.2a). 6 months EDF is high in Nandurbar, Dhule, Jalgaon, Buldana, Akola, Amaravati, Washim, Hingoli, Nashik, Aurangabad, Ahmednagar, Palghar, Thane, Mumbai city,

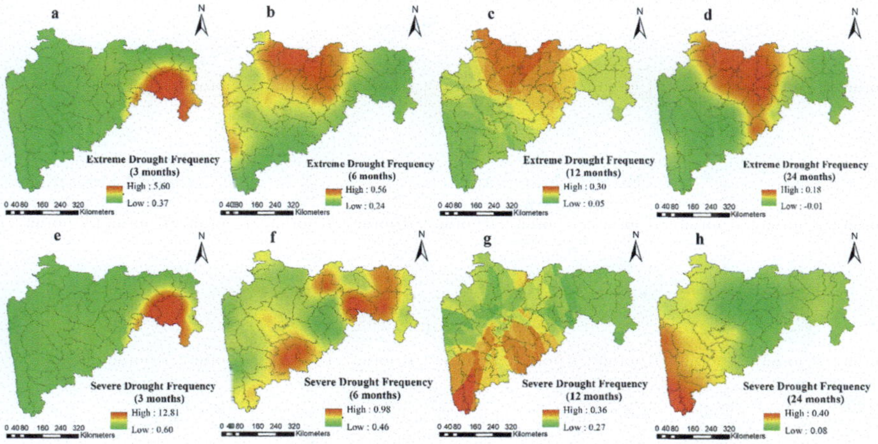

Fig. 8.2 **a** EDF 3 months **b** EDF 6 months **c** EDF 12 months **d** EDF 24 months **e** SDF 3 months **f** SDF 6 months **g** SDF 12 months **h** SDF 24 months

Mumbai suburban, Ratnagiri and Raigad (Fig. 8.2b). 12 months EDF is high in Nandurbar, Dhule, Jalgaon, Buldana, Amaravati, Aurangabad, Jalna, Beed, Latur, Hingoli and Washim (Fig. 8.2c). 24 months EDF is high in Nandurbar, Dhule, Jalgaon, Buldana, Akola, Washim, Hingoli, Aurangabad, Latur, Parbhani and Amaravati (Fig. 8.2d). 3 months Severe Drought Frequency is high in Chandrapur, Gadchiroli and Yavatmal (Fig. 8.2e). 6 months SDF is high in Chandrapur, Yavatmal, Nagpur, Bhandara, Gondia, Gadchiroli, Akola, Solapur, Osmanabad, Latur, Kolhapur, Sindhudurg, Sangli, Ahmednagar, Amaravati, Buldana, Beed, Pune, Nandurbar and Wardha (Fig. 8.2f).12 months SDF is high in Kolhapur, Sindhudurg, Sangli, Satara, Ratnagiri, Solapur, Osmanabad, Latur, Beed, Parbhani (Fig. 8.2g). 24 months SDF is high in Kolhapur, Sindhudurg, Ratnagiri, Raigad, Sangli, Solapur, Osmanabad, Beed, Pune, Ahmednagar, Thane, Mumbai city, Mumbai suburban, Nashik, Dhule and Nandurbar (Fig. 8.2h). The maps are similar for 3 months EDF and SDF signifying that the areas which are in extreme drought conditions are also in severe drought conditions but for other map the situation is not similar. Mainly for EDF of 6, 12 and 24 months the same areas show similar trend of roughly high frequency. Whereas in case of SDF for 12 and 24 months time scale show similar trends of high frequency in south west region.

Drought Magnitude

3 months DM is high in Dhule, Jalgaon, Palghar, Nashik, Mumbai city, Thane, Raigad, Mumbai suburban, Ratnagiri, Satara, Sangli, Kolhapur, Sindhudurg and Pune (Fig. 8.3a). 6 months DM is high in Sindhudurg, Kolhapur, Ratnagiri, Raigad,

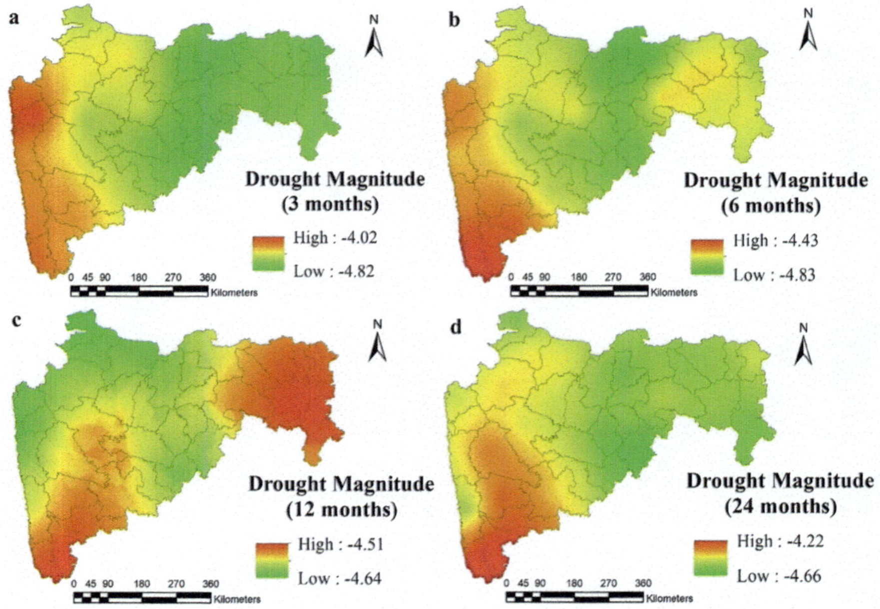

Fig. 8.3 **a** DM 3 months **b** DM 6 months **c** DM 12 months **d** DM 24 months

Mumbai city, Mumbai suburban, Satara, Sangli, Pune, Palghar, Yavatmal, Chandrapur, Gadchiroli and Bhandara (Fig. 8.3b). 12 months DM is high in Gadchiroli, Gondia, Chandrapur, Bhandara, Nagpur, Wardha, Yavatmal, Sindhudurg, Kolhapur, Sangli, Solapur, Satara, Ahmednagar (Fig. 8.3c). 24 months DM is high in Sindhudurg, Kolhapur, Sangli, Solapur, Satara, Pune, Raigad, Mumbai city, Mumbai suburban, Thane, Nashik, Dhule and Ahmednagar (Fig. 8.3d). DM maps for 3, 6 and 24 months show similar trends of high frequency in south west region.

Mean Drought Intensity

3 Months MDI is high in Nandurbar, Dhule, Jalgaon, Nashik, Aurangabad, Palghar, Thane, Sindhudurg, Sangli, Kolhapur, Ratnagiri, Raigad, Ahmednagar, Pune, Beed, Satara, Chandrapur and Gadchiroli (Fig. 8.4a). 6 months MDIis high in Gadchiroli, Chandrapur, Yavatmal, Wardha, Nanded, Parbhani, Hingoli, Washim, Akola, Amaravati, Buldana, Jalgaon, Dhule, Nandurbar, Sindhudurg, Kolhapur, Sangli, Palghar, Nashik, Aurangabad and Jalna (Fig. 8.4b). 12 Months MDI is high in Sindhudurg, Kolhapur, Sangli, Ratnagiri, Satara, Pune, Raigad, Thane, Mumbai city, Mumbai suburban, Palghar, Nashik, Aurangabad, Jalna, Parbhani, Nanded, Hingoli, Washim, Akola, Buldana, Jalgaon, Dhule, Nandurbar (Fig. 8.4c). 24 months MDI is high in Sindhudurg, Kolhapur, Sangli, Satara, Ratnagiri, Raigad, Mumbai city, Mumbai

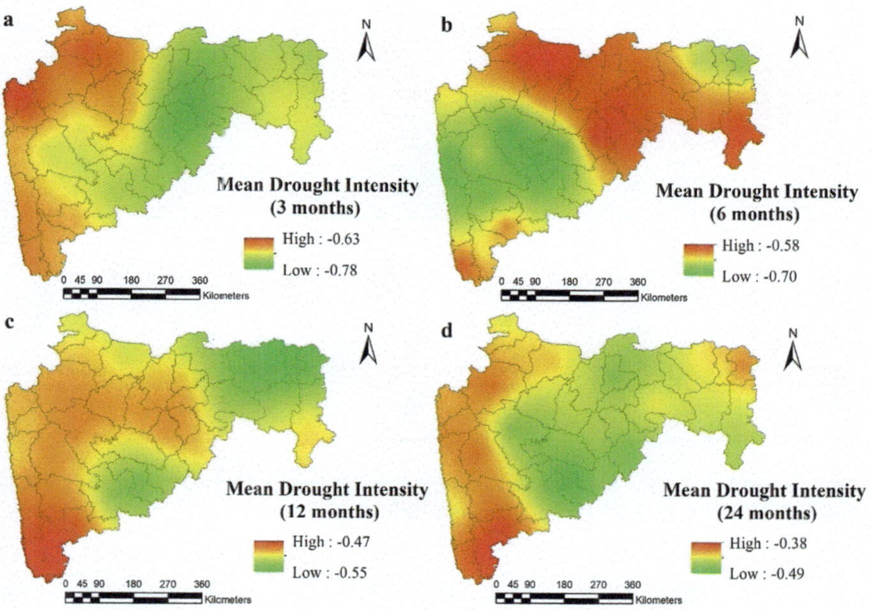

Fig. 8.4 **a** MDI 3 months **b** MDI 6 months **c** MDI 12 months **d** MDI 24 months

suburban, Thane, Pune, Nashik, Dhule, Jalgaon, Gondia, Nagpur, Bhandara, Chandrapur and Yavatmal (Fig. 8.4d). In case of MDI for 3, 12 and 24 months time scale show a roughly similar trend of high frequency in western part of Maharashtra.

Return Period

3 Months RPED is high in Nagpur, Bhandara, Wardha, Pune, Satara, Solapur, Sangli, Kolhapur, Sindhudurg, Ahmednagar and Ratnagiri (Fig. 8.5a). 6 Months RPED is high in Sindhudurg, Kolhapur, Sangli, Satara, Solapur, Osmanabad, Latur, Ratnagiri, Nanded, Yavatmal, Wardha, Chandrapur, Nagpur, Bhandara, Gondia and Gadchiroli (Fig. 8.5b). 12 Months RPED is high in Gondia, Gadchiroli, Sindhudurg, Kolhapur, Raigad and Ratnagiri (Fig. 8.5c). 24 Months RPED is high in Gadchiroli, Gondia, Chandrapur, Bhandara, Nagpur, Wardha, Amaravati and Yavatmal (Fig. 8.5d). 3 Months RPSD is high in Sindhudurg, Kolhapur, Sangli, Solapur, Satara, Ratnagiri, Pune, Mumbai city, Mumbai suburban, Palghar, Nashik, Jalgaon, Gadchiroli and Nandurbar (Fig. 8.5e). 6 Months RPSD is high in Sindhudurg, Ratnagiri, Raigad, Pune, Mumbai city, Mumbai suburban, Thane, Palghar, Nashik, Dhule, Nandurbar, Aurangabad, Satara, Ahmednagar and Solapur (Fig. 8.5f). 12 Months RPSD is high in Kolhapur, Sindhudurg and Ratnagiri (Fig. 8.5g). 24 Months RPSD is high in Gadchiroli, Gondia, Chandrapur, Bhandara, Nagpur, Wardha, Yavatamal,

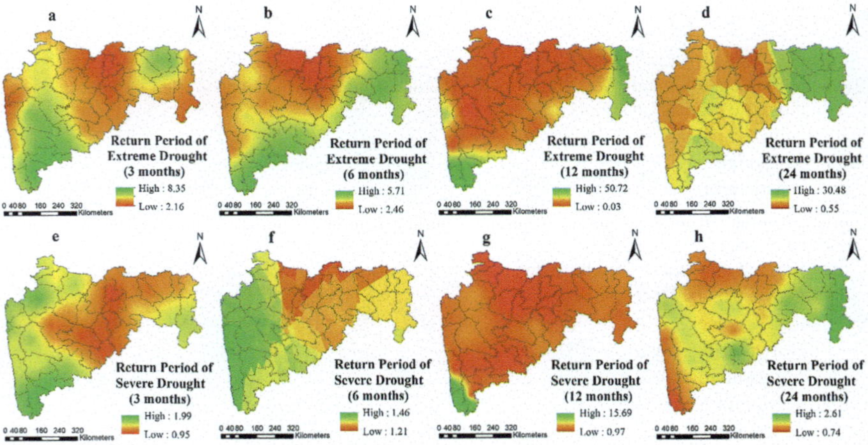

Fig. 8.5 **a** RPED 3 months **b** RPED 6 months **c** RPED 12 months **d** RPED 24 months **e** RPSD 3 months **f** RPSD 6 months **g** RPSD 12 months **h** RPSD 24 months

Osmanabad and Palghar (Fig. 8.5h). The Return Period maps shows roughly a similar sketches for extreme and severe drought conditions for 3, 6, 12 and 24 months time scale mainly signifying that the central Maharashtra is under high return period of both extreme and severe drought.

Conclusion

Due to climate change, weather patterns are changing every day, droughts are very common not only in India but all over the world. The present study assesses the spatio-temporal drought using rainfall data from NASA Power access and then SPI was calculated. Parameters like DF (3, 6, 12, 24 months), DM (3, 6, 12, 24 months), MDI (3, 6, 12, 24 months), RPED and RPSD (3, 6, 12, 24 months) were evaluated and finally spatial layers were prepared using the kriging function in Arc gis software. Using these spatial layers drought vulnerable regions can easily be identified which then can provide us an overview of drought situation in the whole state. Proper mitigation and management policies like using drought resistant crops, rain water harvesting, crop insurance etc., are required to reduce the problems. Drought reduction and supervision is a vital task to overcome the situation of drought in Maharashtra. Though proper fieldwork and grass root level surveying could have been used to improve the study.

References

Bai, X., Shen, W., Wu, X., & Wang, P. (2020). Applicability of long-term satellite-based precipitation products for drought indices considering global warming. *Journal of Environmental Management, 255*, Article 109846.

Banerjee, A., & Bera, S. (2021). GIS-based assessment for agricultural susceptibility and sustainability to meteorological drought in West Bengal. In *Modern cartography series* (Vol. 10, pp. 265–284). Academic Press.

Dikshit, A., Pradhan, B., Huete, A., & Park, H. J. (2022). Spatial based drought assessment: Where are we heading? A review on the current status and future. *Science of the Total Environment, 844*, Article 157239.

Dutta, D., Kundu, A., Patel, N. R., Saha, S. K., & Siddiqui, A. R. (2015). Assessment of agricultural drought in Rajasthan (India) using remote sensing derived Vegetation Condition Index (VCI) and Standardized Precipitation Index (SPI). *The Egyptian Journal of Remote Sensing and Space Science, 18*(1), 53–63.

Guo, B., & Wen, Y. (2019). An optimal monitoring model of desertification in Naiman banner based on feature space utilizing Landsat8 OLI image. *IEEE Access, 8*, 4761–4768.

Hayes, M. J., Svoboda, M. D., Wiihite, D. A., & Vanyarkho, O. V. (1999). Monitoring the 1996 drought using the standardized precipitation index. *Bulletin of the American Meteorological Society, 80*(3), 429–438.

Herrera-Estrada, J. E., Satoh, Y., & Sheffield, J. (2017). Spatiotemporal dynamics of global drought. *Geophysical Research Letters, 44*(5), 2254–2263.

Huete, A., Didan, K., Miura, T., Rodriguez, E. P., Gao, X., & Ferreira, L. G. (2002). Overview of the radiometric and biophysical performance of the MODIS vegetation indices. *Remote Sensing of Environment, 83*(1–2), 195–213.

Kumar, P. V., Osman, M., & Mishra, P. K. (2019). Development and application of a new drought severity index for categorizing drought-prone areas: A case study of undivided Andhra Pradesh state, India. *Natural Hazards, 97*(2), 793–812.

Lan, Y., Chawade, A., Kuktaite, R., & Johansson, E. (2022). Climate change impact on wheat performance—Effects on vigour, plant traits and yield from early and late drought stress in diverse lines. *International Journal of Molecular Sciences, 23*(6), 3333.

McKee, T. B., Doesken, N. J., & Kleist, J. (1993, January). The relationship of drought frequency and duration to time scales. In *Proceedings of the 8th Conference on Applied Climatology* (Vol. 17, No. 22, pp. 179–183).

Nasrollahi, M., Khosravi, H., Moghaddamnia, A., Malekian, A., & Shahid, S. (2018). Assessment of drought risk index using drought hazard and vulnerability indices. *Arabian Journal of Geosciences, 11*(20), 606.

Palmer, W. C. (1965). Meteorological drought. US. *Weather Bureau Research Paper, 45*, 1–58.

Querner, E. P., & van Lanen, H. A. (2001). Impact assessment of drought mitigation measures in two adjacent Dutch basins using simulation modelling. *Journal of Hydrology, 252*(1–4), 51–64.

Saadat, H., Adamowski, J., Bonnell, R., Sharifi, F., Namdar, M., & Ale-Ebrahim, S. (2011). Land use and land cover classification over a large area in Iran based on single date analysis of satellite imagery. *ISPRS Journal of Photogrammetry and Remote Sensing, 66*(5), 608–619.

Udmale, P. D., Ichikawa, Y., Kiem, A. S., & Panda, S. N. (2014). Drought impacts and adaptation strategies for agriculture and rural livelihood in the Maharashtra State of India.

Umran Komuscu, A. (1999). Using the SPI to analyze spatial and temporal patterns of drought in Turkey.

Vicente-Serrano, S. M., Beguería, S., & López-Moreno, J. I. (2010). A multiscalar drought index sensitive to global warming: The standardized precipitation evapotranspiration index. *Journal of Climate, 23*(7), 1696–1718.

Wilhite, D. A., & Glantz, M. H. (1985). Understanding: The drought phenomenon: The role of definitions. *Water International, 10*(3), 111–120.

Wu, D., Li, Y., Kong, H., Meng, T., Sun, Z., & Gao, H. (2021). Scientometric analysis-based review for drought modelling, indices, types, and forecasting especially in Asia. *Water, 13*(18), 2593.

Zargar, A., Sadiq, R., Naser, B., & Khan, F. I. (2011). A review of drought indices. *Environmental Reviews, 19*(NA), 333–349.

Zhu, Y., Wang, W., Singh, V. P., & Liu, Y. (2016). Combined use of meteorological drought indices at multi-time scales for improving hydrological drought detection. *Science of the Total Environment, 571*, 1058–1068.

Chapter 9
Landslide Susceptibility Zonation in Upper Part of Garhwal Himalyan Using Weight of Evidence (WOE) Method

Ruchi Saraswat, Gaurav Bharti, Vaishnavi Bansal, and Suvandan Saraswat

Abstract The landslides that have been taking place in a certain area of Uttarkashi are the subject of this particular piece of literature, which focuses on the topic. Earthquakes and flash floods are two more natural threats that have been brought up in this series of discussions. As a consequence of the increase in the number of activities carried out by humans, the number of landslides that have taken place in these areas has increased. This will make it possible for the required development activities to be carried out in the later stages of the project. It is required to split areas that are prone to risks into various zones in order to establish the amount of the influence that these dangers have. This is done in order to determine the extent of such an impact. Field data, Google Earth, landslip location data, eight Landsat satellite imageries, toposheets prepared by the Survey of India department, and the development of several thematic layers such as slope, altitude, structure, geology, drainage, land use, and land cover have all been utilized in the process of hazard zonation mapping. These hazard layers include slope, altitude, structure, geology, drainage, and land cover. The weight of evidence method was utilized in order to develop LHZ. This was accomplished by combining all of the spatial layers that are regarded to be a part of the GIS domain. A number of separate zones have been established throughout the region in accordance with the landslip threat zonation map. In order to validate the Landslip hazard zonation map, the Landslip Inventory map of the study region was utilized. This allowed for the validation of the map. In addition, the danger zonation map has been validated by mathematical analysis in addition to being tested in the field.

R. Saraswat (✉) · G. Bharti · V. Bansal
Department of Civil Engineering, JSS Academy of Technical Education, Noida, India
e-mail: rsaraswat@jssaten.ac.in

V. Bansal
e-mail: vaishnavibansal@jssaten.ac.in

S. Saraswat
Department of Mechanical Engineering, JSS University, Noida, India
e-mail: suvandan.saraswat@jssaten.ac.in

R. Sarkar et al. (eds.), *Climate Change, Environmental Hazards and Community-Based Resilience*, Disaster Risk Reduction, https://doi.org/10.1007/978-981-95-9485-6_9

Keywords Geographical information system · Landslide hazard zonation · Uttarkashi district

Introduction

Slag the Garhwal Himalaya in the Uttarkashi district is a region highly prone to natural disasters. This mountainous terrain has historically been affected by numerous landslides, earthquakes, and flash floods, making landslides the most frequent natural hazard. Several South Asian countries, including India, are significantly impacted by landslides, which pose severe risks and result in substantial financial losses. Researchers have extensively studied hazard zonation mapping and catastrophe prevention, focusing on mitigating the environmental and economic impacts in the Garhwal Himalaya (Bansal & Sarkar, 2023, Bansal & Sarkar, 2024a, 2024b). These studies often involve geotechnical examinations to assess the stability of landslides and propose corrective strategies for slope stabilization, dependent on the stability findings (Bharti et al., 2022). A weighted overlay technique was employed to incorporate the various thematic layers that were created into an all-encompassing framework. Following the creation of an attribute map, landslide hazard susceptibility zonation was completed, leading to the establishment of a landslide hazard index by incorporating additional necessary weights. Field data facilitated the creation of a zonation map illustrating landslide-prone areas. Numerous geotechnical investigations, employing a wide range of methodologies, have been conducted to distinguish between stable and unstable sites. Cartographic maps were also created to aid data analysis, focusing on landslide hazard potential analysis and the resultant landslide susceptibility potential map, crucial for future development planning.

The landslide susceptibility map was examined using multiple regression analysis, validated through area under the curve (AUC) methods. Geographic Information System (GIS) was employed for landslide hazard zonation, confirmed by comparing landslide rates across different classes. In southern Sikkim, statistical analysis was used for landslide danger zonation, involving the creation of thematic layers in the GIS domain and adjusting weight rating values through trial and error. This process resulted in the creation of a landslide hazard zonation (LHZ) map, which was divided into various zones. Risk assessments using airborne laser scanning data for different threat scenarios were also conducted to determine appropriate actions. The primary goal of this research is to develop a Hazard Hazard Zone (LHZ) map for the Uttarkashi region, classifying steep, landslide-prone terrain into zones based on risk levels.

Literature Review

Landslides are a significant geoenvironmental hazard, especially in mountainous regions such as the Himalayas. Various studies have been conducted to understand the factors leading to landslides and to develop methods for landslide hazard zonation. Anbalagan and Parida (2013) discussed the geoenvironmental problems caused by landslides in the Garhwal Himalayas, highlighting the need for comprehensive risk assessment and management strategies to mitigate the impact of these natural disasters. They emphasized the role of geological and climatic factors in triggering landslides in this region. Shah and Gupta (2002) utilized GIS-based methods to create landslide hazard zonation maps for the Bhagirathi Valley in the Himalayas. Their study demonstrated the effectiveness of remote sensing and GIS technologies in identifying and mapping landslide-prone areas. By analyzing various factors such as slope, aspect, and land use, they provided valuable insights into the spatial distribution of landslide hazards. Varnes (1984) further elaborated on the principles and practices of landslide hazard zonation, providing a foundational framework for subsequent studies. His work emphasized the importance of understanding the underlying mechanisms of landslides to improve hazard prediction and risk management. Deshpande et al. (2009) applied landslide hazard zonation mapping in specific areas of Uttarakhand, including Gopeshwar, Pipalkoti, and Nandprayag. Their research focused on identifying high-risk zones and suggested appropriate mitigation measures. The use of GIS tools allowed for a detailed analysis of the terrain, contributing to more accurate hazard assessments. Similarly, Kumar et al. (2012) conducted a micro-hazard evaluation in the North Western Garhwal Lesser Himalaya, utilizing Geomatics and geosciences techniques to validate landslide occurrences and their contributing factors. In a study by Onagh et al. (2012), multiple linear regression methods were employed to create landslide susceptibility maps in Uttarkashi district. This approach integrated various environmental parameters to predict landslide-prone areas, demonstrating the applicability of statistical methods in landslide studies. Pareta et al. (2012) also used quantitative methods in GIS for landslide hazard zonation, highlighting the importance of integrating multiple data sources to enhance the accuracy of hazard maps. Rawat et al. (2012) conducted a statistical analysis of landslides in South Sikkim, utilizing remote sensing and GIS to understand the spatial distribution and frequency of landslides. Their findings underscored the critical role of advanced geospatial technologies in landslide studies. Abdulwahid and Pradhan (2016) explored the use of airborne laser scanning data (LiDAR) for landslide vulnerability and risk assessment, demonstrating the potential of high-resolution data in improving hazard predictions. Shahinoor and Bayes (2017) combined weights of evidence and spatial multicriteria approaches to model landslide initiation and runout susceptibility in the context of hill cutting and rapid urbanization. Their study provided insights into the complex interactions between human activities and natural processes in triggering landslides. Pham et al. (2015) used the frequency ratio method for landslide susceptibility assessment in Uttarakhand, showing the effectiveness of statistical approaches in hazard mapping. Shit et al. (2016) applied a weighted overlay model for potential landslide susceptibility mapping, demonstrating the integration of various environmental factors in predicting landslide hazards. Lallianthanga et al. (2013) used remote sensing and GIS techniques for landslide hazard zonation in Mizoram, providing a case study of how modern technologies can be applied to landslide risk assessment in different geographic contexts. Lastly, Parkash (2015) and Parkash (2012) provided comprehensive studies on flash floods and landslide disasters in Uttarakhand, emphasizing the need for integrated disaster management approaches to address multiple hazards simultaneously. Their work highlighted the importance of collaboration between various stakeholders, including government agencies, researchers, and local communities, in developing effective risk reduction strategies.

Alimohammadlou et al. (2013) introduce a classification method to better understand landslide processes and their impacts. They advocate for a comprehensive approach to classify landslides based on geomorphological characteristics and triggering factors, enhancing the understanding of different landslide types and their potential impacts. Althuwaynee et al. (2014) explore the estimation of rainfall thresholds for landslide initiation in Kuala Lumpur and its surrounding areas. Ayalew and Yamagishi (2005) apply GIS-based logistic regression for mapping landslide susceptibility in the Kakuda-Yahiko mountains of Japan, demonstrating the effectiveness of GIS in spatial analysis and hazard assessment. Bell and Glade (2004) provide examples of quantitative risk analysis for landslides in Iceland, highlighting the importance of quantifying landslide risks to inform decision-making and develop mitigation strategies. Quantitative risk assessment offers a framework for evaluating potential impacts and prioritizing areas for intervention. Budimir et al. (2015) conduct a systematic review of landslide probability mapping using logistic regression. Their review consolidates various studies, offering a comprehensive overview of logistic regression applications in landslide susceptibility mapping. The findings highlight the robustness of logistic regression as a tool for predicting landslide-prone areas. Can et al. (2005) apply logistic regression analyses to assess the susceptibility of shallow earthflows triggered by heavy rainfall in Turkey, showcasing the use of logistic regression in modeling landslide susceptibility. Das et al. (2010) compare logistic regression with a rock mass classification system along a road section in the northern Himalayas, demonstrating the utility of logistic regression in diverse geological settings. Calvo and Savi (2009) utilize Monte Carlo procedures for debris flow risk assessment, illustrating the application of probabilistic methods in landslide risk evaluation. Their real-world application underscores the importance of incorporating stochastic processes into risk assessment frameworks. Chau et al. (2004) use landslide inventory data and GIS for landslide hazard analysis in Hong Kong. Chung and Fabbri (2003) focus on validating spatial prediction models for landslide hazard mapping. Validation is crucial for ensuring the reliability and accuracy of predictive models. Conforti et al. (2014) evaluate the prediction capability of artificial neural networks for mapping landslide susceptibility in Italy. Their findings demonstrate the potential of machine learning techniques in improving susceptibility mapping. Corominas et al. (2014) provide recommendations for the quantitative analysis of landslide risk. Crozier and Glade (2005) discuss the concepts and approaches to landslide hazard and risk, providing a foundational understanding of the issues involved in landslide risk assessment and emphasizing the need for comprehensive hazard and risk management strategies. Cui et al. (2009) investigate the distribution and risk evaluation of landslide lakes induced by the 2008 Wenchuan earthquake. Dai et al. (2002) provide an overview of landslide risk assessment and management. Their work outlines the key components of risk assessment, including hazard identification, vulnerability analysis, and risk mitigation. This comprehensive overview serves as a valuable reference for understanding the multifaceted nature of landslide risk assessment. Demir et al. (2015) compare frequency ratio and logistic regression methods for landslide susceptibility mapping in Turkey. Devkota et al. (2013) compare certainty factor, index of entropy, and logistic regression models for susceptibility mapping in Nepal, highlighting the importance of selecting appropriate models based on the specific characteristics of the study area.

Study Area

According to Fig. 9.1 and in Table 9.1, the research area extends from Dunda, which is located at a latitude of 30°42″20′ North and a longitude of 78°21″00′, to Bhatwari village, which is located at a latitude of 30°48″30′ North and a longitude of 78°37″15′ East.

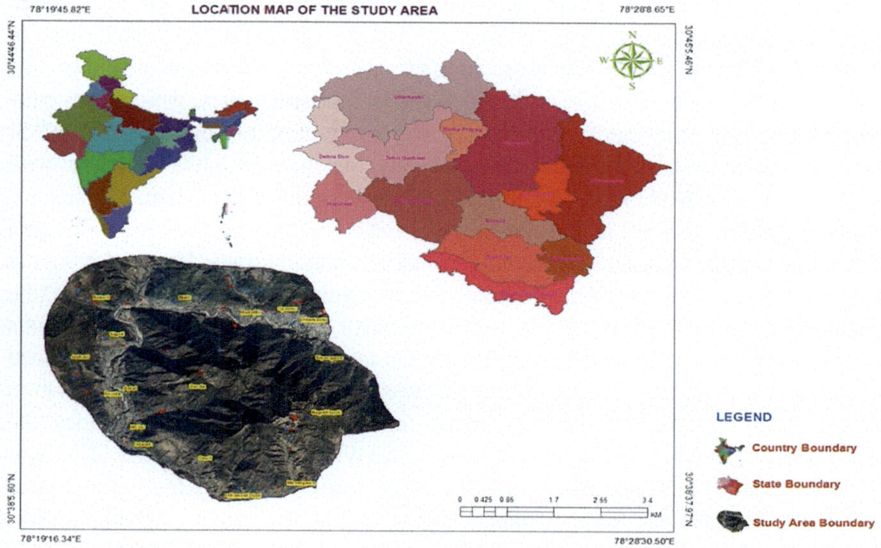

Fig. 9.1 Area of study (Bhatwari to Dunda)

Table 9.1 Specifics of the satellite data analysis used

S. no	1	2
Latitude	30°46′12″ N	30°28′10″ N
Longitude	78°18′48″ E	76°28′34″ E
Yr 2008	28/03/2008	7/4/2008
Yr 2017	17/12/2017	17/12/2017

Bhatwari is a significant town located in the Uttarkashi district, at an elevation of 1680 m, and is situated along the banks of the Bhagirathi River.

Methodology

Smelting Furnace Before beginning the process of implementing the Geographic Information System (GIS), a number of distinct thematic layers were first developed. This was done in order to accomplish the primary goal of generating a Landslip Hazard Zonation map. Next, a weighted grading method was utilized in order to superimpose these layers on top of one another. Utilizing this method, a weight was assigned to each theme layer, with the weight being determined by the relative significance of each component that has an effect on landslides. As can be seen in Fig. 9.2, this research made use of two different methodologies in order to accomplish the task of creating the Landslip Hazard Zonation map.

In order to attain the optimum landslide hazard zonation, it is essential to assign weights to the numerous factors that have an impact on landslides. As can be seen in Table 9.2, which provides the ranking criteria for a number of different attributes, this is accomplished by making use of the pairwise comparison matrix, which takes pairwise comparisons as its input and generates relative weights as its output. In Table 9.3, the weights that have been assigned are presented in a manner that is consistent with the comparison matrix. Table 9.4 contains a presentation of the contrast values for each component, as well as the weights that have been assigned to the individual elements through the use of the weight of evidence methodology. By employing the weight of evidence methodology, one is able to achieve the task of calculating the statistics that are related with the factors that control landslides. A variety of various theme layers, such as slope, slope aspect, precipitation, and geology, were produced and studied with the use of a technology that was based on a Geographic Information System (GIS). For each of the components that have an impact on landslides, the positive weight (Wi+) and the negative weight (Wi–) were found to be. This was accomplished through the process of discovery. You may assess the geographical correlation between predictor variables and landslides by calculating the contrast value (Cw), which is the difference between the two weights. This will allow you to identify the relationship between the two variables. A test of conditional independence was utilised in order to generate a landslide susceptibility map. The results

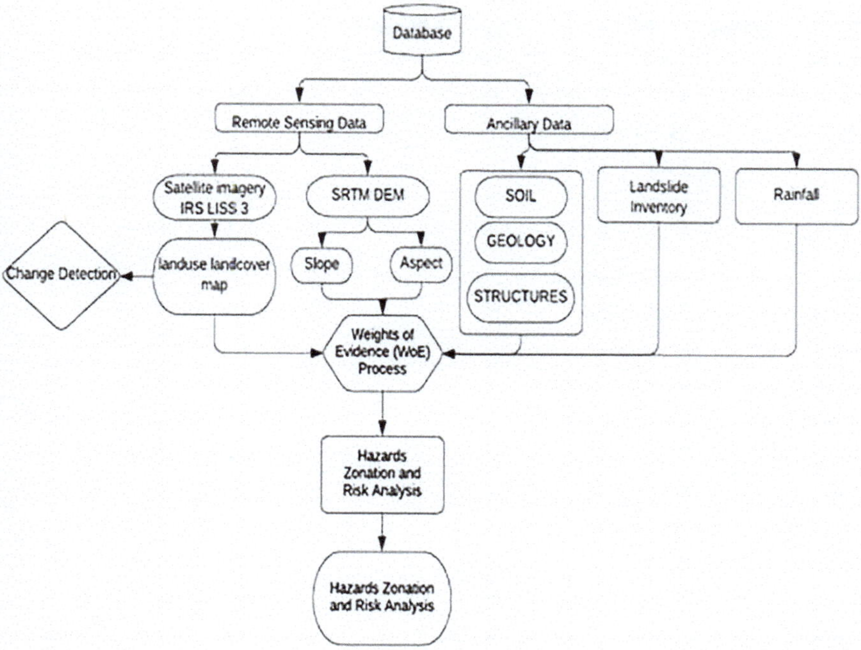

Fig. 9.2 Flowchart of methodology

of this test indicated that ninety percent of the landslides occur in areas that are considered to be high-risk.

This Table 9.3 presents the criteria used in a land suitability assessment, alongside their pairwise matrix comparison values and resulting weights. The criteria include slope, geology, land use, road network, drainage, and landslide risk The Table 9.4 presents a summary of the WOE and contrast factors for various environmental and geological factors. Each factor is broken down into different classes, and for each class, three values are listed: Wi+ (the weight of evidence for the presence of a characteristic), Wi− (the weight of evidence for its absence), and Cw (the contrast factor, calculated as the difference between Wi+ and Wi−).

Preparation of Thematic Data Layer

Slope

Slope, which is measured in degrees when referring to topography, is the gradient. Generally speaking, the slope value falls between 0 and 90°, where 0° represents level areas and 90 degrees represents vertical areas. Critical slope angles are the reason behind the occurrence of landslides. Table 9.5 shows that seventy-five to

Table 9.2 Scoring and ranking criteria according to landscape

Criteria	Class	Level of significance	LHZ appropriateness class
Slope	0–15	8	VHS
Slope	15–30	5	MS
Slope	30–45	3	LS
Slope	45–60	2	VLS
Slope	>60	1	LWS
Geology	Dunda quartzite	9	HGS
Geology	Epideorites	8	VHS
Geology	Pujargaon	7	HS
Geology	Bareti quartzite	6	MHS
Geology	Gamiri quartzite	5	MS
LULC	Forest	9	HGS
LULC	Settlement	8	VHS
LULC	Agriculture	3	LS
LULC	Open	2	VLS
LULC	River	1	LWS
Drainage	<50	1	VLS
Drainage	>50	1	HGS
Road proximity	<50	9	LS
Road proximity	>50	3	HGS
Landslide	<50	9	VLS
Landslide	>50	1	HGS

Table 9.3 Matrix of comparison using computation weights

Criteria	Pairwise matrix comparison						Weights criteria
	Slope (A)	Geology (B)	Land use (C)	Road Network (D)	Drainage (E)	Landslide (F)	(A + B + C + D + E + F)/6
	0.3	.34	0.01	0.59	0.54	0.2	0.33
	0.27	0.15	0.70	0.28	0.50	0.20	0.35
	0.20	0.10	0.10	0.09	0.1	0.07	0.11
	0.16	0.10	0.11	0.12	0.10	0.07	0.11
	0.09	0.06	0.02	0.025	0.025	0.02	0.04
	0.035	0.18	0.002	0.04	0.06	0.043	0.06
Total	1.055	0.93	0.942	1.145	1.325	0.603	1.00

Table 9.4 A synopsis of the classes' weights of evidence (WOE) and contrast considerations

Different factor	Range	Wi+	Wi–	Cw
Rainfall (mm/year)	2540–2640	0.6	–6.3	4.0
	2650–2679	0.0	0.1	–0.1
	2680–2710	0.0	0.0	0.0
	2711–2750	0.0	0.5	–0.5
	2751–2800	0.0	0.0	0.0
Aspect in direction	N	0.7	0.0	0.8
	NE	0.0	0.0	0.0
	E	–0.7	0.0	–0.8
	SE	–0.5	0.0	–0.6
	S	0.2	–0.1	0.3
	SW	0.2	–0.1	0.3
	W	0.2	0.0	0.2
	NW	0.0	0.0	0.0
Slope (Degree)	0–10°	–0.6	1.7	–2.4
	10.1–20°	1.6	–0.4	2.0
	20.1–30°	2.3	–0.1	2.4
	30.1–40°	–9.3	0.0	–9.2
	40.1–60°	–6.4	0.0	–6.3
	60.1–90°	–0.2	0.0	–0.2
Distance to the hill cut (m)	0–100	3.1	–0.5	3.1
	100–200	2.4	–0.2	2.1
	200–300	2.2	–0.2	–0.6
	300–400	0.7	0.0	–0.8
	>400	–2.1	2.5	–2.7
Distance to the road cut (m)	0–100	2.7	–0.5	3.2
	100–200	2.0	–0.1	2.1
	200–300	–0.6	0.0	–0.6
	300–400	–0.8	0.0	–0.8
	>400	–0.9	1.8	–2.8
Permeability of soil	Rapid	0.0	0.0	0.0
	Mixture moderate	0.0	0.0	0.0
	Moderate	0.0	0.6	–0.6
	Slow	0.0	0.4	–0.4
	Very slow	1.7	–6.7	8.2
Stream distance (m)	0–100	0.0	0.1	–0.1
	101–200	0.0	0.1	–0.1
	201–300	–1.3	0.1	–1.2

(continued)

Table 9.4 (continued)

Different factor	Range	Wi+	Wi–	Cw
	301–400	0.8	–0.1	0.9
	>400	0.3	–0.7	0.9
The distance to the drain (m)	0–100	1.6	–0.7	2.2
	101–200	1.1	–0.7	1.2
	201–300	–1.1	–0.2	–1.1
	301–400	0.0	0.1	–0.5
	>400	–1.2	0.1	–2.2
Unit of geology	Dunda quartzite	–2.7	0.9	–3.1
	Bareti quartzite	0.0	0.3	–0.1
	Dharasu formation	0.0	0.1	0.0
	Gamiri quartzite	–1.1	0.0	2.1
	Khattukhal limestone	2.7	–0.3	3.7
	Epidiorites	–1.2	–0.9	–1.7
Land cover	Builtup area	0.6	–0.6	1.3
	Dense and medium dense forest	2.1	2.4	0.0
	Low dense forest	1.1	1.2	–2.4
	Agriculture land	0.7	1.1	–0.7
	Open land	3.1	3.6	–0.1

ninety percent slope angles account for the bulk of landslides. Figure 9.3 displays a slope map of the research region created using a graphic information system (GIS).

This table categorizes landslide potentiality based on the slope gradient of an area. The slope is divided into six percentage ranges, each associated with a specific level of landslide Threat. Areas with a slope of 0–15% are considered to have "No Threat," indicating minimal risk of landslides. As the slope increases, so does the potential Threat: slopes of 15–30% are classified as "Low Threat [LT]," and slopes of 30–45% are categorized as "Low Moderate Threat [LMT]". Moving into higher risk categories, slopes of 45–60% are deemed "Moderate" Threat [MT], while slopes of 60–75% are classified as "High Threat [HT]." The highest risk category, "Very

Table 9.5 Slope variation in relation to the likelihood of landslides

S. no	Slope (%)	Possibility of landslides
1	0–14.99	NT
2	15–29.99	LT
3	30–44.99	LMT
4	45–59.99	MT
5	60–74.99	HT
6	75–90	VHT

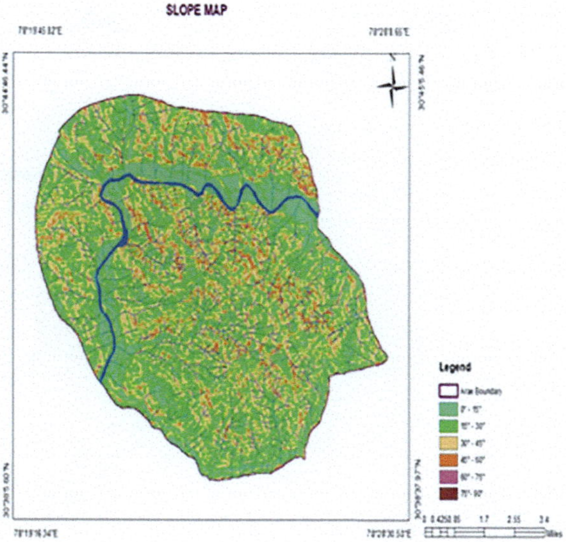

Fig. 9.3 Map of slope

High Threat," applies to areas with slopes of 75–90%. This classification helps in assessing and mitigating landslide risks in terrain with varying slopes.

Slope

When it comes to geology, the region under investigation is mostly made up of significant rock formations such as quartzite, metavolcanics, pujaragon limestone, and so on. In accordance with the information shown in Fig. 9.4, the primary rock formations are epidiorites, augen mylonite, and dunda quartzite taking place in the region under investigation and being more prone to landslides.

Utilizing and Covering Land

Furthermore, one of the main factors influencing the frequency of landslides is land use. This is because landslides are more prone to occur on slopes with less vegetation. The land cover classifications showed that areas that were both heavily forested and exposed to the elements had the highest percentage of landslides. Figure 9.5 shows the research region's land use and land cover map. This map was produced with GIS and originated from a LISS III satellite picture (Fig. 9.6).

Fig. 9.4 Map of geology

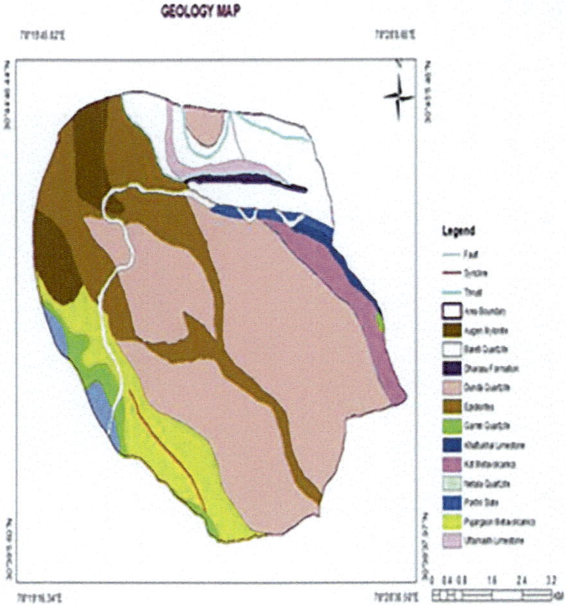

Fig. 9.5 Land use and land cover map

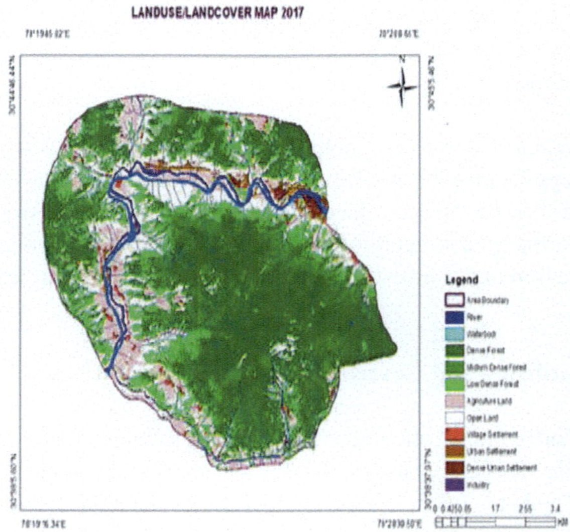

Drainage

When it comes to managing landslides, drainage is an essential component since the density of the soil and the geotechnical features of the soil are mostly determined by drainage. On the basis of stream order categorization, several drainage patterns

Fig. 9.6. Map of drainage

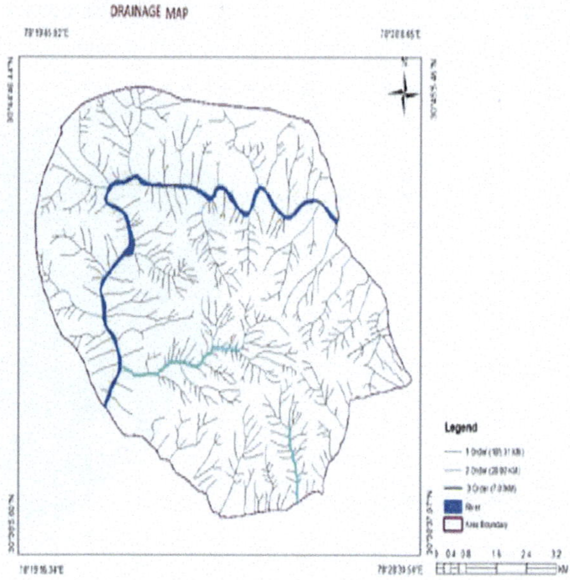

have been observed in the region under investigation. These patterns include the first order pattern consisting of 185.31 kms in length, the second order pattern consisting of 28.92 kms in length, and the third order pattern consisting of 7.07 kms in length.

Drainage Density

A river's drainage density—that is, the length of the stream per unit area—can be used to estimate how much flow it has. The slope angle or relative relief is the main component that decides whether a landslip happens in a particular area since it has a strong correlation with drainage density. Table 9.6 shows that the zone with the highest drainage density is also the one with the highest danger rating. As seen in Fig. 9.7, the zone with the highest drainage density also has the highest hazard rating.

The table classifies landslide potentiality based on varying levels of drainage density and slope angle. It shows that areas with a drainage density of 15 per km and

Table 9.6 Drainage density with slope angle

Sr. no	Density of drainage (in per km)	Angle of slope (in °)	Possibility of landslides
1	15	0–14.99	LT
2	144	15–29.99	LMT
3	215.35	30–44.99	MT
4	284.25	45–59.99	HT
5	359.6	>60	VHT

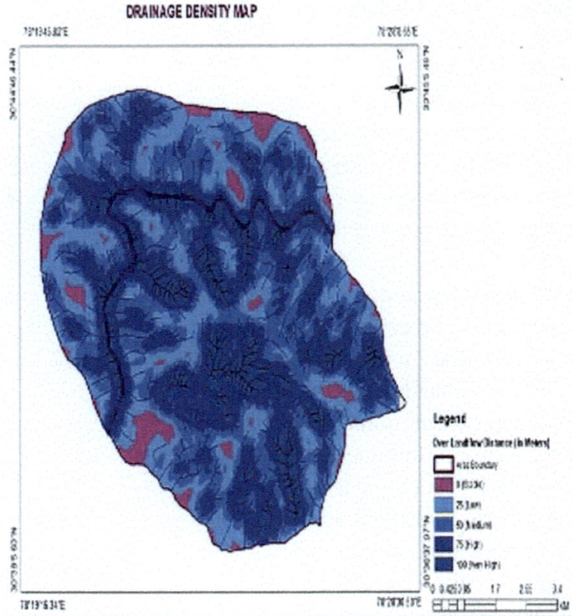

Fig. 9.7 Drainage density

slope angles between 0 and 15° are considered low hazard. As drainage density and slope angle increase, the landslide hazard potential escalates: a drainage density of 144 per km with slope angles of 15–30° falls into the low moderate hazard category, while 215.35 per km with 30–45° is deemed moderate. High hazard areas are those with a drainage density of 284.25 per km and slope angles of 45–60°. Finally, the highest risk, or very high hazard, is assigned to regions with a drainage density of 359.6 per km and slope angles exceeding 60°. This data highlights the correlation between higher drainage densities and steeper slopes with increased landslide risk.

Structure

To determine the rating of the structure, the distance from the major thrusts and faults was taken into consideration. This was due to the fact that there were more slide zones in close vicinity to significant faults. The stability of the slope is significantly impacted by a number of discontinuities, including faults, thrusts, joints, and shear zones, among others. A structure map of the region under investigation is depicted in Fig. 9.8, which demonstrates the existence of a variety of structural elements that are present in that location.

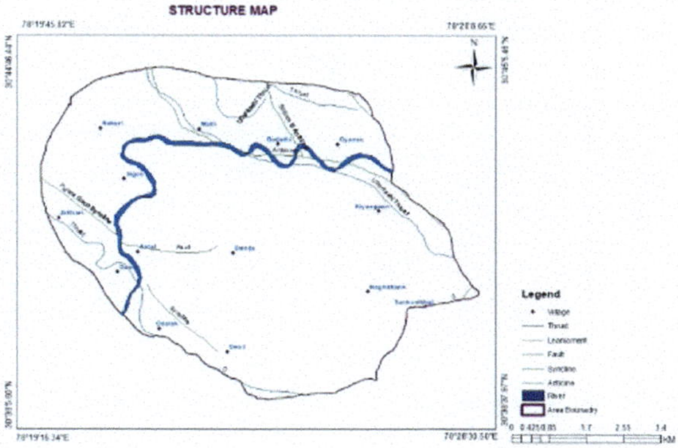

Fig. 9.8 Map depicting structure

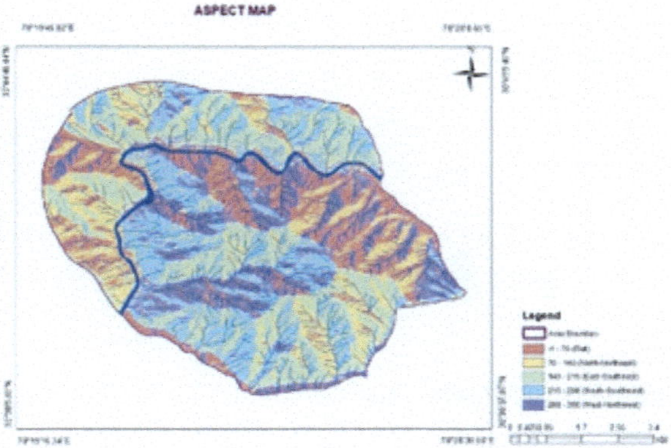

Fig. 9.9 Map depicting aspect

Aspect

Aspect is one of the most important aspects since it shows which way the slope is moving. It's possible that pertinent aspect qualities, such exposure to wind, sunlight, and precipitation, as well as other discontinuities, can regulate the frequency of landslides. The aspect map of the whole region is seen in Fig. 9.9.

Map Depicting Soil

The soil map is one of the most crucial components that must be assembled in order to create a landslip hazard zonation map. The soil map has been found to have been provided by the National Bureau of Soil Survey. The georeferenced soil map allowed for the successful classification of soil into several categories. The many types of soils that are found in the region under investigation are depicted in Fig. 9.10.

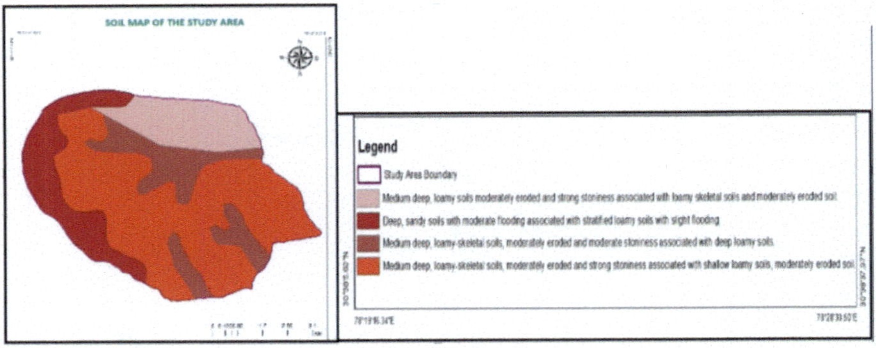

Fig. 9.10. Map depicting soil

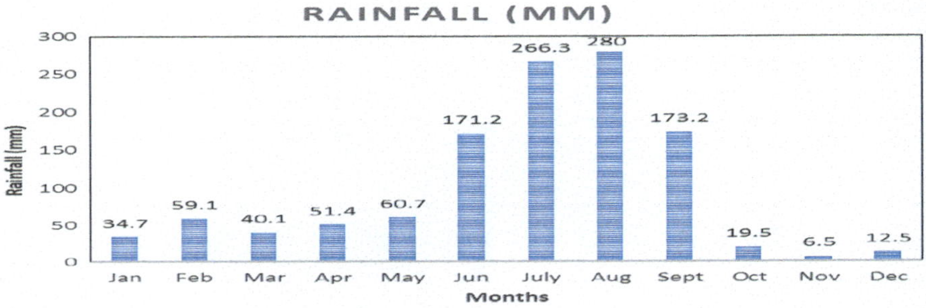

Fig. 9.11 Rainfall data of study area (2013–2017)

Rainfall Data

In light of the fact that the majority of landslides are of the triggered kind, rainfall is one of the most crucial variables that might cause them.

In accordance with the information that is accessible from the IMD 3.1.9 Road Factor, Fig. 9.11 illustrates the monthly variance in rainfall.

Road Factor

The impact that landslides have on roads and railroads is one of the most significant effects that they have, since they have the potential to damage a huge number of people in any given location. The specifications of the roads that are located in this research region are included in Table 9.7, together with the percentage of the land that has been covered by landslides.

The table provides details on various road types, including village roads [VR], national highways [NH], and state highways [SH], outlining their lengths, widths, total covered areas, and the percentage of the total road area they occupy. Village roads have a length of 77.49 km and a width of 4 m, covering an area of 0.310 km^2,

Table 9.7 Specifics of the roads that have landslides

Type of road	Length in km	Width of road	Area covered in total (km^2)	Road area in %
VR	77.49	4	0.310	2.585
NH	21.34	10	0.2135	1.79
SH	23.32	6	0.140	1.17

which is 2.585% of the total road area. National highways are 21.34 km long and 10 m wide, covering 0.2135 km^2, or 1.79% of the road area. State highways with a length of 23.32 km and a width of 6 m, cover 0.140 km^2, accounting for 1.17% of the total road area. This table offers a comparison of the extent and proportion of each road type within the examined region.

Results and Discussions

Map of Landslide Inventory

Preserving the specifics of the complete landslide occurrence, including stable, passive, reactivated, and, most of the time, active landslides, is the main objective of a landslide inventory map. Figures 9.13 and 9.14 demonstrate how useful all of this information is for hazard zonation mapping, vulnerability assessments, and hazard analysis. It also offers details on the landslip that is occurring in that specific location. In the region under evaluation, there are 28 latent landslides and 55 active landslides, according to information gathered during the field study.

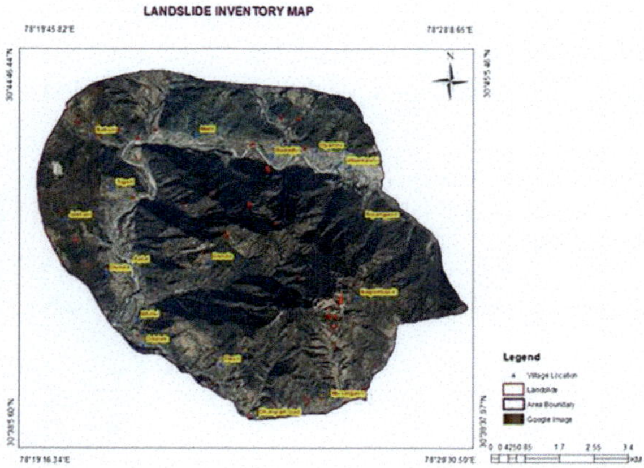

Fig. 9.13 Landslide inventory map

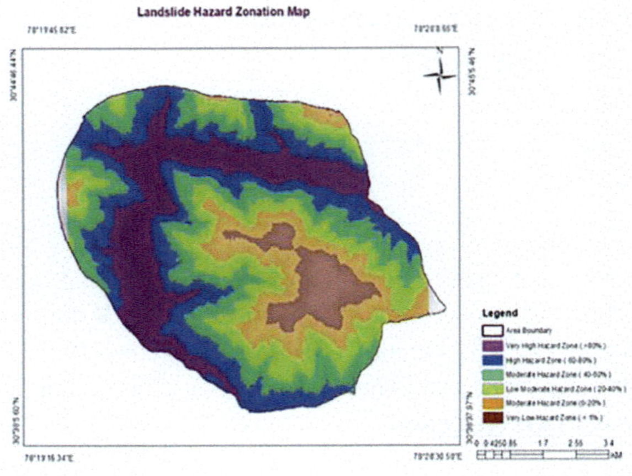

Fig. 9.14 Map of the landslide hazard zone

Zone of Hazard for Landslides

Determining which area is more prone to landslides and how much of a risk they pose at a particular location is the main goal of the Land Hazard Zone (LHZ). The steep terrain that is prone to landslides is divided into several zones on the LHZ map, and each zone is categorized based on how vulnerable it is to landslides. This can only be achieved by identifying the areas affected by landslides and estimating the probability that they will occur within a given time range. Figure 9.15 displays data related to landslide susceptibility and the categorization of locations into different susceptible zones based on areas affected by landslides.

The Table 9.8 classifies different landslide hazard zones, quantifying their areas in square kilometers and as percentages of the total area surveyed. The "Very Low Hazard [VLH]" zone covers 13.4 km^2 (0.28%) and is deemed safe for development, while the "Low Hazard [LH]" zone spans 11.29 km^2 (37.32%), suggesting low risk but with no specific remarks. The "Low Moderate Hazard [LMH]" zone, at 15.88 km^2 (8.94%), indicates local instability. The "Moderate Hazard [MH]" zone covers

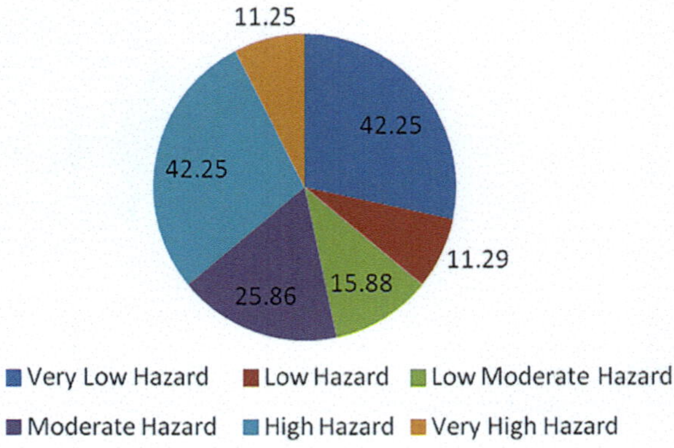

Fig. 9.15 Classification of the study area's landslide threat zones

Table 9.8 Information about the landslide hazard zone

S. no	1	2	3	4	5	6
Zone of landslide hazard	VLH	LH	LMH	MH	HH	VHH
Area (sq.km)	13.4	11.29	15.88	25.86	42.25	11.25
Area (%)	0.28	37.32	8.94	33.13	14.68	5.66
Comments	Secure for use in zone development operations		Instability in the local vulnerable zone		Insecure area for building operations	

25.86 km^2 (33.13%), implying a moderate landslide risk. The "High Hazard [HH]" zone, covering 42.25 km^2 (14.68%), and the "Very High Hazard [VHH]" zone, at 11.25 km^2 (5.66%), are both considered unsafe for development due to high and very high landslide risks, respectively. Zone of Extremely High Threat zone is extremely unstable and faces a persistent risk of landslides, particularly during and after a period of strong rainfall. This zone encompasses a total area of 11.25 km^2. The designation "High Threat Zone" designates areas with steep slopes covered in soil debris and worn rock. When disturbed, these slopes can cause landslides. There is a significant chance of debris sliding in these dangerous areas. This zone covers an area that is 42.25 km^2 in size. Although there may be unstable zones in some areas, this zone is generally steady. This zone is classified as having a moderate level of hazard. Its total size is approximately 15.58 km^2. The Zone of Moderately High Threat zone is comprised of regions that have a moderately high density of flora, a moderately high slope angle, and rocks that are generally compact and extremely hard. Approximately 25.86 km^2 are covered by it. Low Threat Zone refers to regions in which the combination of different regulating parameters is not expected to have a negative impact on the stability of the slope. This zone encompasses an area that is approximately 11.29 km^2 in size. Very Low Hazard Zone is mostly made up of vegetative regions and is situated in a location that is geographically removed from human settlements. Because of this, it is presumed to be free from landslip threats both in the present and in the future. This zone has an area that is approximately 13.4 square.

Conclusions

The present study concludes that the upper Garhwal Himalaya, specifically the Uttarkashi region between Bhatwari and Dunda, is highly prone to landslides due to its steep topography, fragile geological formations, and intense rainfall events. By employing the Weight of Evidence (WoE) method integrated within a Geographic Information System (GIS) framework, a comprehensive landslide hazard zonation (LHZ) map was developed to identify and classify areas according to their susceptibility levels. Multiple thematic layers—including slope, aspect, geology, drainage, land use and land cover, soil type, rainfall, and proximity to roads and hill cuts—were analyzed to assess their spatial correlation with landslide occurrences. The WoE analysis quantitatively evaluated each factor's contribution to landslide initiation by calculating positive and negative weights and contrast values. Results revealed that slope, geology, and drainage density exerted the highest influence, while proximity to roads and structural discontinuities (faults and thrusts) significantly increased susceptibility. Gentle to moderately steep slopes (10–30°) and areas with weak lithological units such as Khattukhal Limestone were identified as the most vulnerable. Approximately 19–20% of the study area falls under high to very high hazard zones, whereas about 33% lies in moderate zones. Low and very low zones are confined to gentle, vegetated terrains and stable geological formations. Validation using the landslide inventory and comparative assessment with the Analytical Hierarchy Process (AHP) model indicated strong spatial agreement—over 80% for the study area and more than 90% for observed landslide locations—confirming the model's robustness. The study establishes that the WoE approach is an efficient, data-driven technique for regional landslide susceptibility mapping. The developed LHZ map provides a scientific basis for land-use planning, infrastructure development, and slope stabilization measures. It also aids policymakers and engineers in prioritizing risk mitigation efforts and promoting sustainable development in the geologically sensitive Himalayan terrain.

References

Abdulwahid, M. W., & Pradhan, B. (2016). *Landslide vulnerability and risk assessment for multi-hazard scenarios using airborne laser scanning data (LiDAR)*. Springer-Verlag

Alimohammadlou, Y., Najafi, A., & Yalcin, A. (2013). Landslide process and impacts: A proposed classification method. *CATENA, 104*, 219–232.

Althuwaynee, O. F., Pradhan, B., & Ahmad, N. (2015). Estimation of rainfall threshold and its use in landslide hazard mapping of Kuala Lumpur metropolitan and surrounding areas. *Landslides, 12*, 861–875.

Althuwaynee, O. F., Pradhan, B., Park, H. J., & Lee, J. H. (2014). A novel ensemble bivariate statistical evidential belief function with knowledge-based analytical hierarchy process and multivariate statistical logistic regression for landslide susceptibility mapping. *CATENA, 114*, 21–36.

Anbalagan, R., & Parida, S. (2013). Geoenvironmental problems due to harmony landslide in Garhwal Himalaya, Uttarakhand, India. *International Journal of Emerging Technology and Advanced Engineering, 3* (Special Issue 3), 553–559. ICERTSD

Ayalew, L., & Yamagishi, H. (2005). The application of GIS-based logistic regression for landslide susceptibility mapping in the Kakuda-Yahiko mountains, Central Japan. *Geomorphology, 65*(1–2), 15–31.

Bansal, V., & Sarkar, R. (2023). Numerical analysis and implications of fluctuating water levels on slope stability. In *International conference on sustainable infrastructure: innovation, opportunities and challenges* (pp. 407–419). Springer

Bansal, V., & Sarkar, R. (2024a). Comparative analysis of slope stability for Kalimpong Region under dynamic loading using limit equilibrium method and machine benchmark learning classifiers. *Iranian Journal of Science Technology, Transactions of Civil Engineering, 48*, 2785–2807.

Bansal, V., & Sarkar, R. (2024b). Prophetical modeling using limit equilibrium method and novel machine learning ensemble for slope stability gauging in Kalimpong. *Iranian Journal of Science Technology, Transactions of Civil Engineering, 48*, 411–430.

Bell, R., & Glade, T. (2004). Quantitative risk analysis for landslides-Examples from Bíldudalur, NW-Iceland. *Natural Hazards and Earth System Sciences, 4*(1), 117–131.

Bharti, G., Shukla, B. K., Srinivasan, V., & Bansal, V. (2022). Experimental study on eccentrically loaded rectangular footing on sand for embedded condition to enhance serviceability of structures. *Materials Today: Proceedings, 61*, 517–522.

Budimir, M. E. A., Atkinson, P. M., & Lewis, H. G. (2015). A systematic review of landslide probability mapping using logistic regression. *Landslides, 12*, 419–436.

Calvo, B., & Savi, F. (2009). A real-world application of Monte Carlo procedure for debris flow risk assessment. *Computers & Geosciences, 35*(5), 967–977.

Can, T., Nefeslioglu, H. A., Gokceoglu, C., Sonmez, H., & Duman, T. Y. (2005). Susceptibility assessments of shallow earthflows triggered by heavy rainfall at three catchments by logistic regression analyses. *Geomorphology, 72*(1–4), 250–271.

Chau, K. T., Sze, Y. L., Fung, M. K., Wong, W. Y., Fong, E. L., & Chan, L. C. P. (2004). Landslide hazard analysis for Hong Kong using landslide inventory and GIS. *Computers & Geosciences, 30*(4), 429–443.

Chung, C. J. F., & Fabbri, A. G. (2003). Validation of spatial prediction models for landslide hazard mapping. *Natural Hazards, 30*, 451–472.

Conforti, M., Pascale, S., Robustelli, G., & Sdao, F. (2014). Evaluation of prediction capability of the artificial neural networks for mapping landslide susceptibility in the Turbolo River catchment (Northern Calabria, Italy). *CATENA, 113*, 236–250.

Corominas, J., van Westen, C., Frattini, P., Cascini, L., Malet, J. P., Fotopoulou, S., & Smith, J. T. et al. (2014). Recommendations for the quantitative analysis of landslide risk. *Bulletin of Engineering Geology and the Environment, 73*, 209–263.

Crozier, M. J., & Glade, T. (2005). Landslide hazard and risk: issues, concepts and approach. *Landslide Hazard and Risk*, 1–40.

Cui, P., Zhu, Y. Y., Han, Y. S., Chen, X. Q., & Zhuang, J. Q. (2009). The 12 May Wenchuan earthquake-induced landslide lakes: Distribution and preliminary risk evaluation. *Landslides, 6*, 209–223.

Dai, F. C., Lee, C. F., & Ngai, Y. Y. (2002). Landslide risk assessment and management: An overview. *Engineering Geology, 64*(1), 65–87.

Das, I., Sahoo, S., van Westen, C., Stein, A., & Hack, R. (2010). Landslide susceptibility assessment using logistic regression and its comparison with a rock mass classification system, along a road section in the northern Himalayas (India). *Geomorphology, 114*(4), 627–637.

Demir, G., Aytekin, M., & Akgun, A. (2015). Landslide susceptibility mapping by frequency ratio and logistic regression methods: An example from Niksar-Resadiye (Tokat, Turkey). *Arabian Journal of Geosciences, 8*, 1801–1812.

Deshpande, P. K., Patil, J. R., Nainwal, D. C., & Kulkarni, M. B. (2009). Landslide hazard zonation mapping in Gopeshwar, Pipalkoti and Nandprayag areas of Uttarakhand. In *Indian geotechnical conference, 16–19 December, Guntur, Andhra Pradesh.*

Devkota, K. C., Regmi, A. D., Pourghasemi, H. R., Yoshida, K., Pradhan, B., Ryu, I. C., & Althuwaynee, O. F., et al. (2013). Landslide susceptibility mapping using certainty factor, index of entropy and logistic regression models in GIS and their comparison at Mugling–Narayanghat road section in Nepal Himalaya. *Natural Hazards, 65*, 135–165.

Kumar, K., Devrani, R., Kathait, A., & Aggarwal, N. (2012). Micro-hazard evaluation and validation of landslide in a part of North Western Garhwal Lesser Himalaya, India. *International Journal of Geomatics and Geosciences, 2*(3), 2012.

Lallianthanga, R. K., Lalbiakmawia, F., & Lalramchuana, F. (2013). Landslide hazard zonation of Mamit town, Mizoram, India using remote sensing and GIS techniques. *International Journal of Geology, Earth and Environmental Sciences*, 3(1), 184–194. ISSN: 2277–2081 (Online).

Onagh, M., Kumra, V. K., & Rai, P. K. (2012). Landslide susceptibility mapping in a part of Uttarkashi district (India) by multiple linear regression method. *International Journal of Geology, Earth and Environmental Sciences* 2(2), 102–120.

Pareta, K., Kumar, J., & Pareta, U. (2012). Landslide hazard zonation using quantitative methods in GIS. *International Journal of Geospatial Engineering and Technology*, *1*(1), 1–9.

Parkash. (2015). *A study on flash floods and landslide disaster on 3rd August 2012 along Bhagirathi valley in Uttarkashi district*. Uttarakhand by National Institute of Disaster Management.

Parkash. (2012). *Training module on comprehensive "Landslide risk management" by National Institute of Disaster Management.*

Pham, T. B., Bui, T. D., Indra, P., & Dholakia, M. B. (2015). Landslide susceptibility assessment at a part of Uttarakhand Himalaya, India using GIS—based statistical approach of frequency ratio method. *International Journal of Engineering Research & Technology (IJERT)*, *4*(11). ISSN: 2278–0181.

Rawat, M. S., Rawat, B. S., Joshi, V., & Kimothi, M. M. (2012). Statistical analysis of Landslides in South district, Sikkim, India using remote sensing and GIS. *IOSR Journal of Environmental Science, Toxicology and Food Technology (IOSR-JESTFT)*, *2* (3), 47–61.

Shah, A. K., & Gupta, R. P. (2002). GIS-based landslide hazard zonation in the Bhagirathi (Ganga) valley, Himalaya. *International Journal of Remote Sensing, 23*, 357–369.

Shahinoor. R. M., & Bayes. A. (2017). Landslide initiation and runout susceptibility modelling in the context of hill cutting and rapid urbanization: A combined approach of weights of evidence and spatial multicriteria. *Journal of Mountain Science-Springer, 14*(10), 1919–1937.

Shit, K. P., Bhunia, S. G., & Maiti, R. (2016). Potential landslide susceptibility mapping using weighted overlay model (WOM) Springer International Publishing Switzerland 2016 Model. *Earth Syst Environ, 2*, 21.

Varnes, D. J. (1984). Landslide hazard zonation: A review of principles and practice. *Natural Hazards*, *3*, 63. UNESCO.

Chapter 10
Assessment of Spatio-temporal Pattern of Drought Using Standardized Precipitation Index in Tamil Nadu, India

Priti Paul and Priyanka Gogoi

Abstract Drought is the largely reoccurring natural hazards because of meteorological imbalances and climate change and impacts every other spheres like agriculture, hydrology and socio-economical. In this research work the rainfall data was collected from 1981 to 2022 from NASA Power Access and here mainly SPI (Standardized Precipitation Index) was used as a drought index. Several parameters like EDF (Frequency of Extreme Drought), SDF (Frequency of Severe Drought), DM (Magnitude of Drought), MDI (Mean Intensity of Drought), RPED (Extreme Drought Return Period) and RPSD (Severe Drought Return Period) for the time scale of 3, 6, 12, 24 months each were calculated from SPI and evaluated for spatio temporal analysis of drought in Tamil Nadu. ArcGIS was used for spatial layer preparation of the parameters stated earlier. The spatio-temporal estimation of drought in Tamil Nadu shows the need and urgency of drought mitigation policies and schemes there to lessen the unlikeable effects of drought.

Keywords Drought magnitude · Intensity of drought · Drought return period · Tamil Nadu

Introduction

Drought is the majorly frequently arising natural risks that indicate a lack of precipitation, resulting to scarcity of water, agricultural loss, and a deteriorated ecosystem. It is a significant problem in the modern world and has numerous detrimental repercussions on human life, agriculture, water resources, society, and the environment (Pei et al., 2018; Rahman & Latch, 2016; Zhang et al., 2015). Drought is the outcome of a substantial time of water scarcity arises because of climatic factors like decline in rain (Mohmmed et al., 2018) or anthropogenic ones, including land-use change (Keesstra, 2007), concretisation, burning fossil fuels, urban expansion, deforestation,

P. Paul · P. Gogoi (✉)
Department of Geography, University of Gour Banga, Malda, West Bengal, India
e-mail: priyankag19161@gmail.com

R. Sarkar et al. (eds.), *Climate Change, Environmental Hazards and Community-Based Resilience*, Disaster Risk Reduction, https://doi.org/10.1007/978-981-95-9485-6_10

pollution etc. According to several reports, climate change is the primary factor that can have a sizeable effect on the environment, agriculture, vegetation, and animals of the influenced region. Droughts are associated with economic losses, altered water supply systems, and food insecurity at the regional level (Ding et al., 2011; Sternberg, 2012). In Saha et al. (2022), it has been found that there are four chief types of droughts: one related to shortage of rain called as meteorological, second related to shortage of water in water resources called hydrological, third being related to insufficient soil moisture called agricultural, and the last one which is the consequence of the first three leading to shortages of water supply called socioeconomic (Wilhite, 2016). If the rain showers deficit ranges between 26 and 50%, it is classified as 'severe drought'; when it reaches 50%, it is classified as 'moderate drought'. Drought vulnerability can be measured using IPCC contributing factors such as exposure, adaptive capacity, and sensitivity (Engstrom et al., 2020). Spatio-temporal drought assessment is a scientific approach to evaluating drought patterns that takes into accounts both their spatial distribution and temporal variability. It involves the systematic evaluation of drought characteristics (Zargar et al., 2011), drought frequency, average drought intensity, drought magnitude, and drought return period. This technique is fundamental for recognizing the evolution of drought occurrences, detecting vulnerability, and guiding drought supervision and mitigation strategies (Mishra & Singh, 2010). Drought is not a uniform phenomenon; it varies with climatic conditions, topography, and land use. Therefore, spatio-temporal analysis enables researchers to spot drought prone regions, monitor drought progression, assess risk, and formulate mitigation strategies. By integrating remote sensing and GIS, researchers can visualize drought patterns at local, national, and international scales. The Standardised Precipitation Index (SPI), Normalized Difference Vegetation Index (NDVI) and Palmer Drought Severity Index (PDSI) are commonly used drought indicators for assessing drought severity and duration throughout time. Traditionally, several drought assessment and monitoring indices rely on rainfall data. These indexes are frequently difficult to get and are not designed to be easily comprehended by policy makers (Ji & Peters, 2003). SPI is the most extensively used and accepted drought index. SPI results at multiple time scales, such as 3, 6, 12, and 24 months, indicate the effects of drought on diverse water availability factors such as soil wetness, water reserves, and ground water. In this research work using parameters like extreme drought frequency, severe drought frequency, drought magnitude, mean drought intensity, return period of extreme drought and return period of severe drought were used for assessing the spatial and temporal patterns of drought in Tamil Nadu using SPI. Tamil Nadu is the second largest economy in India and therefore drought should be properly managed and checked to improve the economic conditions of Tamil Nadu.

Study Area

Tamil Nadu is the eleventh largest state in India, while in terms of people, it ranks sixth (Fig. 10.1). The states that border it are Andhra Pradesh to the north, Kerala and Karnataka to the west. The southernmost point of the Indian Peninsula, according to the 2011 census, there were 72,147,030 people living in Tamil Nadu state and the density of population is 555 people per square kilometer. This state spans 130,058 square kilometers and is flanked by latitudes 8004′ N and 13,035′ N and longitudes 76,018′ E and 80,020′ E. A drought is a threat to Tamil Nadu (Sellaperumal Pazhanivelan et al., 2023). The research area exhibits significant diversity in temperature, rainfall, slope, humidity, soil, geology, and vegetation type. The highest yearly average temperature is 240 °C, which drops to around 100 °C during the winter (Sellaperumal Pazhanivelan et al., 2023).

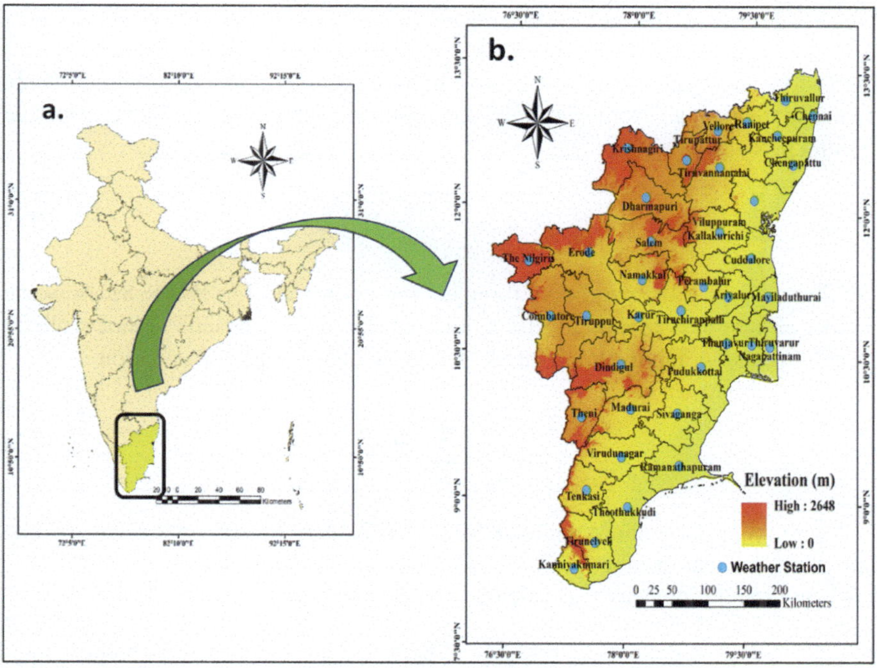

Fig. 10.1 Study area location map

Materials and Methods

Materials

For assessing the spatial and temporal drought in Tamil Nadu twenty five factors based on drought index named SPI were used. Factors were calculated on different time scales of 3 months, 6 months, 12 months and 24 months. The factor like monthly rainfall data was sourced from NASA power access and other factors like extreme drought frequency, severe drought frequency, drought magnitude, mean drought intensity and return period of extreme drought and severe drought are sourced from SPI values.

Methods

The factors listed above used for spatio temporal analysis of drought in Tamil Nadu are calculated using several formulae. Drought frequency was calculated using the formula given below

$$N_{i,100} = \frac{Ni}{i\,.\,n} * 100. \tag{10.1}$$

Where, $N_{i,100}$ = the number of droughts for a timescale I in 100 years, N_i = the number of months with droughts for a timescale I in the n year set, i = time scale (3, 6, 12, 24 months). Drought magnitude was calculated using

$$M_D = \sum_{i=1}^{m} SPI_{ij}. \tag{10.2}$$

Where, M_D= Drought Magnitude, SPI_{ij} = SPI values of a drought event at j time scale, m = number of months.

Mean drought intensity was calculated using formula

$$MI = \frac{MD}{Mean}. \tag{10.3}$$

Where, MI= mean drought intensity and MD = Drought Magnitude and return period of extreme and severe drought was calculated using

$$RP = \frac{n}{Er}. \tag{10.4}$$

Where, RP= return period, n = number of droughts and Er = rank of drought.

Table 10.1 Wet or dry classes according to SPI values

Class	Drought category
2.0+	Extremely wet
1.5–1.99	Very wet
1.0–1.49	Moderately wet
0.99–0	Mildly drought
0 to −0.99	Mild dry
−1.0 to −1.49	Moderate dry
−1.5 to −1.99	Severely dry
−2 and less	Extremely dry

Source World Meteorological Organization, 2012

Standardised Precipitation Index (SPI)

The Standardised Precipitation Index (SPI) is extensively in use which is used to quantify wet or dry conditions (McKee et al., 1993). The values above zero denote wet conditions and below zero denote dry conditions (Table 10.1). In this study the primary index used was SPI for spatio-temporal estimation of drought for districts of Tamil Nadu. SPI can be used for diffrenet time scales of 3, 6, 12 and 24 months. Using SPI values various drought variables were calculated and their logical consideration and directionality of influence is given in Table 10.2. Then these data were standardised using logistic regression model and different maps were generated in Arc GIS software. SPI was calculated as:

$$\mathrm{SPI} = \frac{\mathrm{Pi} - \mathrm{Pm}}{\mathrm{SD}} \tag{10.5}$$

where, P_i = Precipitation, P_m = average precipitation, SD = standard deviation.

Table 10.2 Spatio-temporal assessment of drought vulnerability parameters and their directionality of influence towards drought

Variables	Directionality of influence
DF (Drought Frequency)	positive
DM (Drought Magnitude)	posiitve
MDI (Mean Drought Intensity)	positive
RPD (Return Period of Drought)	negative

Result and Discussion

In the present study, 25 criterias like EDF, SDF, DM, MDI, RPED and RPSD for time scale of 3, 6, 12 and 24 months were chosen for drought mapping. All these parameters are directly or positively related to drought vulnerability. These layers were used for spatio-temporal estimation of drought.

Drought Frequency

The 3 months EDF is high in Tiruchirappalli, Dindigul, Pudukkottai, Karur, Sivagangai, and parts of Madurai (Fig. 10.2a). 6 months EDF is high in the Nilgiris, and parts of Coimbatore (Fig. 10.2b). 12 months EDF is high in Kanyakumari, Tirunelveli, Tenkasi, and Thoothukudi (Fig. 10.2c). 24 months EDF is high in Madurai, Virudhunagar and Ramnathapuram (Fig. 10.2d). High 3 months SDF is found in Ariyalur, Perambalur, Thanjavur, Tiruvarur, Nagapattinam and Mayiladuthurai (Fig. 10.2e). 6 months SDF is high in Ariyalur, Mayiladuthurai, parts of Thanjavur and Tiruvarur (Fig. 10.2f). 12 months SDF is high in the Nilgiris, parts of Madurai, Virudhunagar, and Thoothukudi (Fig. 10.2g). High 24 months SDF is found in Tenkasi, Tirunelveli, Kanyakumari, and Thoothukudi (Fig. 10.2h).

Drought Magnitude

3 months DM is high in Ranipet, Vellore, Tiruvannamalai and Tirupattur (Fig. 10.3a). 6 months DM is high in Salem, Namakkal, Karur, Tiruchirappalli, Perambalur, Ariyalur, Nagapattinam, Mayiladuthurai, Thanjavur, Pudukkottai, Tiruvarur, parts of Tiruppur, Sivagangai and Dindigul (Fig. 10.3b). 12 months DMis high in Tiruvallur, Ranipet, Vellore, Kancheepuram, Chennai, and Chengapattu (Fig. 10.3c). 24 months DM is high in Kancheepuram, Chengapattu, Viluppuram and Chennai (Fig. 10.3d).

Mean Drought Intensity

3 months MDI is high in Ranipet, Vellore, Tirupattur and Tiruvannamalai (Fig. 10.4a). 6 months MDI is high in Sivagangai, Namakkal and Salem (Fig. 10.4b). 12 months MDI is high in Theni, Virudhunagar, Tenkasi, Madurai, Ramnathapuram, Tirunelveli and Kanyakumari (Fig. 10.4c). 24 months MDI is high in Tiruvallur, Chennai, Chengapattu, and parts of Viluppuram (Fig. 10.4d).

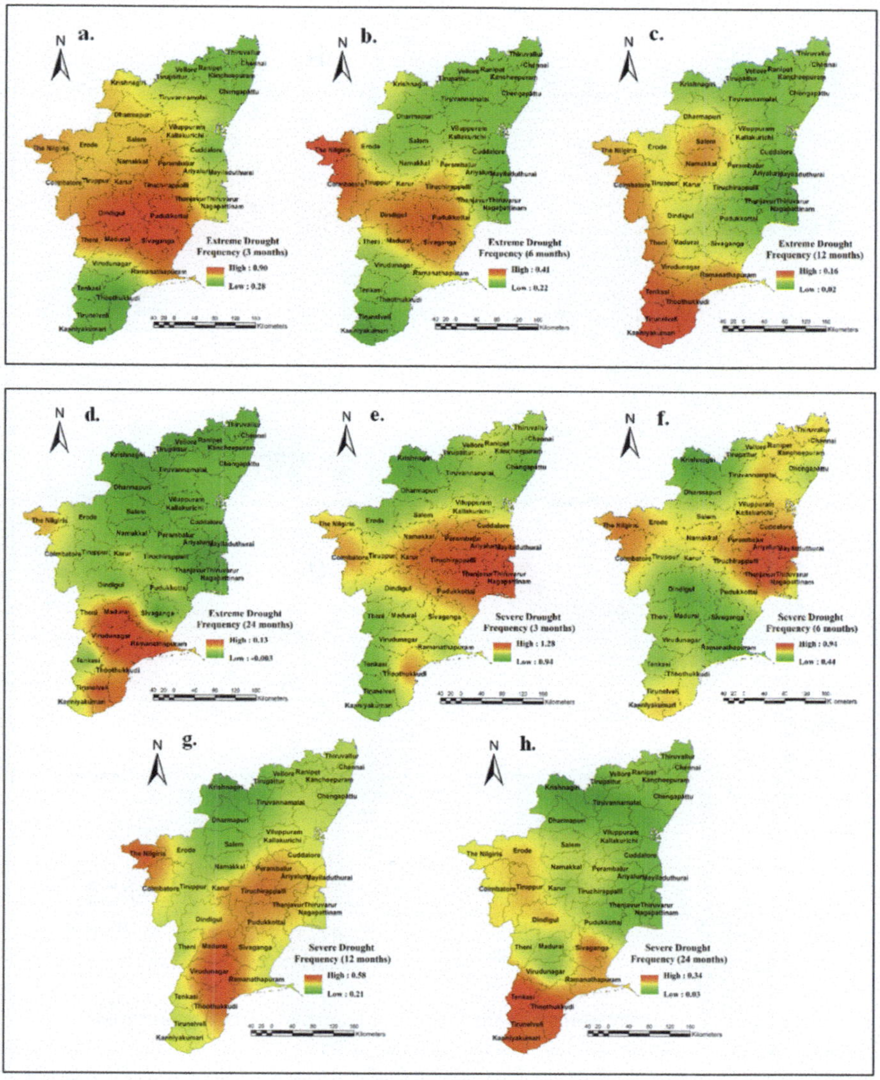

Fig. 10.2 **a** EDF 3 months **b** EDF 6 months **c** EDF 12 months **d** EDF 24 months **e** SDF 3 months **f** SDF 6 months **g** SDF 12 months **h** SDF 24 months

Drought Return Period

3 months RPED is high in Thoothukudi, and parts of Tirunelveli (Fig. 10.5a). 6 months RPED is high in Ramnathapuram and parts of Sivagangai (Fig. 10.5b). 12 months RPED is high in Mayiladuthurai, parts of Thanjavur and Tiruvarur (Fig. 10.5c). 24 months RPED is high in Tiruvallur, Chennai, Kancheepuram, Kallakurichi, parts of Namakkal and Salem (Fig. 10.5d). 3 months RPSD is high

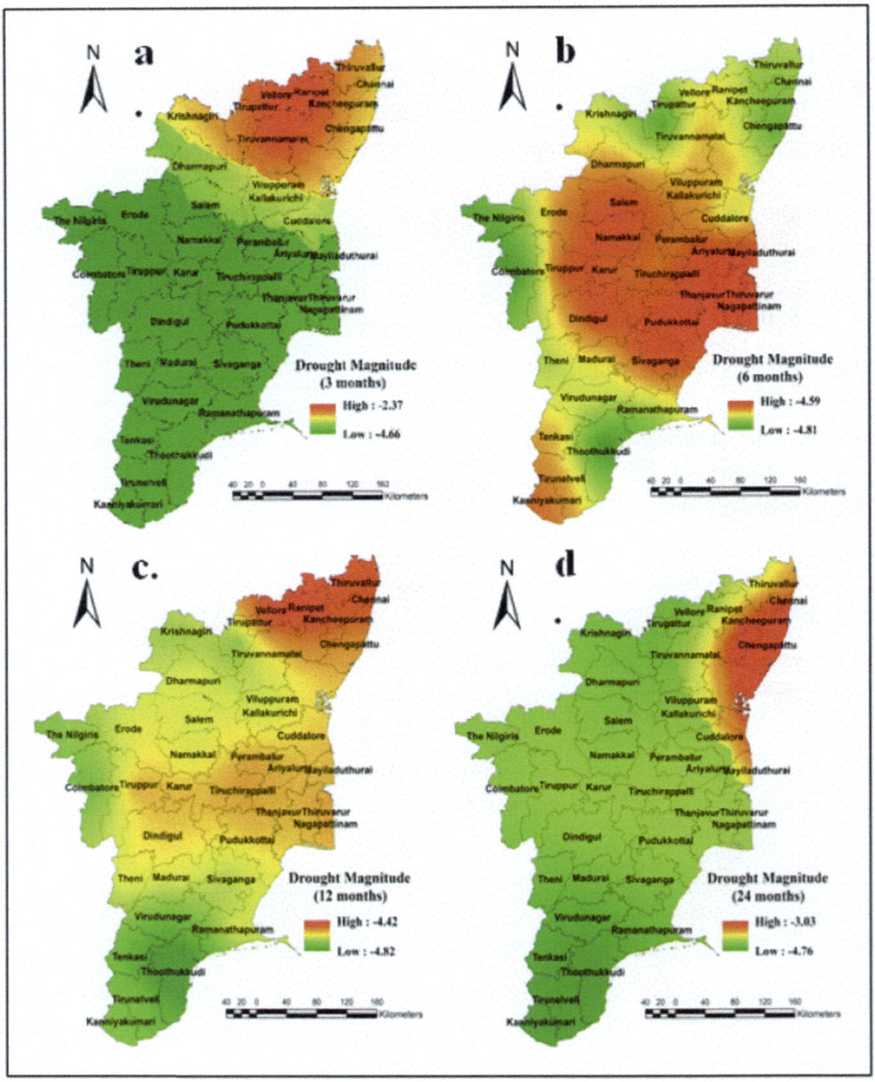

Fig. 10.3 **a** DM 3 months **b** DM 6 months **c** DM 12 months **d** DM 24 months

in Dharmapuri and Chengapattu (Fig. 10.5e). 6 months RPSD is high in Tiruvallur, parts of Ranipet and Kancheepuram (Fig. 10.5f). 12 months RPSD is high in Krishnagiri, Tirupattur, Vellore and parts of Dharmapuri (Fig. 10.5g). 24 months RPSD is high in Krishnagiri, Tirupattur, parts of Vellore and Dharmapuri (Fig. 10.5h).

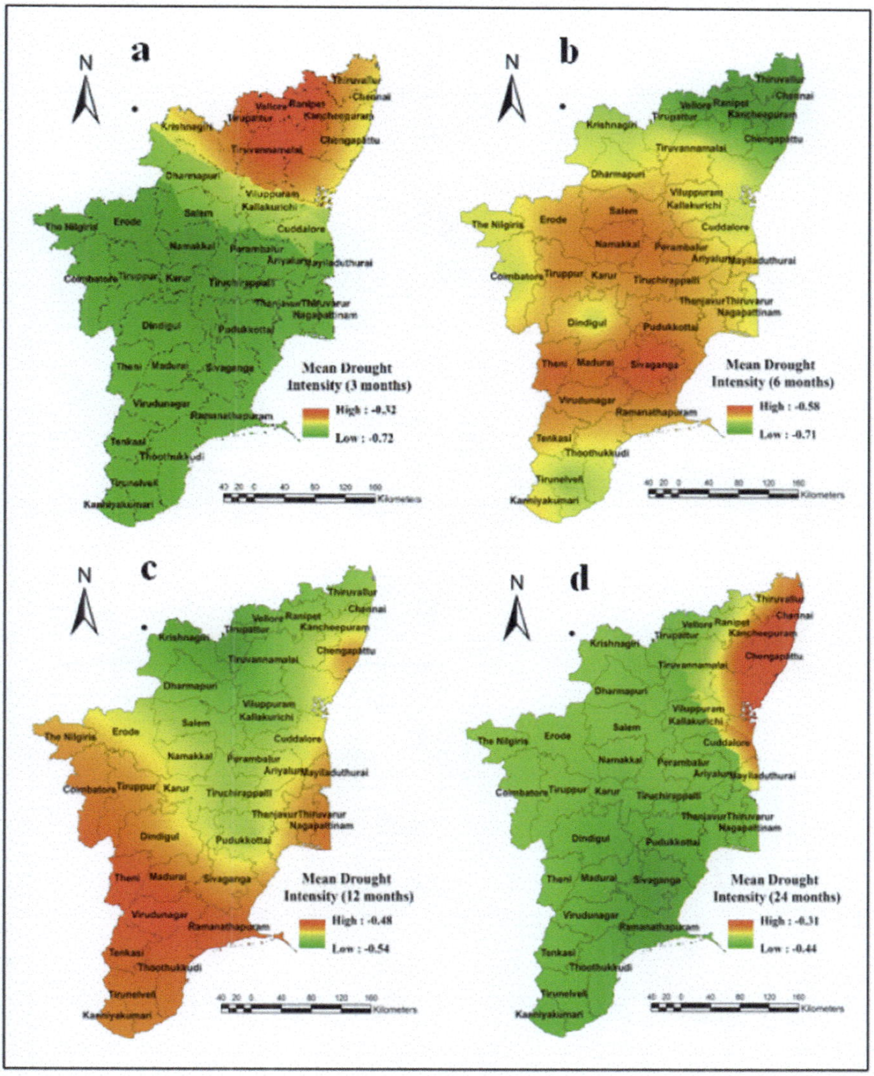

Fig. 10.4 **a** MDI 3 months **b** MDI 6 months **c** MDI 12 months **d** MDI 24 months

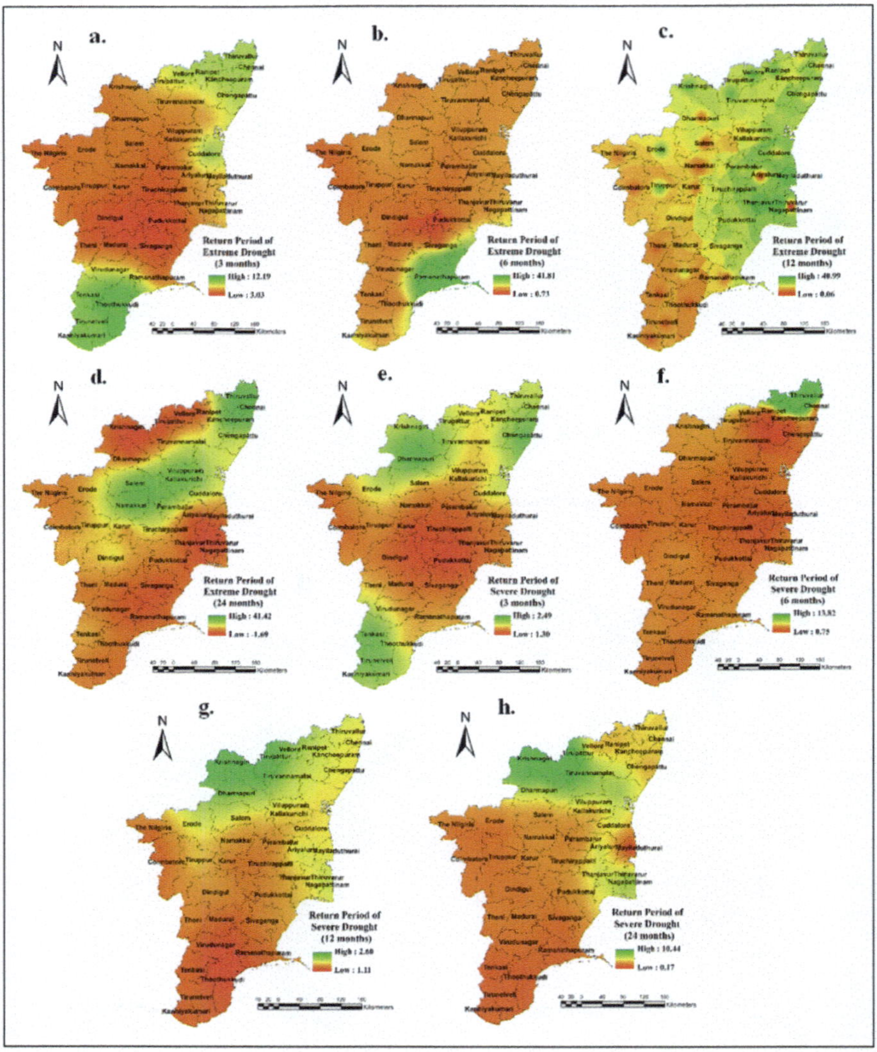

Fig. 10.5 **a** RPED 3 months **b** RPED 6 months **c** RPED 12 months **d** RPED 24 months **e** RPSD 3 months **f** RPSD 6 months **g** RPSD 12 months **h** RPSD 24 months

Conclusion

In conclusion it can be said that due to changing weather and climatic patterns, irregular rainfall and dry droughts are common nowadays. Different indicators like drought frequency, drought magnitude, mean drought intensity and return period of extreme drought and severe drought were selected for the time scale of 3, 6, 12 and 24 months to better reflect the drought scenario in Tamil Nadu both spatially and temporally. Tamil Nadu being the front face of northeast monsoon still faces drought like situation in the june to september period. Since field work could have improved the study but limited resource and time was the constraint. So drought mitigation and drought reduction policies are needed to better manage and improve the drought like conditions for Tamil Nadu.

References

Ding, Y., Hayes, M. J., & Widhalm, M. (2011). Measuring economic impacts of drought: A review and discussion. *Disaster Prevention and Management: An International Journal, 20*(4), 434–446.

Engström, J., Jafarzadegan, K., & Moradkhani, H. (2020). Drought vulnerability in the United States: An integrated assessment. *Water, 12*(7), 2033.

Ji, L., & Peters, A. J. (2003). Assessing vegetation response to drought in the northern Great Plains using vegetation and drought indices. *Remote Sensing of Environment, 87*(1), 85–98.

Keesstra, S. D. (2007). Impact of natural reforestation on floodplain sedimentation in the Dragonja basin, SW Slovenia. *Earth Surface Processes and Landforms: The Journal of the British Geomorphological Research Group, 32*(1), 49–65.

McKee, T. B., Doesken, N. J., & Kleist, J. (1993, January). The relationship of drought frequency and duration to time scales. In *Proceedings of the 8th Conference on Applied Climatology* (Vol. 17, No. 22, pp. 179–183).

Mishra, A. K., & Singh, V. P. (2010). A review of drought concepts. *Journal of Hydrology, 391*(1–2), 202–216.

Mohmmed, A., Zhang, K., Kabenge, M., Keesstra, S., Cerdà, A., Reuben, M., & Ali, A. A. (2018). Analysis of drought and vulnerability in the North Darfur region of Sudan. *Land Degradation & Development, 29*(12), 4424–4438.

Pazhanivelan, S., Geethalakshmi, V., Samykannu, V., Kumaraperumal, R., Kancheti, M., Kaliaperumal, R., & Yadav, M. K. (2023). Evaluation of SPI and rainfall departure based on multi-satellite precipitation products for meteorological drought monitoring in Tamil Nadu. *Water, 15*(7), 1435.

Pei, W., Fu, Q., Liu, D., Li, T. X., Cheng, K., & Cui, S. (2018). Spatiotemporal analysis of the agricultural drought risk in Heilongjiang Province, China. *Theoretical and Applied Climatology, 133*(1), 151–164.

Rahman, M. R., & Lateh, H. (2016). Meteorological drought in Bangladesh: Assessing, analysing and hazard mapping using SPI, GIS and monthly rainfall data. *Environmental Earth Sciences, 75*(12), 1026.

Saha, S., Gayen, A., Gogoi, P., Kundu, B., Paul, G. C., & Pradhan, B. (2022). Proposing novel ensemble approach of particle swarm optimized and machine learning algorithms for drought vulnerability mapping in Jharkhand, India. *Geocarto International, 37*(25), 8004–8035.

Sternberg, T. (2012). Chinese drought, bread and the Arab Spring. *Applied Geography, 34*, 519–524.

Wilhite, D. A. (2016). Drought as a natural hazard: Concepts and definitions. In *Droughts* (pp. 3–18). Routledge.

Zargar, A., Sadiq, R., Naser, B., & Khan, F. I. (2011). A review of drought indices. *Environmental Reviews, 19*(NA), 333–349.

Zhang, Q., Sun, P., Li, J., Xiao, M., & Singh, V. P. (2015). Assessment of drought vulnerability of the Tarim River basin, Xinjiang, China. *Theoretical and Applied Climatology, 121*(1), 337–347.

Chapter 11
Integrated Physical–Numerical Modelling for Determining Rainfall Thresholds of Slope Failure in the Pithoragarh District, Uttarakhand Himalayas

Nishant Kumar, Gaurav Bharti, Vaishnavi Bansal, and Chetna

Abstract Rainfall-induced landslides are among the most recurrent and destructive geohazards in the Himalayan region, where steep slopes, fragile lithology, and intense monsoonal precipitation collectively contribute to frequent slope failures. This study aims to establish rainfall thresholds responsible for slope instability through an integrated physical–numerical modeling framework in the Pithoragarh district of Uttarakhand, India. Five representative slope sites (S1–S5) were selected to capture lithological and geomorphological variability. Laboratory-scale physical modeling was conducted using a rainfall simulator to replicate progressive infiltration and failure behavior under controlled rainfall intensities. Concurrently, numerical simulations were performed using GeoStudio's SEEP/W and SLOPE/W modules to quantify pore-pressure variation and Factor of Safety (FoS) reduction under saturated conditions. Results revealed a significant decline in FoS from 1.548–2.010 (dry) to 0.748–0.981 (saturated), confirming that rainfall infiltration leads to increased pore-water pressure and loss of matric suction, ultimately triggering slope failure. The integrated results identified a critical rainfall threshold of 45–50 mm cumulative rainfall at an intensity of approximately 28 mm h^{-1}, corresponding to the onset of instability where FoS drops below unity. The close agreement between physical and numerical outcomes validates the robustness of the hybrid modeling approach. The derived intensity–duration (I–D) threshold aligns with global empirical ranges and provides a reliable basis for early warning and hazard mitigation in Himalayan

---

N. Kumar · G. Bharti (✉) · V. Bansal · Chetna
Department of Civil Engineering, JSS Academy of Technical Education, Noida, India
e-mail: gauravbharti000@gmail.com

N. Kumar
e-mail: nk7838331@gmail.com

V. Bansal
e-mail: vaishnavibansal@jssaten.ac.in

Chetna
e-mail: chetna18032012@gmail.com

R. Sarkar et al. (eds.), *Climate Change, Environmental Hazards and Community-Based Resilience*, Disaster Risk Reduction, https://doi.org/10.1007/978-981-95-9485-6_11

terrain. The study underscores the importance of coupling experimental validation with computational analysis for realistic prediction of rainfall-induced slope failures in data-scarce mountain environments.

Keywords Rainfall-induced landslides · Slope stability · Physical modeling · Numerical simulation · Rainfall threshold · Pithoragarh Himalayas

Introduction

Landslides are among the most devastating natural hazards in mountainous regions worldwide, posing a significant threat to lives, infrastructure, and the environment. Globally, rainfall-induced landslides account for the majority of slope failures, especially in tropical and subtropical regions characterized by steep terrain, fragile lithology, and intense seasonal precipitation. The phenomenon is particularly critical in the Himalayas, one of the world's youngest and most tectonically active mountain systems, where orographic precipitation, steep slopes, and heterogeneous geology converge to create highly unstable geomorphic conditions. In India, rainfall-triggered landslides frequently disrupt transportation corridors, destroy settlements, and threaten hydropower infrastructure, particularly during the southwest monsoon season. The delicate balance between rainfall intensity, soil moisture, and slope stability makes it imperative to identify rainfall thresholds that can act as reliable precursors for early warning and risk mitigation. Rainfall acts as the primary triggering agent by infiltrating the soil mass, increasing pore-water pressure, and decreasing matric suction, thereby reducing the effective stress and shear strength of slope-forming materials. The failure mechanism typically initiates when infiltration exceeds the soil's hydraulic conductivity, leading to saturation of the subsurface layers and a consequent reduction in cohesion. In mountainous regions like the Himalayas, where geomaterials are highly weathered and fragmented, even moderate rainfall events can initiate slope failure. Also reported that shallow failures in the Chamoli–Joshimath region of Uttarakhand were strongly associated with rainfall thresholds, whereas Vikram Gupta et al., (2016) highlighted similar mechanisms during the Balia Nala landslide in Nainital. These examples emphasize the need for regional threshold estimation and slope-specific analysis.

Modeling of rainfall-induced landslides provides an effective framework for quantifying these interactions between hydrological and geotechnical parameters. Broadly, three categories of modeling approaches are employed: physical, numerical, and empirical. Physical modeling replicates the actual slope geometry and hydrological conditions at reduced scale, enabling direct visualization of failure mechanisms under controlled laboratory conditions. Such approaches are particularly useful for examining pore-pressure evolution and deformation processes, as seen in the works of Li et al., (2016). However, the limitations of scaling laws and material similarity often restrict their direct field applicability. Numerical models, on the other hand, employ analytical formulations and computational tools—such as limit equilibrium, finite element, and finite difference methods—to simulate slope response under varying rainfall scenarios. These models are efficient in quantifying safety factors and identifying potential slip surfaces but rely heavily on accurate input parameters and boundary conditions. Empirical and statistical models, including intensity–duration (I–D) and cumulative rainfall thresholds, are derived from historical landslide inventories and rainfall records. While statistically robust at regional scales, their predictive reliability decreases when applied to smaller catchments with distinct geomorphological characteristics.The Himalayan terrain exhibits a particularly complex rainfall–landslide relationship. Seasonal monsoons produce spatially variable rainfall patterns that often exceed 2000 mm annually in certain subcatchments, causing rapid infiltration in colluvial deposits and saprolitic soils. Field observations and hydrological analyses have revealed that soil composition, structure, and permeability exert a decisive influence on infiltration and failure onset. It was also emphasized that the fragile lithology of Nepal's Lesser Himalaya results in rapid pore-pressure development during monsoon events and the role of antecedent moisture in progressive failure in Uttarakhand. The interaction between rainfall and soil hydrodynamics hence remains central to understanding slope instability in Himalayan conditions.The study of rainfall thresholds is vital for developing operational early warning systems. These thresholds define the critical rainfall intensity or cumulative depth that may initiate failure under existing slope conditions. Two main approaches exist for their determination: (i) empirical derivation from rainfall–landslide databases and (ii) process-based derivation through physically or numerically simulated models. Empirical thresholds provide region-specific relationships between rainfall parameters and landslide occurrence. Nevertheless, in areas with sparse data or variable microclimatic conditions, physically based modeling offers a more reliable alternative by explicitly linking rainfall infiltration, pore pressure, and stability conditions.

Recent advances in computational geomechanics and sensor-based monitoring have further strengthened threshold modeling. Real-time monitoring systems integrating tiltmeters, piezometers, and rainfall gauges have demonstrated strong potential for early warning. Yet, such systems remain cost-intensive and require continuous maintenance, making them impractical for many high-risk Himalayan slopes. Consequently, laboratory-scale physical modeling complemented by numerical analysis emerges as a pragmatic compromise—providing both visual insight and quantitative validation. The combination of these approaches allows for calibration of threshold rainfall values with realistic soil properties and hydrological responses. Despite numerous landslide incidents, only a limited number of studies have attempted integrated physical–numerical modeling for rainfall-induced failures. The scarcity of long-term rainfall and landslide data further complicates empirical threshold development. Hence, there is a pressing need to establish slope-specific thresholds through laboratory-controlled modeling supported by numerical simulation. The district of Pithoragarh in eastern Uttarakhand exemplifies this necessity: its steep terrain, complex lithology, and recurrent monsoon-triggered failures pose continuous challenges to infrastructure and local communities (Khanduri, 2022).The present study builds upon this motivation by employing both physical and numerical modeling to establish rainfall thresholds for a representative Himalayan slope. The research aims to overcome the limitations of purely empirical or numerical approaches by combining laboratory simulation with computational verification. The specific objectives are:

1. To construct a semi-similar physical model of a rainfall-prone slope and simulate varying rainfall intensities to determine failure initiation mechanisms.
2. To perform complementary numerical analyses using GeoStudio (SLOPE/W and SEEP/W) for validation of physical model results.
3. To define a rainfall threshold corresponding to slope failure.

Literature Review

Rainfall-induced landslides have been extensively studied across the world through empirical, statistical, physical, and numerical methods. Early efforts focused on defining rainfall thresholds capable of predicting the onset of slope failure. Berti et al., (2012) introduced probabilistic rainfall thresholds using a Bayesian approach, enabling the estimation of failure probability under given intensity–duration conditions. Similarly, Melillo et al., (2016) derived automatic rainfall event reconstructions for Sicily, identifying threshold curves applicable to regional early warning systems. In tropical regions such as Honduras, García-Urquia (2016) established rainfall frequency contour lines to delineate susceptible areas, while Posner & Georgakakos (2015) incorporated soil-moisture variability to refine thresholds for real-time prediction in Central America. These foundational studies demonstrated that cumulative rainfall and antecedent conditions jointly influence the likelihood of failure, shifting emphasis from rainfall intensity alone to combined hydro-geotechnical controls. Within Asia, several researchers have examined rainfall thresholds in landslide-prone regions characterized by intense monsoonal precipitation. Althuwaynee et al., (2015) estimated rainfall thresholds for Kuala Lumpur using statistical correlation and applied them to hazard mapping, while Saadatkhah et al., (2015) employed the TRIGRS model for dynamic rainfall–infiltration simulation in Malaysia. In Vietnam, Long & De Smedt (2018) analyzed rainfall-induced susceptibility using GIS-based modeling, emphasizing terrain curvature and lithology as key influencing factors. These works collectively highlight the diversity of rainfall thresholds across climatic and geomorphic settings. Tiranti & Rabuffetti (2010) developed rainfall-duration thresholds for shallow landslides in northern Italy, integrating hydrological parameters into operational early warning systems. Rossi et al., (2017) compared satellite rainfall estimates with gauge measurements, underscoring the utility of remote-sensing data in landslide modeling. Lepore et al., (2012) produced a rainfall-induced susceptibility zonation for Puerto Rico, demonstrating the adaptability of rainfall-based indices across scales. Together, these studies provide valuable methodological references for threshold derivation using both in-situ and remotely sensed data.

In India, landslide research has evolved from inventory-based mapping to process-driven modeling. Dahal et al., (2008a) pioneered the use of weights-of-evidence (WOE) and GIS for susceptibility mapping of rainfall-triggered landslides in the Lesser Himalaya (Saraswat et al., 2020). Subsequent studies extended this framework by integrating hydrological and geomorphological variables. Panchal & Shrivastava (2022) applied the Analytic Hierarchy Process (AHP) for hazard assessment along the NH-5 corridor, while Kincal & Kayhan (2022) combined multiple methods to map susceptibility in Turkey, illustrating cross-regional methodological parallels. Recent advances emphasize machine learning and data-driven models for landslide prediction. Nanda et al., (2024) employed ensemble machine-learning algorithms—including Random Forest and Support Vector Machine—to predict rainfall-induced landslides along the Bandipora–Gurez highway, achieving significant improvement over traditional logistic regression. Li et al., (2022) demonstrated the feasibility of global dynamic rainfall-induced susceptibility mapping using machine-learning integration of meteorological and terrain parameters, marking a shift toward scalable predictive modeling. Similarly, Wu (2019) explored extreme rainfall inventories to simulate subsequent landslide evolution, thereby coupling temporal and spatial prediction aspects. From a geomatics perspective, Mondini et al., (2011) and Ardizzone et al., (2007) utilized optical and LiDAR imagery for semi-automatic recognition of shallow landslides triggered by rainfall, reinforcing the importance of remote-sensing and DEM-based inputs for calibration of hydrological models. Yang (2017) further refined GIS-based slope-unit approaches for rainfall-induced susceptibility assessment, integrating both geomorphic and hydrological variables. These remote-sensing and GIS-driven methodologies are increasingly employed for rapid hazard assessment in data-sparse regions. Probabilistic and physically based numerical models have also been explored. Dolojan et al., (2021) simulated infiltration and slope-stability processes during Typhoon Hagibis in Japan, successfully replicating observed failures. Lin et al., (2021) employed the CF-SINMAP model to evaluate rainfall-induced hazards in Hunan, China, combining hydrological modeling with stability indices. Zhao et al., (2019) analyzed grid-size effects on logistic-regression-based susceptibility mapping, revealing that spatial resolution critically affects prediction accuracy. These approaches underline the growing sophistication of numerical and GIS-integrated hazard modeling. Within the Indian Himalayan context, rainfall-threshold analysis remains limited despite increasing landslide occurrences. Paswan & Shrivastava (2022) established threshold rainfall depth for slope failure through combined physical and numerical modeling, determining that a cumulative rainfall of $\approx$80 mm at an intensity of 30 mm h^{-1} triggered failure in their modeled slope, corroborated by safety-factor reduction from 1.23 to 0.63. Their subsequent study (Paswan & Shrivastava, 2023) enhanced this framework through a tilt-based monitoring system, validating the predictive reliability of laboratory-derived thresholds. Complementary field evidence from Khanduri (2022) in Pithoragarh confirms that heavy monsoon rainfall leads to widespread slope instability, suggesting strong regional applicability of threshold-based approaches. Comparative reviews such as Reichenbach et al., (2018) and Dikshit et al., (2020) synthesized statistically based susceptibility models, concluding that hybrid frameworks combining empirical thresholds with physical process simulation yield superior reliability. Dai et al., (2003) characterized rainfall-induced landslide mechanisms from a remote-sensing perspective, integrating soil-moisture dynamics and slope-angle criteria. Collectively, these studies demonstrate a progressive transition from empirical rainfall–landslide correlation toward integrated, data-driven, and physically constrained modeling paradigms. Despite these advances, several research gaps remain evident. Most existing rainfall thresholds are region-specific and cannot be universally applied due to variations in lithology, slope morphology, and hydrological response. Limited long-term datasets in Himalayan terrain hinder the derivation of statistically stable thresholds. Moreover, while machine-learning and GIS-based frameworks improve prediction accuracy, they often lack physical interpretability, restricting their operational deployment for real-time warnings. Integration of laboratory-derived thresholds with numerical validation offers a promising direction for reconciling empirical reliability with process understanding.Thus, the cumulative body of research suggests that rainfall threshold modeling, when supported by both experimental validation and computational simulation, provides the most reliable pathway for developing site-specific early warning systems. Building on these foundations, the present study seeks to refine threshold determination for rainfall-induced landslides in Himalayan conditions by combining controlled physical modeling, field-based parameterization, and numerical validation.

Study Area

Pithoragarh as shown in Fig. 11.1, located in the eastern part of Uttarakhand, India, is characterized by rugged Himalayan terrain, steep slopes, and complex geological structures, making it highly susceptible to landslides, particularly during the monsoon season. The district has experienced numerous rainfall-induced landslides, leading to significant losses of life, livestock, and infrastructure. For instance, in 2020, the Bangapani tehsil witnessed devastating landslides resulting in 17 fatalities and the loss of 23 cattle. Over the past five decades (1971–2020), landslides in Pithoragarh have claimed approximately 433 human lives and 248 cattle, with rainfall and cloudbursts identified as the primary triggers (Paswan & Shrivastava, 2022). The region's climatic conditions, combined with its geological and structural frameworks, contribute to its high landslide susceptibility. Notably, heavy rains in 2016 caused large-scale slope failures in the areas around Didihat, Bastari, and Naulra villages, resulting in 21 casualties (Khanduri, 2022). During the 2020 monsoon, intense precipitation between July 19 and 29 triggered numerous landslides and debris flows, leading to substantial loss of life and property across the district. A total of 28 deaths were recorded during this period, with villages such as Tanga and Gaila Malla Tok of Pattharkot being among the worst affected. The rainfall regime in Pithoragarh follows a pronounced monsoonal pattern, with approximately 77.4%

of the total annual precipitation concentrated between June and September. Historical records indicate significant variability in rainfall trends over the past century, characterized by an increasing pattern up to around 1935, followed by a gradual decline in subsequent decades. Such fluctuations have considerable implications for agriculture, water resource management, and disaster preparedness in the region. Given the district's hydroclimatic instability and fragile geomorphic setting, it is imperative to establish region-specific rainfall thresholds to anticipate and mitigate rainfall-induced landslide hazards effectively. Integrating Geographic Information System (GIS)-based statistical modeling with remote-sensing datasets can substantially enhance the spatial precision of susceptibility mapping, thereby supporting sustainable land-use planning and hazard mitigation strategies tailored to the Pithoragarh district (Paswan & Shrivastava, 2023). Five experimental slope locations were selected within the Pithoragarh district of Uttarakhand, India, to represent varying geomorphological and lithological conditions along the eastern Himalayan terrain. The selected sites include S1 (29.959120° N, 80.607103° E), S2 (29.956460° N, 80.648224° E), S3 (30.049511° N, 80.755127° E), S4 (30.055365° N, 80.776051° E), and S5 (30.090445° N, 80.805731° E). These coordinates span the central and northeastern sectors of the district, capturing diverse slope orientations, soil textures, and hydrological characteristics.

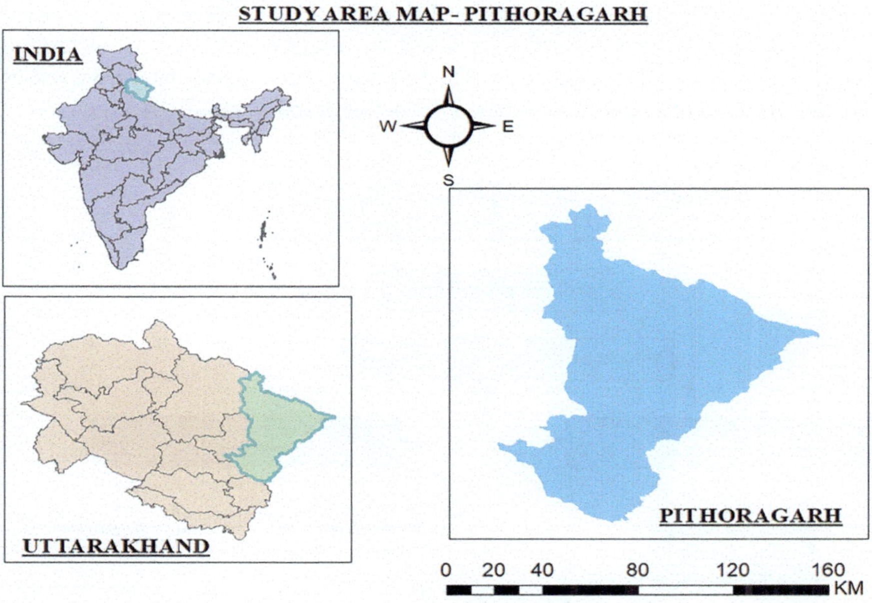

Fig. 11. 1 Study area (Pithoragarh)

Methodology

The methodology adopted in this study integrates both physical and numerical modeling techniques to simulate and analyze rainfall-induced landslides through a threshold-based approach (Paswan & Shrivastava, 2022) as shown in Fig. 11.2. The overall workflow, as illustrated in Fig. 11.3, begins with comprehensive soil investigation and culminates in the identification of rainfall thresholds for slope instability. The methodology comprises three major components: (i) soil investigation and model preparation, (ii) physical modeling and threshold determination, and (iii) numerical simulation for validation and analysis.

Soil Investigation

The first stage involves detailed geotechnical characterization of slope-forming materials collected from the Pithoragarh region of Uttarakhand. Laboratory tests were conducted to determine soil properties such as bulk density, specific gravity, cohesion, angle of internal friction, permeability, and grain size distribution. These parameters play a decisive role in defining the infiltration behavior and shear strength of the slope materials (Dahal et al., 2008b; Lin et al., 2021). The soil samples were oven-dried, sieved, and subjected to standard compaction and direct shear tests in accordance with IS: 2720 guidelines. The resulting geotechnical dataset served as the input for both physical and numerical modeling frameworks.

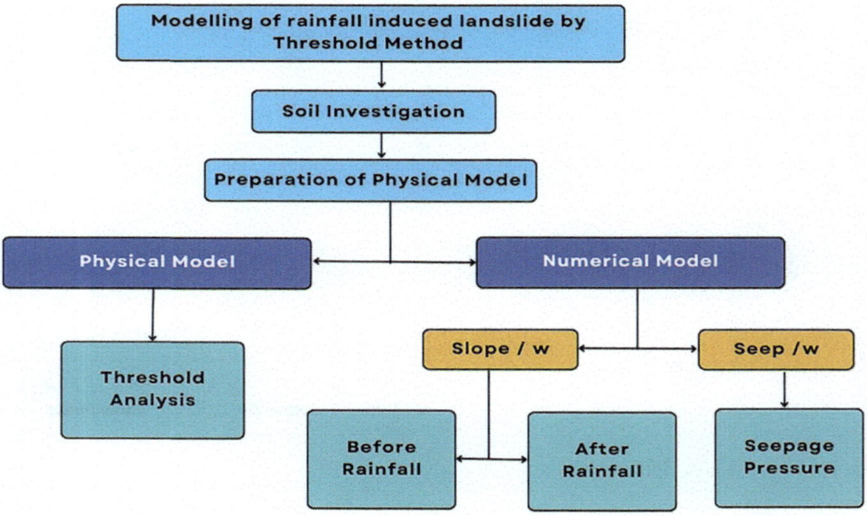

Fig. 11.2 Methodology

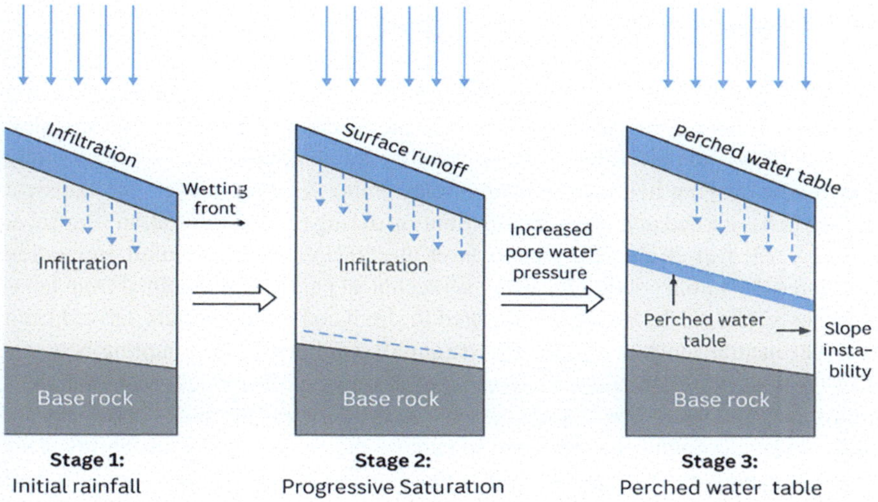

Fig. 11.3 Sequential evolution of rainfall in slopes

Preparation of the Physical Model

A scaled-down physical model was prepared within a transparent flume of dimensions optimized to replicate the natural slope geometry of the Pithoragarh terrain. The geometric scale was selected such that the slope angle, material density, and permeability closely resemble the prototype conditions under field rainfall scenarios (Li et al., 2022). The soil was compacted layer by layer to achieve uniform density, and the slope face was inclined at a representative angle derived from local topographic measurements. A rainfall simulator equipped with nozzles was mounted above the flume to uniformly distribute rainfall intensity, thereby imitating natural monsoonal precipitation patterns observed in the study area (Paswan & Shrivastava, 2023). The rainfall intensities were varied incrementally to observe the progressive development of pore-water pressure, seepage, and surface deformation. Real-time measurements were recorded using tilt sensors and piezometers installed at different depths within the slope mass. These instruments provided continuous monitoring of the hydromechanical response of the slope under controlled rainfall conditions. The physical model thus enabled direct visualization of slope failure initiation and allowed quantification of the rainfall intensity and duration at which the first signs of instability occurred (Melillo et al., 2016).

Numerical Modelling

Parallel to the physical experiment, numerical simulations were performed using GeoStudio software, specifically employing the SLOPE/W and SEEP/W modules. The SLOPE/W model evaluates slope stability by computing the factor of safety (FoS) under varying hydrological conditions, while SEEP/W models the transient seepage behavior and pore-pressure distribution during rainfall infiltration (Bansal & Sarkar, 2023; Lin et al., 2021; Dolojan et al., 2021). The pre-rainfall simulation considered dry initial conditions, using geotechnical parameters obtained from laboratory tests. The model was then subjected to simulated rainfall inputs derived from the experimental setup to represent post-rainfall conditions. The coupling between SEEP/W and SLOPE/W allowed dynamic assessment of the reduction in the FoS as infiltration progressed. The seepage pressure distribution and the rate of pore-pressure buildup were continuously computed to capture the onset of slope instability. This two-stage modeling framework ensured that both mechanical (shear strength) and hydraulic (pore pressure, seepage flux) factors were incorporated into the analysis. Such an integrated approach has been widely applied in similar studies worldwide (Bansal & Sarkar, 2024a, Bansal & Sarkar, 2024b, Berti et al., 2012; Tiranti & Rabuffetti, 2010; García-Urquia, 2016).

Threshold Analysis

The results from the physical and numerical models were combined to determine the rainfall threshold responsible for slope failure. The critical threshold was defined as the rainfall intensity–duration (I–D) combination at which a rapid reduction in slope stability occurred, evidenced by a marked drop in the factor of safety (Paswan & Shrivastava, 2022). In this study, the physical model demonstrated failure when cumulative rainfall approached approximately 80 mm under an intensity of 28 mm h^{-1}, which corresponded to a reduction in FoS from 1.23 (pre-rainfall) to 0.63 (post-rainfall). The correlation between rainfall intensity and failure initiation was further validated through the numerical analysis. The post-rainfall SEEP/W simulations indicated a steep rise in pore-water pressure near the toe and mid-slope regions, aligning with the experimentally observed failure zones. Such agreement between the two models underscores the robustness of the hybrid threshold determination method (Nanda et al., 2024).

Integration of Physical and Numerical Models

The combined use of physical and numerical modeling facilitates a comprehensive understanding of rainfall-induced slope failure mechanisms. The physical model provides direct insight into the deformation process and failure pattern, while the numerical model enables parametric analysis and generalization under varying conditions (Bharti et al., 2022; Rossi et al., 2017; Reichenbach et al., 2018). This methodological integration bridges the gap between experimental observation and theoretical prediction, yielding threshold values that are both physically interpretable and practically applicable for early warning systems.

Application and Relevance

The threshold values derived from this study are directly applicable to rainfall-induced landslide risk assessment in Himalayan regions such as Pithoragarh, where steep slopes, heterogeneous geology, and intense monsoonal rainfall prevail. These results can be incorporated into GIS-based hazard zonation and early warning frameworks (Panchal & Shrivastava, 2022; Dahal et al., 2008a). Moreover, the combination of physical validation and numerical simulation enhances the reliability of landslide prediction systems by capturing both hydrological and mechanical triggers.

Results and Discussions

Overview of Experimental Conditions

The experimental model was designed to analyze the rainfall-induced slope failure process under controlled laboratory conditions. The total slope area used in this study measured 0.61 m^2, representative of the geometric and hydraulic characteristics of the Pithoragarh terrain. A fixed rainfall intensity of 28 mm h^{-1} was applied using a calibrated rainfall simulator to replicate high-intensity monsoonal precipitation typical of the region. Rainfall depth was increased incrementally by 9.33 mm every 20 minutes, enabling stepwise evaluation of hydrological response and slope deformation through progressive wetting. This incremental design provided a realistic simulation of natural rainfall events where cumulative infiltration, rather than instantaneous intensity, governs slope failure mechanisms (Paswan & Shrivastava, 2022). The rainfall sequence produced a systematic evolution of surface wetting, infiltration, and eventual slope instability. Figure 11.3 depicts the progression of these events, highlighting distinct stages of initiation, development, and ultimate failure. During the initial phase, the slope surface exhibited only minor erosion and darkening, signifying superficial wetting. With continued rainfall, pore-water pressures increased, leading to internal weakening of the soil structure and visible deformation near the mid-slope and toe sections.

Table 11. 1 F.O.S For all 5 sites in dry and saturated condition

S.NO	Study region site	Cohesion KPa	Density KN/m3	Friction angle (Degree)	FOS
1	S1(dry)	15	15	30	1.84
2	S1(sat)	15	17	30	0.978
3	S2(dry)	18	16	35	2.010
4	S2(sat)	18	18	35	0.981
5	S3(dry)	20	16	29	1.725
6	S3(sat)	20	19	29	0.948
7	S4(dry)	15	13	30	1.727
8	S4(sat)	15	15	30	0.748
9	S5(dry)	16	16	28	1.548
10	S5(sat)	16	18	28	0.831

Factor of Safety (FoS) Variation under Dry and Saturated Conditions

The Factor of Safety (FoS) was determined for five slope sections (S1–S5) under both dry and saturated states using SLOPE/W numerical simulations. The computed results, summarized in Table 11.1 and Fig. 11.4, illustrate the significant reduction in slope stability following saturation.

Across all sites, FoS values under dry conditions ranged between 1.548 and 2.010, indicating stable slopes. However, upon saturation, FoS values sharply declined to between 0.748 and 0.981, reflecting failure or near-failure states. The highest stability under dry conditions was observed in S2 (FoS = 2.010), corresponding to higher cohesion (18 kPa) and friction angle (35°), whereas the lowest stability under saturated conditions occurred in S4 (FoS = 0.748), due to its lower density and cohesion (15 kPa). This reduction in FoS across all slope sections demonstrates the critical role of rainfall infiltration and pore-pressure buildup in destabilizing slope-forming materials. The results are consistent with findings by Dahal et al., (2008b) and Melillo et al., (2016), who reported that infiltration-driven loss of matric suction significantly reduces the shear strength of residual soils, thereby triggering shallow slope failures.

Physical Model Observations

The physical rainfall simulation provided direct visual evidence of slope behavior under progressive saturation. During the first 20 min (9.33 mm rainfall), no deformation was observed; however, the slope surface became darker, indicating initial infiltration. Between 20–40 minutes (18.66 mm cumulative rainfall), hairline cracks appeared near the crest, marking the onset of micro-shear displacement. By 60 min

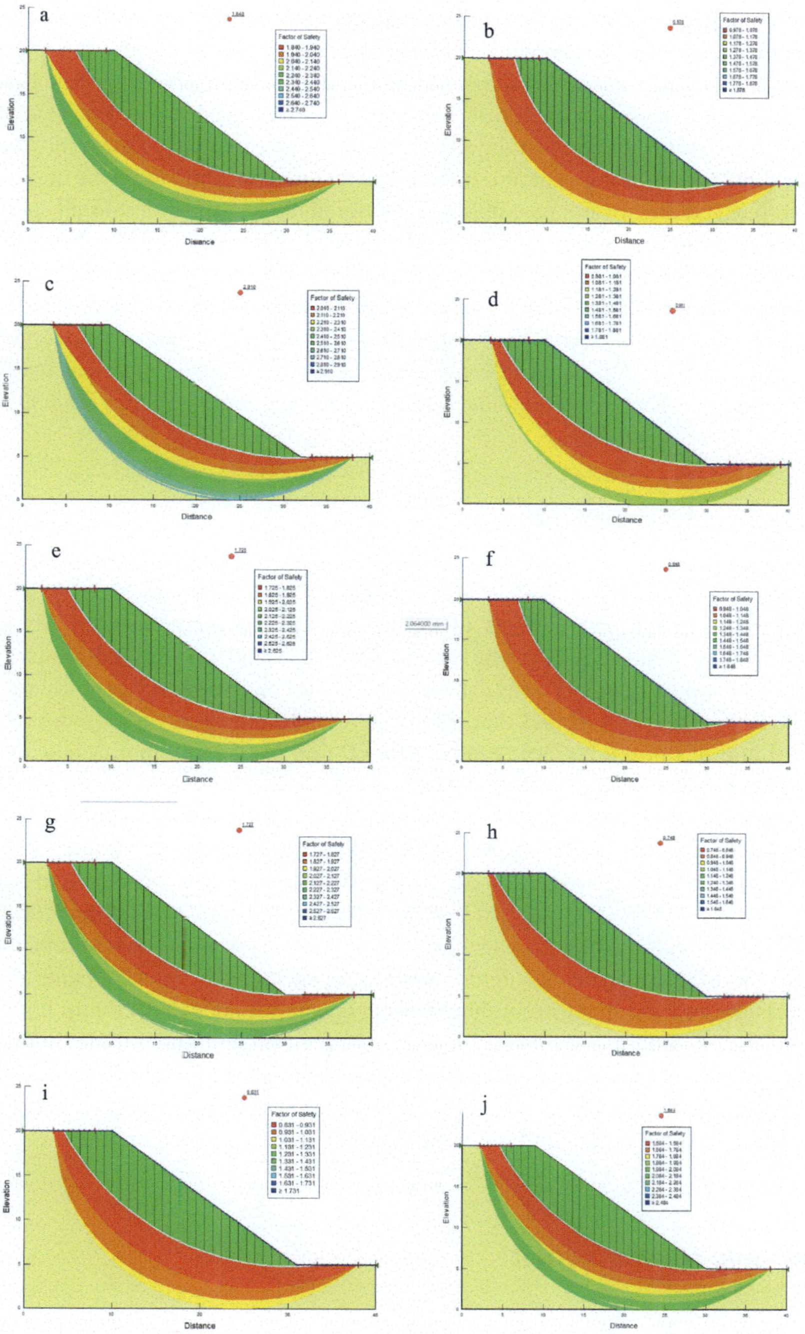

Fig. 11. 4 F.O.S for S1:a (dry), b(saturated), F.O.S for S2:c (dry), d(saturated), F.O.S for S3:e (dry), f(saturated), F.O.S for S4:g (dry), h(saturated), F.O.S for S5:i (dry), j(saturated)

(28 mm cumulative rainfall), the cracks extended toward the mid-slope, and localized subsidence was observed. Failure was initiated between 80–90 min (37 mm cumulative rainfall) when the slope exhibited visible surface bulging near the toe and tension cracking near the crest. The critical failure occurred at approximately 100 min (46.6 mm cumulative rainfall), when the slope experienced mass movement and shallow sliding along a curved surface. The failure plane initiated at the mid-slope and propagated downward, resembling a rotational mode typical of rainfall-induced shallow landslides (Paswan and Shrivastava, 2023; Lin et al., 2021). Piezometric readings revealed a steady increase in pore-water pressure during rainfall, peaking just before the observed failure. The rise in pore pressure directly corresponded to increased rainfall depth, confirming that the slope reached a near-saturated condition at the time of collapse. These observations align with the numerical findings, confirming the strong hydraulic–mechanical coupling governing slope instability.

Numerical Model Results

Numerical analysis using SEEP/W as shown in Fig. 11.5 and SLOPE/W was performed to quantify the effects of infiltration and corresponding stability variations. For each site (S1–S5), the pre-rainfall (dry) and post-rainfall (saturated) states were simulated under identical slope geometry and boundary conditions. The SEEP/W simulation indicated that infiltration depth increased linearly during the first 40 min of rainfall but accelerated after 60 min as the soil approached saturation. The pressure head near the toe reached +3.6 kPa, while negative suction near the crest decreased from −6.19 kPa to−1.09 kPa, signifying near-complete loss of matric suction. Correspondingly, the SLOPE/W results showed a continuous decline in FoS, with the average reduction exceeding 55% between dry and saturated conditions. Spatial distribution maps of FoS revealed that instability first developed in the mid-slope region where seepage flux converged, followed by failure propagation toward the toe. This behavior agrees with experimental deformation patterns and supports the concept that the combined effect of rainfall intensity (28 mm h^{-1}) and limited slope drainage capacity accelerates pore-pressure buildup. Similar hydraulic responses were reported in infiltration-based slope models by Dolojan et al.,(2021) and García-Urquia (2016), emphasizing that the transition from negative to positive pore pressure is the dominant failure mechanism under sustained rainfall.

Threshold Determination

Integrating the physical and numerical results, the critical rainfall threshold for slope failure was identified as approximately 45–50 mm cumulative rainfall at an intensity of 28 mm h^{-1}**.** This threshold corresponds to the rainfall depth and duration at which FoS values dropped below 1.0 and visible deformation began. The derived rainfall

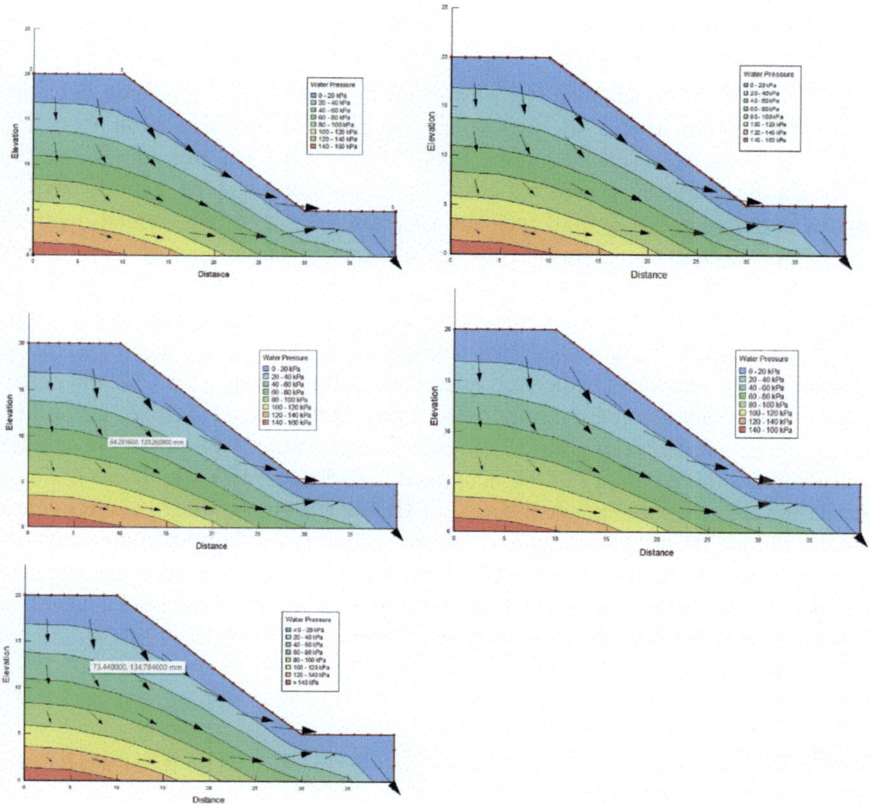

Fig. 11.5 (a) Seep/W analysis of site S1, (b) Seep/W analysis of site S2, (c) Seep/W analysis of site S3, (d) Seep/W analysis of site S4 and (e) Seep/W analysis of site S5 for water pressure

threshold is slightly lower than the 80 mm value reported by Paswan and Shrivastava (2022), owing to differences in slope geometry and soil density. Nonetheless, both studies confirm that rainfall-induced failures in Himalayan terrain typically initiate within a narrow cumulative rainfall range of 40–80 mm. Expressed in intensity–duration (I–D) form, the threshold can be represented as: where I denotes rainfall intensity (mm h^{-1}) and D represents duration (hours). This empirical relation fits well within the global threshold envelope for shallow landslides later refined by Melillo et al., (2016).

Conclusion

The present study systematically investigated rainfall-induced landslide behavior in the Pithoragarh district through integrated physical and numerical modeling approaches. Five representative slope sites (S1–S5) were analyzed to evaluate the effect of rainfall intensity, soil properties, and saturation on slope stability. Experimental simulation at a rainfall intensity of 28 mm h^{-1} and incremental depth of 9.33 mm per 20 min successfully replicated natural slope failure conditions. Results revealed a substantial reduction in the Factor of Safety (FoS) from 1.548–2.010 under dry conditions to 0.748–0.981 upon saturation, confirming that rainfall infiltration significantly decreases soil shear strength by elevating pore-water pressure and reducing matric suction. The derived critical rainfall threshold of 45–50 mm corresponds to the point where FoS drops below unity, indicating the onset of instability. Strong agreement between physical observations and SLOPE/W–SEEP/W simulations validates the reliability of the integrated modeling framework. These findings highlight the threshold-dependent nature of Himalayan slope failures and provide a scientific basis for early warning and risk mitigation in rainfall-prone areas. Incorporating these threshold values into GIS-based landslide hazard models can enhance regional prediction accuracy and contribute to sustainable slope management in Pithoragarh and similar Himalayan terrains.

References

Althuwaynee, O. F., Pradhan, B., & Ahmad, N. (2015). Estimation of rainfall threshold and its use in landslide hazard mapping of Kuala Lumpur metropolitan and surrounding areas. *Landslides, 12*, 861–875.

Ardizzone, F., Cardinali, M., Galli, M., Guzzetti, F., & Reichenbach, P. (2007). Identification and mapping of recent rainfall-induced landslides using elevation data collected by airborne Lidar. *Natural Hazards and Earth System Sciences, 7*(6), 637–650.

Bansal, V., & Sarkar, R. (2023). Numerical analysis and implications of fluctuating water levels on slope stability. In *International conference on sustainable infrastructure: Innovation, opportunities and challenges* (pp. 407–419). Springer.

Bansal, V., & Sarkar, R. (2024a). Comparative analysis of slope stability for Kalimpong region under dynamic loading using limit equilibrium method and machine benchmark learning classifiers. *Iranian Journal of Science Technology, Transactions of Civil Engineering, 48*, 2785–2807.

Bansal, V., & Sarkar, R. (2024b). Prophetical modeling using limit equilibrium method and novel machine learning ensemble for slope stability gauging in Kalimpong. *Iranian Journal of Science Technology, Transactions of Civil Engineering, 48*, 411–430.

Berti, M., Martina, M. L. V., Franceschini, S., Pignone, S., Simoni, A., & Pizziolo, M. (2012). Probabilistic rainfall thresholds for landslide occurrence using a Bayesian approach. *Journal of Geophysical Research: Earth Surface, 117*(F4).

Bharti, G., Shukla, B. K., Srinivasan, V., & Bansal, V. (2022). Experimental study on eccentrically loaded rectangular footing on sand for embedded condition to enhance serviceability of structures. *Materials Today: Proceedings, 61*, 517–522.

Dahal, R. K., Hasegawa, S., Nonomura, A., Yamanaka, M., Dhakal, S., & Paudyal, P. (2008a). Predictive modelling of rainfall-induced landslide hazard in the lesser Himalaya of Nepal based on weights-of-evidence. *Geomorphology, 102*(3–4), 496–510.

Dahal, R. K., Hasegawa, S., Nonomura, A., Yamanaka, M., Masuda, T., & Nishino, K. (2008b). GIS-based weights-of-evidence modelling of rainfall-induced landslides in small catchments for landslide susceptibility mapping. *Environmental Geology, 54*, 311–324.

Dai, F. C., Lee, C. F., & Wang, S. J. (2003). Characterization of rainfall-induced landslides. *International Journal of Remote Sensing, 24*(23), 4817–4834.

Dikshit, A., Sarkar, R., Pradhan, B., Segoni, S., & Alamri, A. M. (2020). Rainfall induced landslide studies in Indian Himalayan region: A critical review. *Applied Sciences, 10*(7), 2466.

Dolojan, N. L. J., Moriguchi, S., Hashimoto, M., & Terada, K. (2021). Mapping method of rainfall-induced landslide hazards by infiltration and slope stability analysis: A case study in Marumori, Miyagi, Japan, during the October 2019 typhoon Hagibis. *Landslides, 18*, 2039–2057.

Garcia-Urquia, E. (2016). Establishing rainfall frequency contour lines as thresholds for rainfall-induced landslides in Tegucigalpa, Honduras, 1980–2005. *Natural Hazards, 82*, 2107–2132.

Gupta, V., Bhasin, R. K., Kaynia, A. M., Tandon, R. S., & Venkateshwarlu, B. (2016) Landslide hazard in the nainital township kumaun himalaya India: The case of september 2014 balia nala landslide. *Natural Hazards 80(2),* 863-877. https://doi.org/10.1007/s11069-015-2002-5

Khanduri, S. (2022). Rain-induced slope instability: Case study of monsoon 2020 affected villages in Pithoragarh District of Uttarakhand, India. *International Journal of Earth Sciences Knowledge and Applications, 4*(1). 1–18.

Kincal, C., & Kayhan, H. (2022). A combined method for preparation of landslide susceptibility map in Izmir (Türkiye). *Applied Sciences, 12*(18), 9029.

Lepore, C., Kamal, S. A., Shanahan, P., & Bras, R. L. (2012). Rainfall-induced landslide susceptibility zonation of Puerto Rico. *Environmental Earth Sciences, 66*, 1667–1681.

Li, B., Liu, K., Wang, M., He, Q., Jiang, Z., Zhu, W., & Qiao, N. (2022). Global dynamic rainfall-induced landslide susceptibility mapping using machine learning. *Remote Sensing, 14*(22), 5795.

Lin, W., Yin, K., Wang, N., Xu, Y., Guo, Z., & Li, Y. (2021). Landslide hazard assessment of rainfall-induced landslide based on the CF-SINMAP model: A case study from Wuling Mountain in Hunan Province, China. *Natural Hazards, 106*, 679–700.

Long, N. T., & De Smedt, F. (2018). Analysis and mapping of rainfall-induced landslide susceptibility in a Luoi District, Thua Thien Hue Province, Vietnam. *Water, 11*(1), 51.

Melillo, M., Brunetti, M. T., Peruccacci, S., Gariano, S. L., & Guzzetti, F. (2016). Rainfall thresholds for the possible landslide occurrence in Sicily (southern Italy) based on the automatic reconstruction of rainfall events. *Landslides, 13*, 165–172.

Mondini, A. C., Guzzetti, F., Reichenbach, P., Rossi, M., Cardinali, M., & Ardizzone, F. (2011). Semi-automatic recognition and mapping of rainfall induced shallow landslides using optical satellite images. *Remote Sensing of Environment, 115*(7), 1743–1757.

Nanda, A. M., Lone, F. A., & Ahmed, P. (2024). Prediction of rainfall-induced landslide using machine learning models along highway Bandipora to Gurez road, India. *Natural Hazards*, 1–29.

Panchal, S., & Shrivastava, A. K. (2022). Landslide hazard assessment using analytic hierarchy process (AHP): A case study of National Highway 5 in India. *Ain Shams Engineering Journal, 13*(3), 101626.

Paswan, A. P., & Shrivastava, A. K. (2022). Modelling of rainfall-induced landslide: A threshold-based approach. *Arabian Journal of Geosciences, 15*(8), 795.

Paswan, A. P., & Shrivastava, A. K. (2023). Evaluation of a tilt-based monitoring system for rainfall-induced landslides: Development and physical modelling. *Water, 15*(10), 1862.

Posner, A. J., & Georgakakos, K. P. (2015). Soil moisture and precipitation thresholds for real-time landslide prediction in El Salvador. *Landslides, 12*(6), 1179–1196.

Reichenbach, P., Rossi, M., Malamud, B. D., Mihir, M., & Guzzetti, F. (2018). A review of statistically-based landslide susceptibility models. *Earth-Science Reviews, 180*, 60–91.

Rossi, M., Kirschbaum, D., Valigi, D., Mondini, A. C., & Guzzetti, F. (2017). Comparison of satellite rainfall estimates and rain gauge measurements in Italy, and impact on landslide modeling. *Climate, 5*(4), 90.

Saadatkhah, N., Kassim, A., & Lee, L. M. (2015). Hulu Kelang, Malaysia regional mapping of rainfall-induced landslides using TRIGRS model. *Arabian Journal of Geosciences, 8*, 3183–3194.

Saraswat, R., Parkash, S., Pal, S. (2020). Landslide susceptibility assessment due to construction of buildings in Garhwal area using GIS. In: Ahmed, S., Abbas, S., Zia, H. (eds) Smart cities—opportunities and challenges. Lecture notes in civil engineering, vol. 58. Springer,, https://doi.org/10.1007/978-981-15-2545-2_70.

Tiranti, D., & Rabuffetti, D. (2010). Estimation of rainfall thresholds triggering shallow landslides for an operational warning system implementation. *Landslides, 7*, 471–481.

Wu, C. (2019). Landslide susceptibility based on extreme rainfall-induced landslide inventories and the following landslide evolution. *Water, 11*(12), 2609.

Yang, S. R. (2017). Assessment of rainfall-induced landslide susceptibility using GIS-based slope unit approach. *Journal of Performance of Constructed Facilities, 31*(4), 04017026.

Zhao, Y., Wang, R., Jiang, Y., Liu, H., & Wei, Z. (2019). GIS-based logistic regression for rainfall-induced landslide susceptibility mapping under different grid sizes in Yueqing. *Southeastern China. Engineering geology, 259*, 105147.

Chapter 12
Assessing the Forest Fire Susceptibility Using Machine Learning Algorithms in Tehri Garhwal District of Uttarakhand

Subhendu Jana, Nirmal Biswas, Anik Saha, and Sunil Saha

Abstract Forest fire seriously damage the Himalayan region's socio-ecological stability, particularly in Uttarakhand's Tehri Garhwal district. This study aimed to analyze the forest fire susceptibility (FFS) utilizing three different modelling approaches: Random Forest (RF), Bagging, and Logistic Regression (LR). Using R studio and SPSS, the models were run considering 900 fire sites and 22 geo-environmental drivers. Susceptibility was divided into five classes using Jenks natural breaks, which outperformed the other interval approaches in decreasing intra-class variance. The results indicate that RF models provided the most reliable spatial distribution, identifying 17.84% of the district as very high susceptibility, while the LR model tended to over-predict very high susceptibility zone (19.86%). Model validation via receiver operating characteristic (ROC) curves demonstrated exceptional predictive accuracy; the RF model achieved the highest Area Under Curve (AUC) of 0.977 (97.74%), followed by Bagging (0.966) and LR (0.866). Notably the inquiry revealed that 46% of the total area situated in high to very high susceptibility zones, particularly within the Devprayag, Ghansali, Jakhani-Dhar and Narendra Nagar regions. In order to safeguard the endangered tribes and ecosystems of the Tehri Garhwal, forest agencies and disaster management authorities can use the produced FFS maps as a high-precision decision-support tool to prioritize resources, improve early warning system and execute focused conservation strategies.

Keywords Forest fire susceptibility · Tehri Garhwal · Machine learning models · GIS and remote sensing · Random forest

---

S. Jana · N. Biswas · S. Saha (✉)
Department of Geography, University of Gour Banga, Malda, West Bengal, India
e-mail: sunilgeo.88@gmail.com

S. Jana
e-mail: subhendudeula1234@gmail.com

N. Biswas
e-mail: nirmalofficial69@gmail.com

A. Saha
Department of Geography, Harishchandrapur College, Malda, West Bengal, India
e-mail: aniksahamalda@gmail.com

R. Sarkar et al. (eds.), *Climate Change, Environmental Hazards and Community-Based Resilience*, Disaster Risk Reduction, https://doi.org/10.1007/978-981-95-9485-6_12

Introduction

Forest fires are the unsupervised and unplanned scorching of vegetation in an ecosystem such as woodlands, grazing land, bushland, or tundra which utilizes the fuel from nature and expands due to environmental factors e.g., landscape and wind (Sharma & Kumar, 2025). Forests serve as essential for sustaining ecological equilibrium, water resource preservation, plants, wildlife, and extensive kinds of plants and animals (Chandra & Arora, 2006; Raj et al., 2025). Natural origins of forest fires involve lightning (Yadav & Devi, 2019), eruptions of volcanoes (Cai et al., 2021; Zhao, 2015), the frictional force caused by twirling stones, drying bamboo stalks rubbing across each other (Satendra, 2014), and in particularly severe cases, a fallen meteorite or asteroid (Morgan et al., 2013). The fire activity is becoming prolonged as a manifestation of climate change and the growing period of high temperatures that appears along with them (Sinha et al., 2021). With time the regularity involving forest fires is increases as because of human involvement in the natural forest environment (FSI, 2022). As a physical event, a forest fire has the capability to jeopardize a society's, economy's, and environment's functions (Boonchut et al., 2005). It also exerts an enormous effect on the atmosphere's chemistry (Randerson et al., 2006), bio-geochemical balance (Baker & Schimel, 2002), and ecosystem architecture (Cochrane, 2003). According to research, the annual worldwide concentrations of carbon and nitrogen from fires are 3.53 Pg and 8 Tg N (Crutzen & Lelieveld, 2001), (Page = 1 trillion kilos). The intensity of condensation nuclei of clouds in upper atmosphere is also raised due to Aerosol concerning fires that have potentiality to effects the magnitude, impacting cloud microphysics, precipitation, and circulation through the atmosphere (Andreae et al., 2004; Ramanathan et al., 2001). In present context forest fire becomes a common form of horrific hazard worldwide. Globally there were 61,410 forest fires on average a year from 2013 to 2022, affecting 7.2 million acres on average (Baker, 2023). Approximately 70 people perished as a result of wildfires in the world in 2020. Worldwide more than 2400 people lost their life due to wildfires in past three decades (Statista, 2023). According to 2020 research by WWF (World Wildlife Fund) International, humans are responsible for 75% of all wildfires worldwide, and human activity may be largely to blame for the increasing trend in forest fires in 2022. India covers an area of around 328 million hectares, of which 68 million hectares are forests (Forest Survey of India, 2003; Sharma & Kumar, 2025). According to national forest inventory assessments, roughly one-third (about 35.71%) of the country's forested areas have little to no history of major fire incidents. In contrast, more than half (around 54.45%) experience occasional fire occurrences, about 7.49% fall into a moderately fire-prone category, and approximately 2.40% are classified as highly vulnerable to frequent fires., human played a significant role to conducting forest fire with 95 percentage (FSI, 2021). The state of Uttarakhand is endowed with an extensive diversity of vegetation, from alpine to tropical deciduous. Forest fires are quite common in Uttarakhand, and they have become more prevalent in recent years as temperatures have risen (Chauhan et al., 2018). In Uttarakhand, forest cover declined by 1.11 kha due to fires and 16.8 kha in response to all other drivers, equivalent to 428 kt of CO_2 emissions between 2001 and 2021. As per Global Forest Watch only in the year 2022, 340 ha of land in Uttarakhand was burn down. A total of 1,278,343.90 United States dollars (or 10.53 crore Indian rupees) were destroyed as a result of the forest fires that collapsed in Uttarakhand in May 2019 (Kukreti, 2019). As a district of Uttarakhand, TehariGharwal is the worst affected district in the aspect of the forest fire (Singh et al., 2025). Fires were responsible for 22% of tree cover loss in Tehri Garhwal, comparable to CO_2 emissions of 6.34kt between 2001 and 2021. Between 2001 and 2021, this district lost 574 hectares of vegetation owing to all other causes of loss and 162 hectares due to forest fires (Statista, 2022). Based on the research type, employed strategies and techniques to monitor susceptibility of may be grouped into three categories: Statistical, probabilistic, and machine learning techniques (Bui et al., 2016). Through the use of particular mathematical functions and formulae, probabilistic (mechanistic) techniques model and forecast potential forest fires behavior (Pastor et al., 2003; Mishra et al., 2025). These techniques can thus simulate and forecast how the fire will behave in space and time (Gigovi et al., 2019). The literature describes BEHAVE (Andrews, 1986) and LANDIS-II (Sturtevant et al., 2009) as among the most frequently utilized mechanistic forest fire models. Whenever the research area is extensive statistical modeling of forest fires is superior to probabilistic modeling (Gigovi et al., 2019). Meanwhile, logistical regression (Chuvieco et al., 2010; De Vasconcelos et al., 2001; Pourghasemi, 2016, 2016b), Monte Carlo simulations (Conedera et al., 2011), weights-of-evidence (Amatulli et al., 2007), logistic generalized additive model (Vilar et al., 2010), evidential belief function (Pourghasemi, 2016, 2016b), and geographically weighted regression (Oliveira & Oehler, 2012) are all statistical strategies and approaches used for forest fire modeling. Due to the valuable accuracy of assessing forest fires, machine-learning technologies have been designed and deployed such as the support vector machine (Sakr et al., 2010; Pourtaghi et al., 2016), Random Forest (A. Kernel Logistic Regression (Bui et al., 2016), Entropy Maximum (Arpaci et al., 2014), Artificial Neural Networks (ANN), and Arpaci et al., 2014; Oliveira and Oehler, 2012) are among the commonly applied modelling techniques in this field (Satir et al., 2016; De Vasconcelos et al., 2001; Maeda et al., 2009). Hybrid machine learning models are used to eliminate the error that generated superior outcomes in comparison to traditional and single models (Pham et al., 2016). In reality, modeling and predicting forest fires remains hard due to many complicated interactions among influence and ignition components at regional scale (Bui et al., 2016). This study's primary goal is to develop maps of FFS maps for the Tehri Garhwal district of Uttarakhand using a set of overseen and flexible algorithms for machine learning, including their group forms, and to assess the precision and effectiveness of these models. In this investigation, a total of, a total of 900 incidence of forest fires locations are obtained from Forest Fire Survey of India. From these, for training, 630 locations (70%) were chosen at random, while the remainder 270 (30%) locations were reserved for confirmation. The training dataset, together with 22 conditioning variables, served like the input various models for machine learning, to generate FFS maps using Logistic Regression (Chuvieco et al., 2010; De Vasconcelos et al., 2001; Pourghasemi, 2016, 2016b) as a statistical model, Machine learning models like Random Forest (Oliveira & Oehler, 2012) and Bagging model (Yariyanet al., 2020) as a hybrid machine learning model. After constructing those models, their accuracy has been evaluated by implementing the ROC curve. The primary goal of this research work is to produce a comprehensive forest fires inventory of Tehri Garhwal district of Uttarakhand with detailed records of past fire incidents. The study also contemplates critical examination of different conditioning factors contributing towards the incidences of forest fire, namely topography, vegetation, climatic condition, and human-induced activities. It also aims to delineate specific forest fire susceptibility zones within the district and deliver essential insights that can support risk assessment, prevention planning, and mitigation efforts, and effective management of future fire hazards.

Study Area

The district of Tehri Garhwal serves as the focal point for this study, situated in the Western Himalayan agro-climatic zone. Encapsulated by the Yamuna and Ganga river systems, the region covers a vast 3,642 sq. km expanse characterized by dramatic elevation changes (Fig. 12.1). These vertical gradients create a unique mosaic of sub-tropical and cold temperate environments. The local community, totaling nearly 619,000 residents as of 2011, is spread across 1,862 villages and major urban centers like Chamba and Devprayag. Despite a significant slowdown in population growth over the last two decades, the area remains agriculturally driven, though farmers rely heavily on rainfed crops like jowar and barley due to limited irrigation infrastructure. With a vast majority of the district (over 3,200 sq. km) designated as forest land, the threat of wildfires is a critical concern. This study aims to delineate high-risk zones to provide a framework for proactive forest fire mitigation.

Materials and Methods

Forest Fire Inventory Map

The Spatial inventory of forest fire performs an indispensable part since the backbone of Forest Fire Susceptibility Mapping depends on an association between the geographical coordinates from previous forest fires and the local ignition parameters (Bui et al., 2017). As a result, using precise inventory data can result in an adequate prediction model and useful evidence widely recognized as an essential aspect of formulating decisions for disaster avoidance and reduction. The map of forest fire inventory which was prepared use a mix of primary and secondary sources of data. This study's main sources of information are journals, reports, case studies,

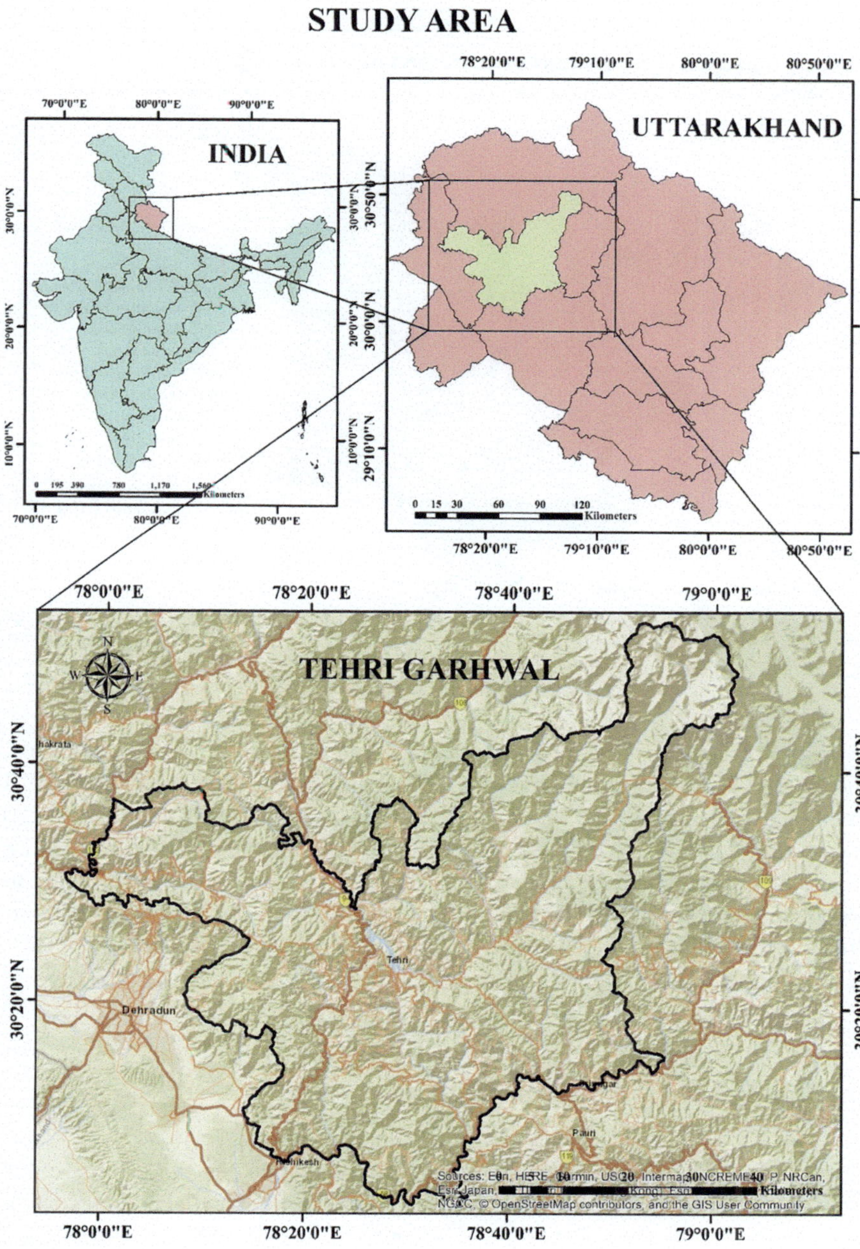

Fig. 12.1 Study area

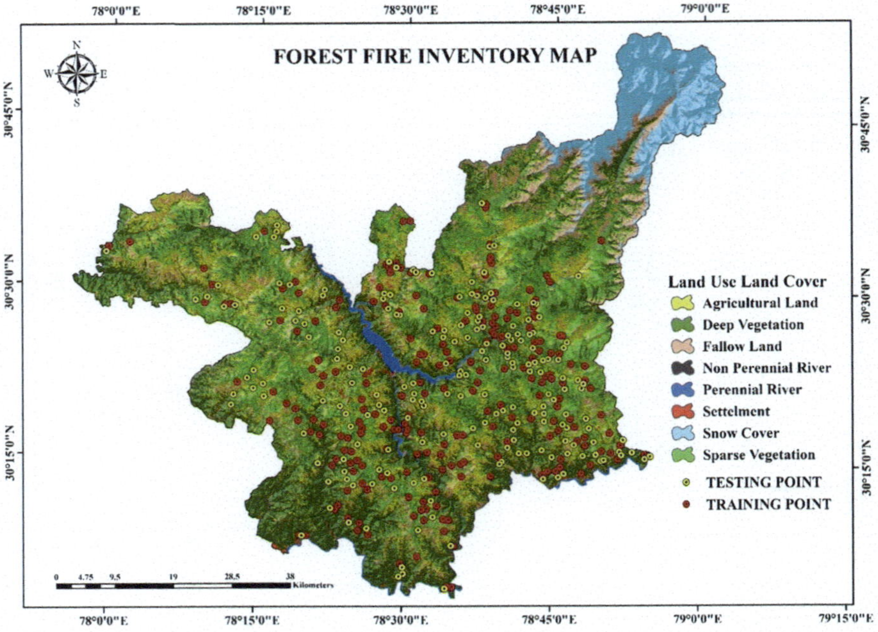

Fig. 12.2 Forest fire inventory map

and newspaper articles. The inventory map of forest fire was developed in this study, a total of 900 forest fire sites listed by India's Forest Survey between 2010 and 2023 were compiled. Of these, 630 locations (70%) were used for model training and 270 locations (30%) were reserved for testing and performance evaluation (Fig. 12.2). Additionally, ArcGIS was employed to generate 900 random non-fire points to support the analysis and to simulate non-Forest Fire regions. To minimize overfitting concerns in modeling, non-Forest Fire points were generated at randomness and plotted on the forest fire inventory map.

Conditioning Factors for Forest Fires

The accuracy of the parameters is considered for prediction has an immense effect on the quality of the prediction. The ignition parameters that are taken into consideration for predicting the forest fires and it represent the range of forest fire sources. Additionally, when the elements that would have contributed to the wildfire's growth and escalation as well as its origins will determine its size. This study incorporated 22 variables to model forest fire occurrence (Fig. 12.3). Elevation, slope, aspect, curvature, and topographic location index were among these variables. (TPI), topographic wetness index (TWI), NDVI, land surface temperature (LST), NDWI, water ratio index (WRI), NDBI, land use/land cover, average annual temperature, average

Table 12.1 Forest fire conditioning factors (FFCFs), specific information utilized, along with the sources for each

FFCFs	Data used	Sources
Altitude	SRTM DEM	USGS earth explorer (https://earthexplorer.usgs.gov/)
Slope		
Aspect		
TWI		
Curvature	LANDSAT 8 OLI IMAGE (30*30)	
LULC		
LST		
NDWI		
WRI		
NDBI	Reference district soil map	National Bureau of soil survey and land use planning
Soil texture	Open street map	https://extract.bbbike.org/
Distance from river		
Distance from settlement		
Distance from road	Nasa power data access viewer	https://power.larc.nasa.gov/data-access-viewer/
Air temperature		
Wind speed		
Relative humidity		
Average rainfall	WRIS	https://indiawris.gov.in/wris/

temperature per year, wind speed, relative humidity, and yearly rainfall, and distances from roads, rivers, religious centres, and settlements (Table 12.1). The detailed descriptions of these factors are provided below.

Elevation

Elevation strongly affects precipitation patterns, wind exposure, and fuel moisture, all of which influence fire behavior. Lower elevations warm earlier and receive less rainfall, causing fuels to dry more quickly, while higher zones experience different fuel conditions and more frequent lightning-related ignitions. Elevation in the study region varies from 333 m to 6,728 m (Fig. 12.3A). Data from the Digital Elevation Model (DEM) USGS (SRTM 30 m) were used and processed in ArcGIS.

Slope

Slope is a critical factor influencing forest-fire behaviour, as fires tend to advance more rapidly uphill where rising heat more effectively warms and dries the fuels above. In Tehri Garhwal, the slope values range from 0° to 75° reflecting the slope from gentle to very steep (Fig. 12.3B). In this investigation, the slope was obtained using ArcGIS from the DEM. slope analysis tools.

Aspect

Slope orientation influences both vegetation patterns and the amount of solar radiation received. In lower-elevation forests, south- and slopes that face west usually have sparse the vegetation and lighter fuel loads because they receive more sunlight, causing fuels to dry faster under higher temperatures and lower humidity. North-facing slopes, however, retain more vegetation and heavier fuel loads. Their cooler, shaded conditions delay fuel drying, though the higher accumulation of vegetation can lead to more severe wildfires. In this study, aspect was extracted from 30-m USGS (SRTM) DEM data and analyzed in ArcGIS (Fig. 12.3C).

Curvature

Curvature is a crucial topographic parameter that influences the pattern and rate of fire spread. It reflects how quickly the slope angle changes, distinguishing between surfaces that are concave (negative curvature) and surfaces that are convex (positive curvature) (Tiwari et al., 2021; Jaafari, 2018). This investigation's curvature for the area was produced utilizing the curvature analysis tool in ArcGIS, derived from DEM data. Within the Tehri Garhwal district, curvature values range from 36 to 33 (Fig. 12.3F).

Topographic Position Index (TPI)

The discrepancy between the elevation and the Topographic Position Index of a particular cell and the mean height of the in and around cells (Pawluszek & Borkowski, 2017). TPI influences forest fire indirectly. Flat or low-lying areas tend to have higher settlement density compared to ridges or valleys, which increases human activity and, consequently, elevates the likelihood of forest fire incidents in these zones.

The Topographic Wetness Index (TWI)

The Index of Topographic Wetness in this case symbolizes the degree of surface moisture. Or the wetness level across terrain features. for the reduce in topographic moisture, combined with a higher requirement of water to saturate the land surface, which is increases both the likelihood of forest fires and their potential rate of the spread (Jaafari, 2018). TWI plays a critical role to determining how fires propagate

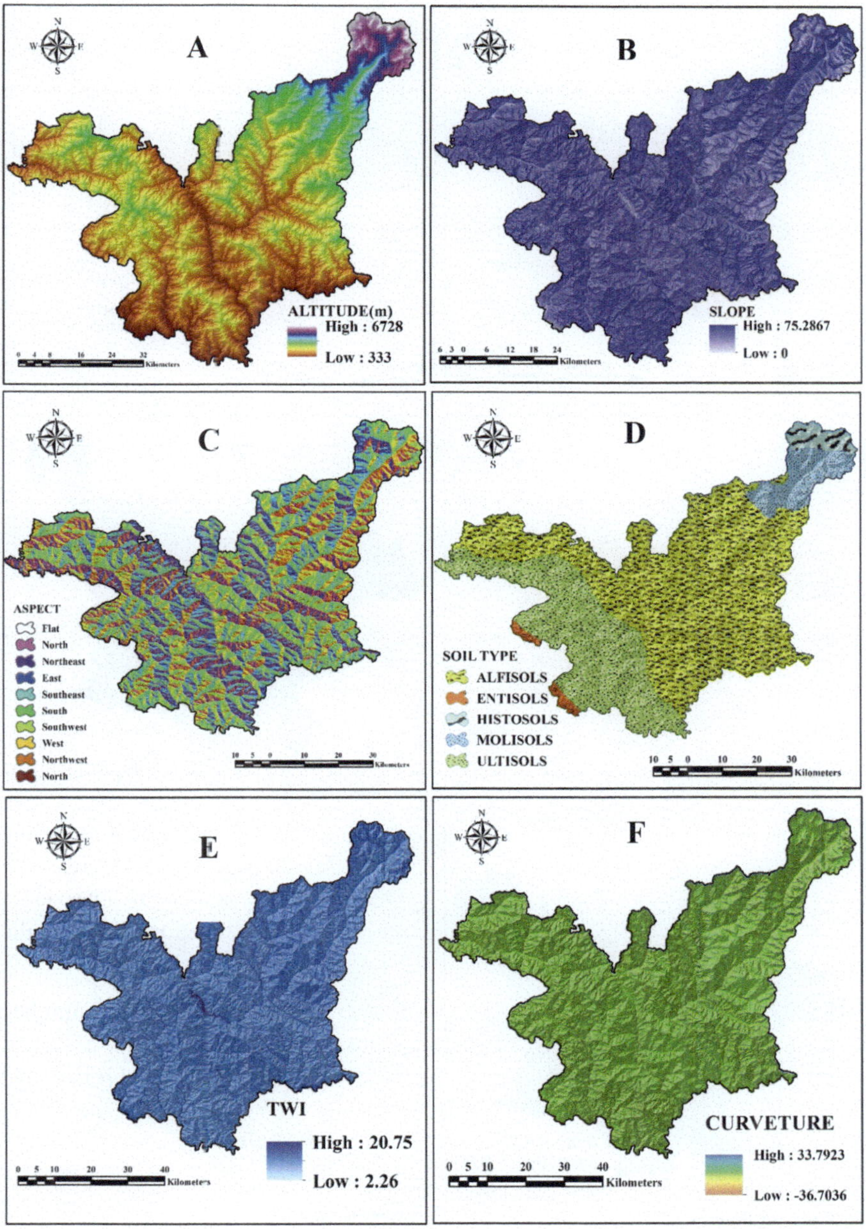

Fig. 12.3 Forest fire conditioning factors

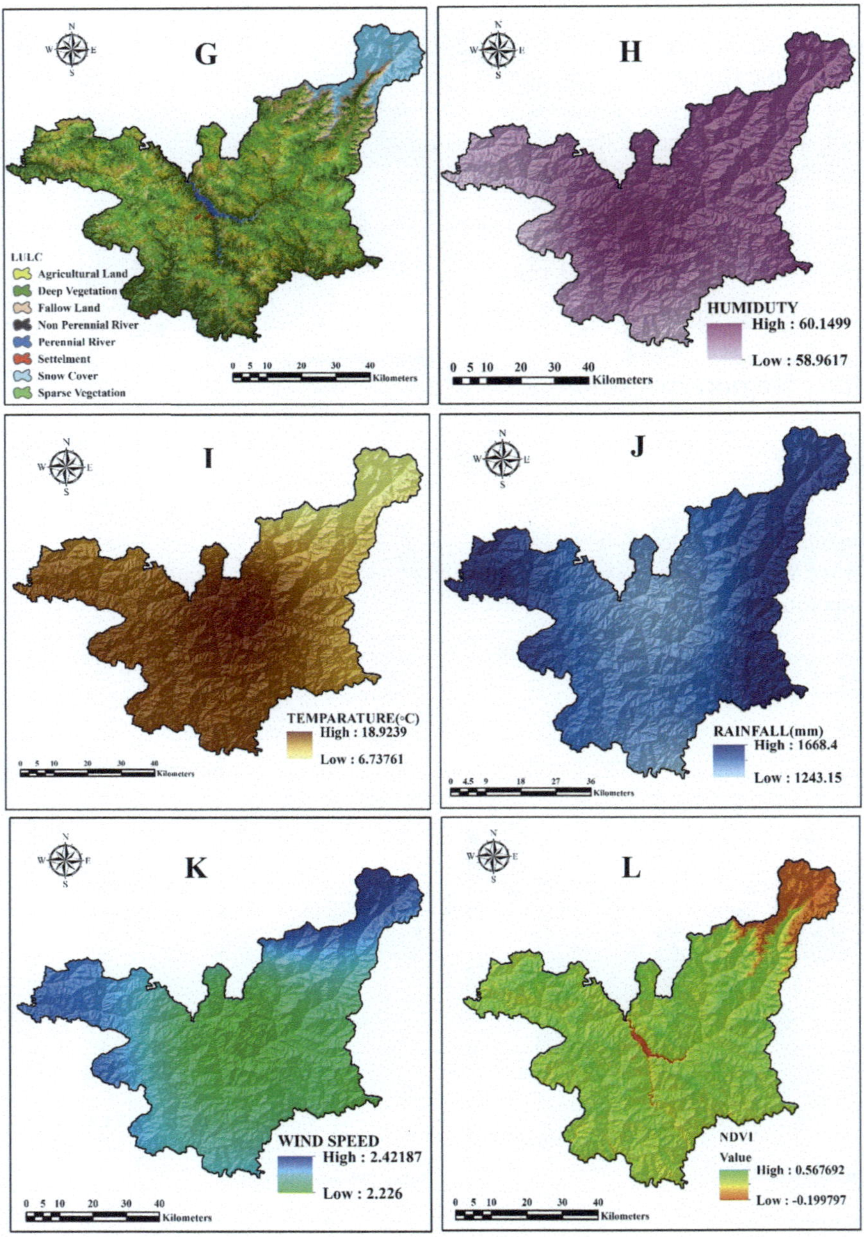

Fig. 12.3 (continued)

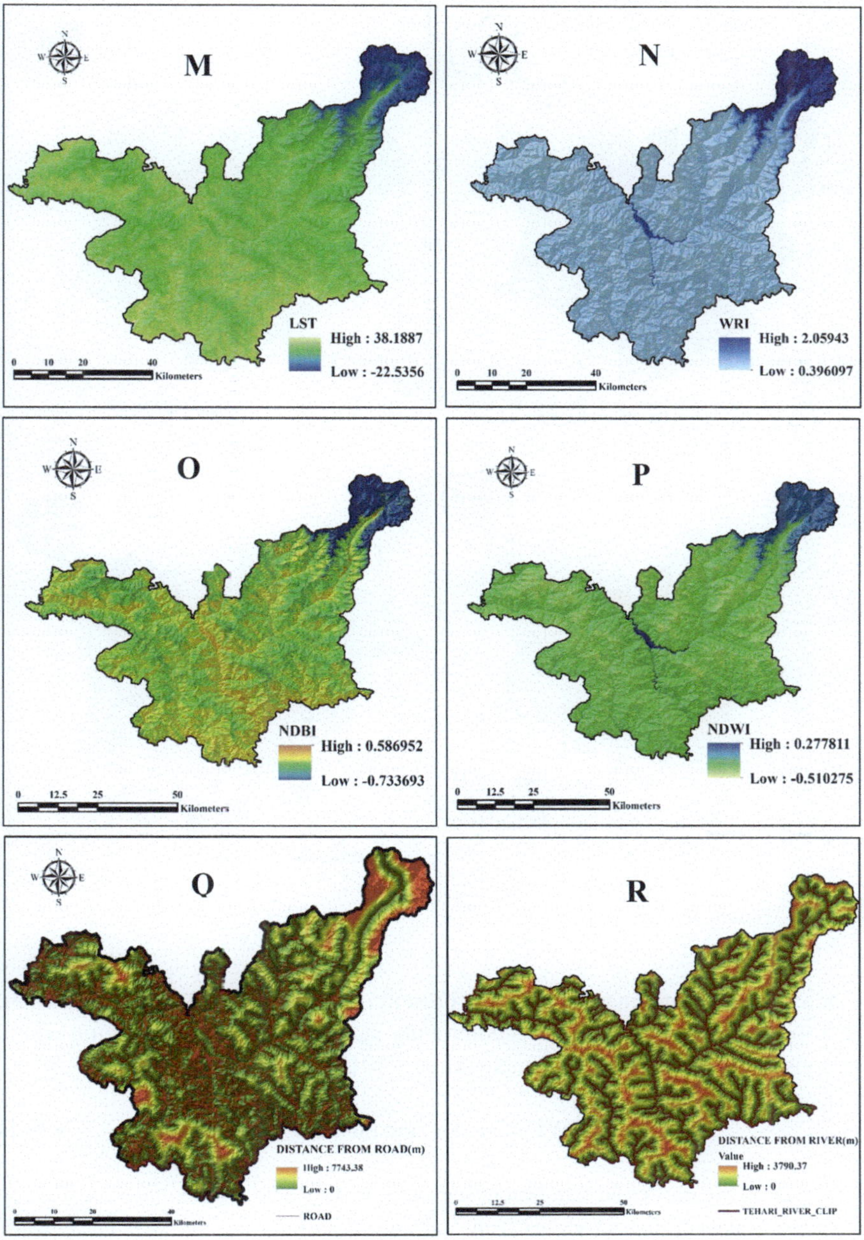

Fig. 12.3 (continued)

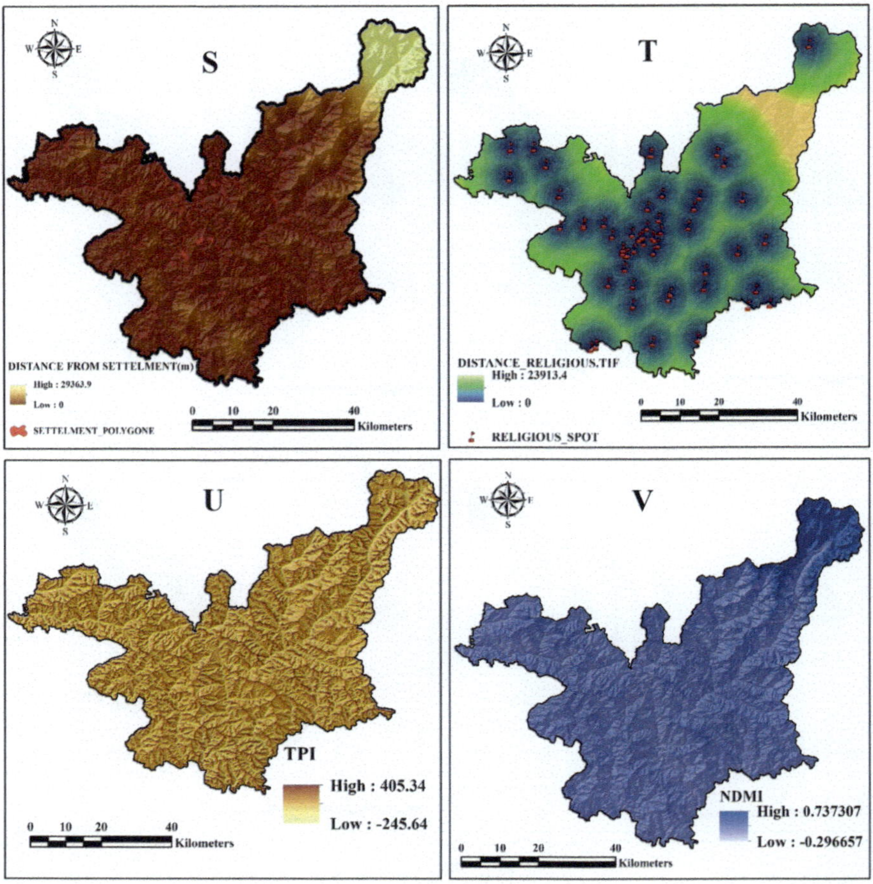

Fig. 12.3 (continued)

by indicating the amount of moisture is present on the surface (Jaafari et al., 2019). The Topographic Wetness Index is commonly calculated using the equation which is proposed by Nihat Ay and Shunichi Amari (2015).

$$\text{TWI} = \ln \ \text{As}/(\tan(\text{slope})) \tag{12.1}$$

where 'As' is the particular catchment region in meters The greater the. The TWI in the Tehari Gharwal district varies from 2.26 to 20.75 km (Fig. 12.3E). The greater the TWI worth, the lower the chance of a Forest fire.

Rainfall

Rainfall is is among the most important climatic variables influencing forest fire behavior. Variations in precipitation directly affect fuel moisture levels and the degree of surface wetness (Abedi Gheshlaghi, 2019; Kayet et al., 2020). Reduced the rainfall generally results in drier fuels which is increasing the probability of fire incident, while higher precipitation in this area tends to lower fire risk (Saha et al., 2023). In this study, the average rainfall distribution was generated using the IDW interpolation technique based on data collected from WRIS stations located around the Tehri Garhwal district. Over the past five years (2017–2022), annual rainfall in the district varied between 1240 and 1670 mm—classified as moderately high—yet forest fire events still occur during the summer season (Fig. 12.3J).

Relative Humidity

The part that climate and weather play an essential role in determining the moisture content of vegetation or on relative humidity and, thereby affecting fuel availability. Extended dry periods when reduce fuel moisture, and often causing forest fires to occur more frequently and with greater intensity (Oregon State University). Low relative humidity facilitates easier ignition and rapid fire spread, whereas vegetation freshly moistened by rainfall resists ignition. In the Tehri Garhwal district, average annual relative humidity during the last five years (2017–2022) ranged from 58 to 60% (Fig. 12.3H). These values were obtained from NASA POWER data and later processed and visualised in ArcGIS.

Temperature

Temperature is a critical climate element that greatly influences both the incidence and severity of forest fires (Satir, 2016). Higher temperatures contribute to reduced fuel moisture content and the formation of local winds, which indirectly affect fire behaviour. As temperature increases, fuel moisture generally decreases, making vegetation more susceptible to burning, and several studies highlight the strong geographical correlation between temperature and the incidence of forest fires (Tiwari et al., 2021). In the Tehri Garhwal district, the average annual temperature over the past five years (2017–2022) ranged from 6 °C to 18 °C (Fig. 12.3I). These data were acquired from NASA POWER and mapped using the ArcGIS environment.

Wind Speed

Wind speed is a major factor which is influencing forest fire behaviour because it is affects vegetation, topography, and surface moisture conditions (Wang et al., 2021). In the dry season, stronger winds enhance the capacity of fires to spread quickly across the landscape (Kanga et al., 2017) for Sudden shift in wind direction can also widen the fire front, making suppression efforts that increasingly the challenging. In the Tehri Garhwal district, the average annual wind speed over the last five years (2017–2022) are ranged between 2.2 and 2.5 mile per second (Fig. 12.3K). These data were retrieved from NASA POWER and mapped within the ArcGIS environment.

Soil Texture

Soil texture is regarded as a crucial component because it influences the environmental conditions under which wildfires occur and indirectly affects fire behaviour (Pourghasemi, 2016, 2016b). Land use practices further modify fire vulnerability, as various land cover types display differing degrees of ignition potential. For this reason, land cover characteristics have been incorporated as an additional ignition-related parameter (Bui et al. 2016b, 2017; Pourghasemi, 2016, 2016b; Eugenio et al., 2016).

NDVI

A popular tool is the Normalized Difference Vegetation Index (NDVI) to measure vegetation vigour and greenness, making it valuable for evaluating plant density and detecting variations in vegetation health. Mapping the spatial distribution of vegetation helps in recognizing the scope and seriousness of potential forest effects of fire NDVI reflects The amount of moisture in plants and soil—factors directly linked to the risk of a fire starting and spread—because it is derived from the photosynthetic activity of vegetation (Saloum and Abdou, 2016; Tiwari et al., 2021). The NDVI value is determined by applying the subsequent formula:

$$\text{The NDVI} = \frac{\text{NIR} - \text{RED}}{\text{NIR} + \text{RED}} \tag{12.2}$$

where NIR stands for Near Infra-Red (not equal to 0) electromagnetic reflectance, RED for Red (not equal to 0), and NDVI for Normalized Difference Vegetation Index. Reflectance in Near Infra-Red (not equal to 0); RED = Electromagnetic reflectance in Red (not equal to 0).

The NDVI in the Tehari Gharwal district varies from –0.2 to 0.6 (Fig. 12.3L). NDVIs closer to 1 indicate denser vegetation. High NDVI means a high chance of Forest Fire, and Low NDVI means a low chance of Forest Fire.

Normalized Difference Moisture Index (NDMI)

The propagation of fires due to plant water stress has a high correlation with soil and plant moisture, As stated by several research (Jaafari, 2018). Furthermore, the possibility of forest fires is influenced more by soil wetness than by the predominant climatic factors. One of the most used indices in research on Forest Fire vulnerability is the NDMI, which measures plant water stress (Fornacca et al., 2018). In Tehri Gharwal district the NDMI differs from –0.7 to 0.5 (Fig. 12.3V). The ratio is computed using the equation:

$$\mathrm{NDMI} = \frac{(\mathrm{NIR} - \mathrm{SWIR})}{(\mathrm{NIR} + \mathrm{SWIR})} \tag{12.3}$$

where SWIR stands for Short-Wave Infra-Red band and NIR for Near-Infra Red band.

WRI

The WRI, or Water Ratio Index, is designed to identify surface water by utilizing the characteristic spectral reflectance patterns of water bodies. However, the index may also respond to elements like shadows, moist soil, or certain types of vegetation, which can create false indications of water presence. In general, WRI values greater than 1 are interpreted as areas likely representing water (Shen et al., 2010). In Tehri Gharwal district the WRI differs from 0.4 to 2.05 (Fig. 12.3N). The high WRI value indicates that likelihood of a forest fire, in vice versa. The ratio is computed using the equation:

$$\mathrm{WRI} = \frac{(Green + Red)}{(NIR + SWIR)} \tag{12.4}$$

where *SWIR* stands for Short-Wave Infra-Red band and *NIR* for Near-Infra Red band.

NDBI

The NDBI helps differentiate built-up surfaces from other land features based on their spectral characteristics. Negative NDBI scores typically indicate water bodies or regions with relatively high moisture content of the soil. Increased positive NDBI values correspond to built-up regions, whereas reduced positive values generally represent vegetated surfaces. Since vegetation with lower moisture content is more susceptible to burning, areas showing lower positive NDBI values often exhibit a higher likelihood of fire occurrence. The NDBI is determined using the formula that follows value lies between –1 and +1. In Tehri Gharwal district the WRI differs from –0.7 to 0.6 (Fig. 12.3O). So, the higher values NDBI indicate that high probability of forest fire. The following formula is used to calculate the ratio:

$$\mathrm{NDBI} = \frac{(\mathrm{SWIR} - \mathrm{NIR})}{(\mathrm{SWIR} + \mathrm{NIR})} \ldots\ldots \tag{12.5}$$

where SWIR stands for Short-Wave Infra-Red band and NIR for Near-Infra Red band.

NDWI

NDWI stands for Normalized Difference Water Index designed to highlight and extract bodies of open water from satellite imagery by enhancing the spectral response of water Water features often have NDWI values higher than 0.5., while vegetation produces much lower values, allowing water bodies to be clearly distinguished from vegetated surfaces. In Tehri Gharwal district the NDWI differs from –0.51 to 0.28 (Fig. 12.3P). The higher values NDWI indicate that low probability of forest fire, and the other way around. The ratio is computed using the equation McFeeters (1996).

$$\text{NDWI} = \frac{(Green - NIR)}{(Green + NIR)} \tag{12.6}$$

where NIR = Near-Infra Red band.

Distance from Settlement (DFS)

Local communities living near forested areas may ignite fires for various cultural and livelihood-related activities, including burning agricultural residue, deterring wildlife, clearing pathways, collecting forest products, producing charcoal, resin tapping, cooking, or even unintentionally through discarded cigarette butts (Satendra, 2014). Forested zones located close to near the human settlements which are widely recognised as being under continuous anthropogenic pressure and making them more susceptible to fires, resource exploitation, and deforestation in this area (Roman-Cuesta et al., 2004; Cayuela et al., 2006). In the Tehri Garhwal district, DFS values range from 0 km to 2.9 km (Fig. 12.3S). Areas with greater DFS values that indicate a greater chance of forest incident of a fire whereas regions closer to settlements show relatively lower fire probability in respect to other.

Distance from Road (DFR)

The Road networks is a critical component of forest infrastructure, and it is influencing both management and conservation efforts in this region (Veena et al., 2017). However, the presence of roads also increases the direct human interaction on forest ecosystems and thereby elevating the likelihood of fire outbreaks. Activities associated with road construction—such as excavation, vegetation removal—and the movement of travellers or tourists can further contribute to fire ignition (Jaffari et al., 2019). In the Tehri Garhwal district, DFR values range from 0 to 7.7 km (Fig. 12.3Q).

Lower DFR values indicate the areas with a higher potential for forest fires, whereas greater distances correspond to lower forest fire probability on any area.

Distance from River

According to Zhang et al. (2019) and Jaffari et al. (2021), rivers and drainage networks frequently serve as a natural barrier to fire, which spreads by generating areas with higher moisture content and lower flammability. The distance from river features in the study area ranges from 0 to 3.7 km (Fig. 12.3R). The likelihood of a forest fire developing is lower in places near river channels and higher in locations farther away.

Distance from Religious Centre

Large tourist sites and religious or pilgrimage centres normally attract huge crowds, increasing the chances of accidental or negligent ignition of fire. Fires lit for cooking, rituals, or recreation could spiral into a wildfire in favourable conditions (Satendra, 2014) In the study area, distances from religious centres range from 0 to 24 km (Fig. 12.3T). which Regions located closer to these sites exhibit a greater probability of forest fire, while which areas at greater distances show lower risk.

LST

LST is a fundamental input to wildfire analysis, as it directly influences fuel moisture and therefore potential ignition, fire spread, and overall fire behaviour. Past research on such topics has been conducted (Pourtaghi et al., 2016; Sharma & Kumar, 2025), highlights LST as a key variable in determining vegetation sensitivity and a vital component in wildfire risk modelling. In the Tehri Garhwal district, LST values range from –22 °C to 38 °C (Fig. 12.3M). Generally, a higher LST shows lower fuel moisture and hence points towards a higher potential for forest fire incidents.

LULC

Wildfire behavior is largely a product of its environment; while vegetation acts as the primary fuel, the specific type and arrangement determine how a fire evolves (Nikhil et al., 2021). Research indicates that fire risk is highest in areas with dense, dry biomass, often exacerbated by human activity in non-wooded plantation lands (Veena et al., 2017). These anthropogenic risks include unintentional ignitions and the traditional farming practice of using fire to clear fallen bark and encourage grass regrowth (Satendra et al, 2021; Veena et al., 2017). If left unmanaged, these forest plantations can become hazards to human populations (Gómez-González et al., 2018). Spatial analysis using 2023 Landsat 8 imagery reveals a diverse landscape ranging from agricultural land (14.3%) to snow cover (7%) but highlights that the vast majority of fire activity (90%) is confined to areas categorized as deep or sparse vegetation (Fig. 12.3G).

Forest Fire conditioning factors: (A) Elevation, (B) Slope steepness, (C) Slope aspect, (D) Soil Type, (E) TWI, (F) Curvature, (G) Land use land cover, (H) Humidity, (I) Temperature, (J) Rainfall (K) Wind Speed, (L) NDVI, (M) LST, (N) WRI, (O) NDBI, (P) NDWI, (Q) Distance from Roads, (R) Distance from River, (S) Distance from Settlement, (T) Distance from Religious Centre, (U) TPI, (V) NDMI.

Methods

Random Forest Model

The algorithm Random Forest (RF), at first proposed by Breiman (1996), the Random Forest (RF) algorithm is a machine-learning, non-parametric method of multivariate classification. Since its formulation, applications of RF have been favored in many research fields, including psychology (Strobl et al., 2009) and a image from remote sensing analysis (Ham et al., 2005). Several works have underlined its good predictive performance for both regression and classification problems (Youssef et al., 2015; Pourghasemi et al., 2017; Trigila, 2015). The RF method works as an ensemble system built by bootstrap aggregation of multiple decision trees created on resampled datasets (Micheletti et al., 2014). Every tree is grown using random subsets of the training samples and predictor variables, while the remaining samples are used as out-of-bag information for internal model validation (Breiman, 1996). The model assigns each case to a final class based on combined decisions by considering outputs from a large set of individual trees which have been grown on samples that were bootstrapped, together with Each node's random feature selection (Saha et al., 2022; Aditian et al., 2018). In this work, the RF algorithm was run by means of the 'Random Forest' package in R Studio, supported by spatial processing in ArcGIS.

Bagging

Among the pioneering techniques of machine learning, bagging, propounded by Breiman in 1996, is based on the principle of bootstrap sampling. Because of its high predictive power, this method has seen a wide range of applications in landslide, flood, forest fire, and other hazard susceptibility studies during recent years. This technique is found particularly useful because it generates several Trees of decisions from different bootstrapped portions of a dataset, followed by combines the output to form a more stable and reliable final model. The technique improves the overall accuracy of model prediction by removing any fluctuations that might be caused by the alignment error. According to Breiman (1996) and later supported (Yariyan et al.,

2020), a typical classifier using Bagging operates through a three-stage process. In such a process, resampling is done repeatedly to generate a diverse subset for training, which acts like bootstrap samples. When these subsets are combined into full feature space at the time of classification, they help strengthen the predictive stability and accuracy of the model in an ensemble for various applications where Bagging has been used in machine learning applications.

Logistic Regression (LR)

Logistic Regression applies to a problem where an event outcome is binary, and the dependent variable has only two possible states either the forest fire could be present or absent as outlined (Akgun, 2012). This approach examines the association of more than one independent factor, represented by V1, V2…Vn, with a single dependent variable, denoted as Y, in order to determine the probability via logit of the recurrence of an event transformation. The model appraises the contribution of these predictor variables to the likelihood of forest fire occurrence as well as its absence in values of 0 or 1. Although the model of LR does not directly provide a susceptibility map, the probability outputs obtained from it can be considered as susceptibility levels. In this study, the model was calibrated and fitted using Eq. 12.7. The relationship could be expressed as the following form:

$$Y = Logit(p) = ln\left\{\frac{p}{1-p}\right\} = C0 + C1V1 + C2V2 + \ldots + CnVn \qquad (12.7)$$

Here, the likelihood of a forest fire occurring (the dependent variable Y) is represented by p. The term **p/(1 − p)** denotes the odds ratio. $\mathbf{C_0}$ refers to the intercept, while $\mathbf{C_1}$, $\mathbf{C_2}$ **… Cn** are the regression coefficients for assessing the impact of the independent variables $\mathbf{V_1}$, $\mathbf{V_2}$ **… Vn** on the occurrence of the event.

Validation Technique

The ROC Curve

The ROC curve is a commonly used for evaluating and comparing probabilistic prediction models, and it has been frequently applied in hazard-related studies (Saha et al., 2024). It provides that a graphical illustration of a model's capacity to differentiate between event and non-event conditions in many situation, offering a dependable means of assessing prediction accuracy (Ajin et al., 2025). In hazard analysis—such as landslides, forest fires, or floods—the ROC graph displays sensitivity (the proportion of correctly identified event locations) along the Y-axis, while specificity (the correctly identified non-event locations) is shown on the X-axis (Saha et al., 2023). To comprehensively evaluate model performance, and both success-rate curves (using training data) and forecast-rate curves (using independent validation data) should be generated (Ajin et al., 2025).

AUC stands for Area under the Curve, measures a model's overall predictive capability for both event and non-event classes in susceptibility modelling. AUC values range between 0.5 and 1.0. Values close to 0.5 indicate poor discriminatory power (Ajin et al., 2025; Khatun et al., 2024), whereas higher values denote more robust and reliable predictive performance (Saha et al., 2022). All the aforementioned parameters, namely sensitivity, specificity, precision, and AUC, were calculated by the following equations:

$$\text{Sensitivity} = \text{AR} / (\text{AR} + \text{EB}) \tag{12.8}$$

$$\text{Specificity} = \text{AB} / (\text{PF} + \text{ED}) \tag{12.9}$$

$$\text{Precision} = \text{AR} / (\text{AR} + \text{PF}) \tag{12.10}$$

$$\text{AUC} = (\Sigma\text{AR} + \Sigma\text{ED}) / (\text{L} + \text{Q}) \tag{12.11}$$

where AB and AR are true negatives and true positives, PF and EB are False negative and false positive results, L is the total number of event pixels (landslide, flood, or forest fire), and Q is the number of non-event pixels.

Result and Discussion

Forest fire susceptibility maps (FFSMs) were created utilizing a single-model-based evaluation method in the first phase of the investigation. Three distinct FFSMs for the Tehri Garhwal district or study area were created following the use of logistic regression models as a statistical model, bagging as a hybrid machine learning approach, and random forest as a machine learning methodology. Using R programming and SPSS, all models were run in R Studio. Each model's results were categorized into five groups based on susceptibility levels: Very High (VH), High (H), Moderate (M), Low (L), and Very Low (VL). According to Jenks' natural breaks, the Forest Fire Susceptibility Maps (FFSMs) were classified using four different methods: quantile, equal interval, geometric interval, and natural breaks. Because of its impartiality, minimum variance among classes, and maximum distance between each class, Jenks' natural breaks performed better than the others. Similar results were also published by Pal and Mandal (2021), who remarked how well it distributed detailed flood-related data.

There are several climatic factors which jointly affects on the intensity of forest fires. The main climatic factors are such as wind speed, rainfall, humidity, and temperature etc. The extended dry weather (late April to late May) accelerates the leaf fall of virtually all old leaves, and as the fallen leaves pile up, they generate a carpet of particles on the forest ground that has a high surface-to-volume ratio and a low bulk fuel density that is easily ignited by subsurface fire (Verma et al., 2019). Along with this due to climate change, the amount of rainfall became decreased which is driven by the rise of relative humidity and temperature are the primary reason for forest fires. Forest fires thrive in situations of extreme air temperatures and aridity (low humidity) driven on by intense radiation from the sun over southern and western aspects (Verma et al., 2019). A transparent, dense, sticky liquid known as the resin is a byproduct of secondary metabolism that contains hydrocarbons, terpenoids, and additional substances that increase its flammability (Ormeno et al., 2009) found in mountainous regions (Khatun et al., 2024). The change in conventional land management practices in Tehri Gharwal district is one of the reasons for additional forest fires. Forests are now less often utilized for grazing, collecting grasses, animal bedding, and dry wood, as well as for controlled burning to remove forest trash, as a result of natives moving to cities in quest of better employment opportunities. In certain places, accumulated fuel loads raise the possibility of more serious and devastating forest fires (Chuvieco et al., 2010; Saha et al., 2022).

Forest Fire Susceptibility Area by RF, Bagging, LR

FFS was assessed using three machine learning models—Random Forest (RF), Bagging, and Logistic Regression (LR) (Figs. 12.4, 12.5 and 12.6). The percentage of land that falls in a particular class differs among the models, indicating some differences in how each interprets and maps fire-susceptible areas. Of these, the Random

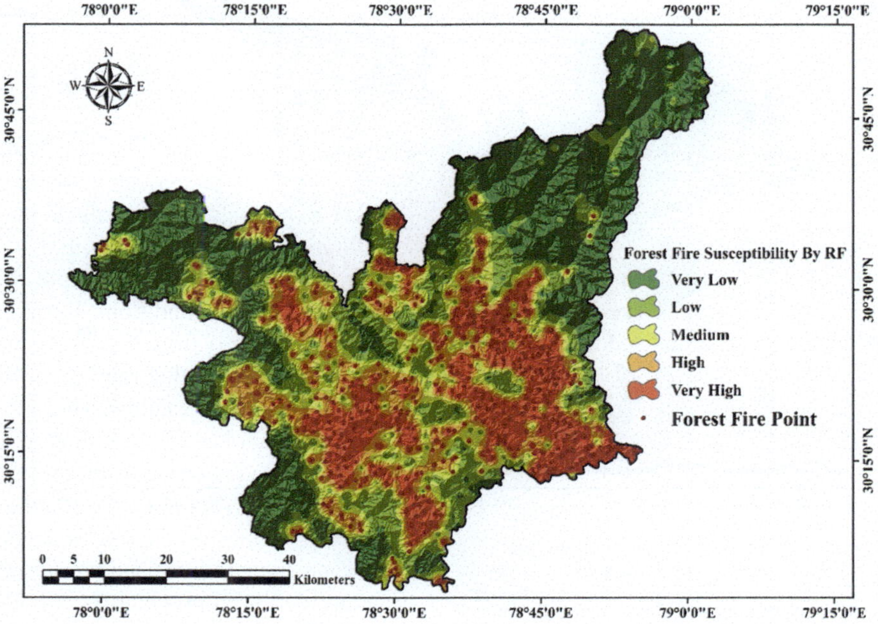

Fig. 12.4 Forest fire susceptibility map by RF

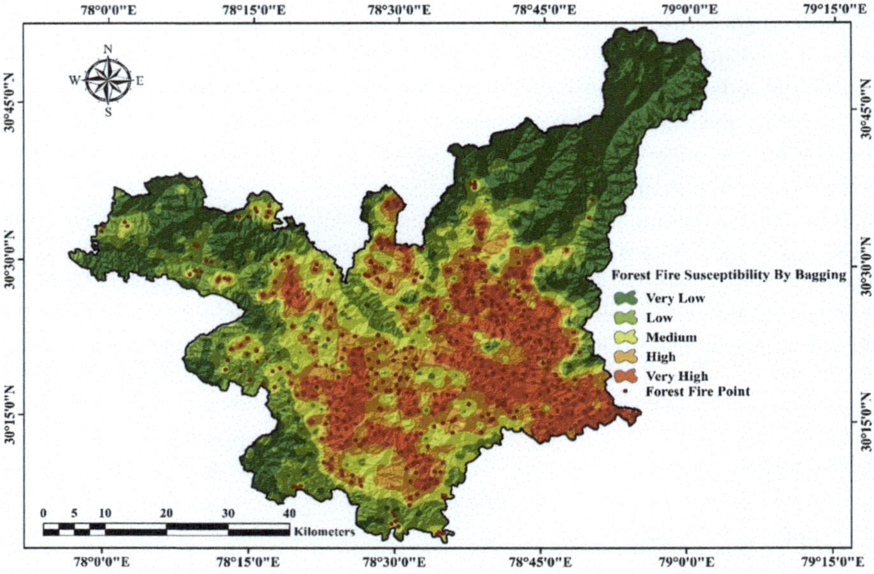

Fig. 12.5 Forest fire susceptibility by Bagging

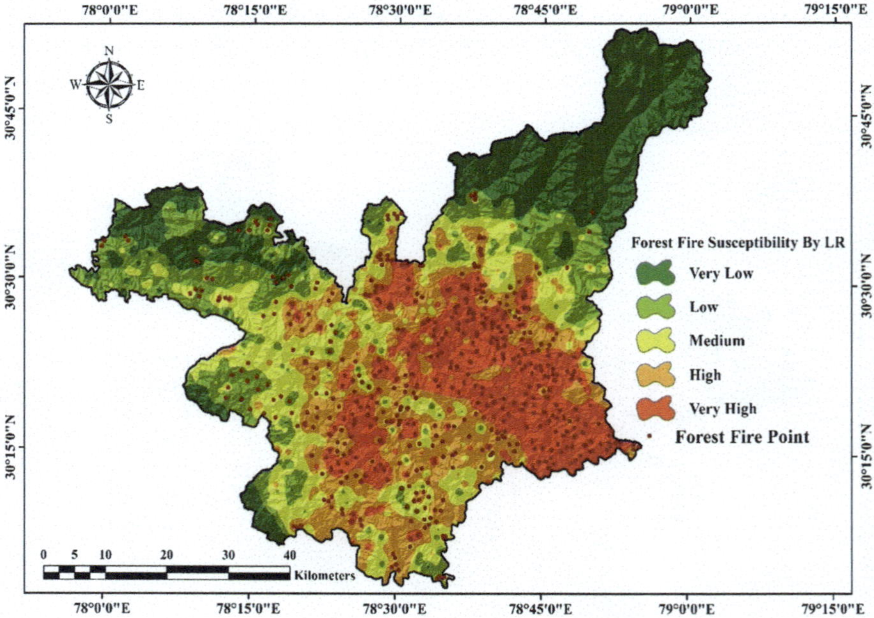

Fig. 12.6 Forest fire susceptibility by LR

Table 12.2 Forest fire susceptibility area by RF, Bagging, LR

Forest fire susceptibility zone	RF		Bagging		LR	
	Effected area (sq. km)	Effected area (%)	Effected area (sq. km)	Effected area (%)	Effected area (sq. km)	Effected area (%)
Very low	1443.3	37.904	1179.06	30.96	864.94	22.72
Low	684.104	17.966	720.089	18.91	648.77	17.04
Medium	472.871	12.419	577.838	15.18	722.17	18.97
High	527.563	13.855	665.007	17.46	815.70	21.42
Very high	679.297	17.84	665.527	17.48	756.07	19.86

Forest model assigns the largest share of the district to the Very Low susceptibility zone susceptibility zone, covering 1443.3 sq. km (37.904%). This is followed by Low with 684.104 sq. km (17.966%) and Medium with 472.871 sq. km (12. 419%). From the analysis of data it can be found that the High and Very high FFS zones covered 527.563 sq. km (13.855%) and 679.297 sq. km (17.84%), of the study area. These numbers show that RF successfully distinguishes between high-risk and low-risk locations and offers a well-balanced categorisation. In case of Bagging model the Very Low zone covered (30.96%) less area than RF model, while the Low category covered 720.089 sq. km (18.91%). The Medium and High zones covered 577.838 sq. km (15.18%) and 665.007 sq. km (17.46%), and the Very High zone covered 665.527 sq. km (17.48%). This distribution suggests that, the Bagging tends to classify more areas into moderate and high-risk levels compared to RF (random forest). LR produced the most risk-oriented classification. The Very Low zone is covered significantly lower area in compare to RF model (22.72%), while the Low class covered 648.77 sq. km (17.04%). The Medium category covered 722.17 sq. km (18.97%), and the High and Very High classes covered 815.70 sq. km (21.42%) and 756.07 sq. km (19.86%), respectively. These results show that LR (logistic regression) assigned a larger share of the landscape to higher susceptibility zones and reflecting its limited ability to capture complex and non-linear fire-related factors. Overall analysis and the comparison between 3 methodology it indicated that RF provided the most reliable and detailed susceptibility map and Bagging offers a smoother and more generalized distribution and on other hand LR is useful but tends to overestimate high-risk areas.

Model Validation

The model's performance was evaluated in this research utilizing ROC analysis based on both training and validation datasets. AUC stands for Area under the Curve was generated for each dataset, and the results are presented in Fig. 12.7. A higher AUC value indicates stronger predictive capability. According to the success-rate

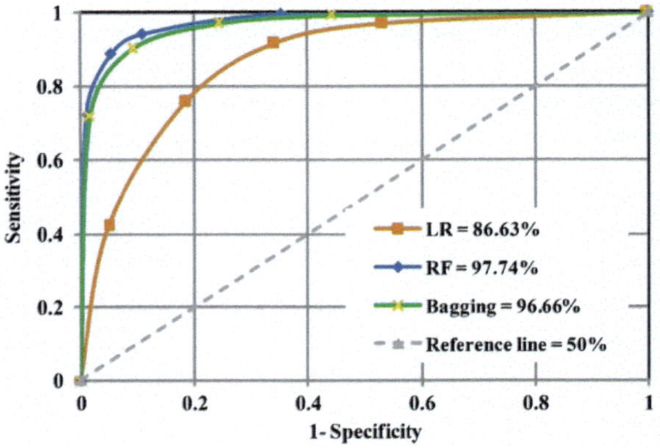

Fig. 12.7 Validation of forest fire ROC curve (success rate curve using training dataset) used to susceptibility maps

curves, the Random Forest (RF) model illustrated the highest accuracy. Out of all the models used for FFS assessment. The value of RF (AUC) is 0.977 (97.74%) followed by Bagging (AUC) 0.966 (96.66%), LR (AUC) 0.866 (86.63%) respectively which shows high performance level of this models (Fig. 12.7).

Conclusion

Forest fire susceptibility in Tehri Garhwal has been most accurately predicted through ensemble machine-learning models integrating 900 fire locations and 22 environmental drivers. While Logistic Regression achieved an AUC of 0.866, the Random Forest and Bagging models reached near-perfect accuracy with AUCs of 0.977 and 0.966, respectively. These high-precision maps highlight an alarming level of risk across the district. Specifically, the regions of Narendra Nagar, Jakhani-Dhar, Ghansali, and Devprayag emerge as high-danger areas where the combination of human proximity and challenging terrain creates a "perfect storm" for wildfire outbreaks. The Strong winds, decreased humidity, increasing temperatures, and pre-monsoon dryness are examples of climate stressors that hasten the drying and ignite potential of fuel. Since around 90% of previous fires have occurred in areas with sparse, deep vegetation, land-use patterns also have a significant influence in this circumstance. The importance of human-related ignition sources, which are crucial to forest fires, is highlighted by proximity to communities, places of worship, and transportation networks. Overall, the study shows that machine-learning approaches are useful tools for understanding how fire behaves in the Himalayan region in our study area. They also offer crucial direction for focused forest fire management and prevention. Using the susceptibility maps made here and this map, forest agencies and disaster management organisations may prioritise high-risk regions in this region, enhance early warning systems, and mitigate the consequences of future fires in Tehri Garhwal. To assist the Tehri-Garhwal tribes and improve the condition of the woods, the development of advanced ecological modeling must be given high attention. Future studies will enable researchers to choose the most accurate instruments for predicting changes in the environment. Implementing the strong conservation policies necessary to protect our natural environment requires this scientific approach.

References

AbediGheshlaghi, H. (2019). Using GIS to develop a model for forest fire risk mapping. *Journal of the Indian Society of Remote Sensing, 47*(7), 1173–1185.

Aditian, A., Kubota, T., & Shinohara, Y. (2018). Comparison of GIS-based landslide susceptibility models using frequency ratio, logistic regression, and artificial neural network in a tertiary region of Ambon, Indonesia. *Geomorphology, 318*, 101–111.

Akgun, A. (2012). A comparison of landslide susceptibility maps produced by logistic regression, multi-criteria decision, and likelihood ratio methods: A case study at İzmir, Turkey. *Landslides, 9*(1), 93–106.

Ajin, R. S., Saha, A., Saha, S., & Kanungo, D. P. (2025). Improving the spatial prediction of machine learning-based landslide susceptibility models by integrating the particle swarm optimization algorithm. *Stochastic Environmental Research and Risk Assessment, 39* (12), 5827–5854. https://doi.org/10.1007/s00477-025-03101-1

Amatulli, G., Peréz-Cabello, F., & De La Riva, J. (2007). Mapping lightning/human-caused wildfires occurrence under ignition point location uncertainty. *Ecological Modelling, 200*(3–4), 321–333.

Andreae, M. O., Rosenfeld, D., Artaxo, P., Costa, A. A., Frank, G. P., Longo, K. M., & Silva-Dias, M. A. F. D. (2004). Smoking rain clouds over the Amazon. *Science, 303*(5662), 1337–1342.

Andrews, P. L. (1986). *BEHAVE: Fire behavior prediction and fuel modeling system: BURN subsystem, Part 1* (Vol. 194). US Department of Agriculture, Forest Service, Intermountain Research Station.

Arpaci, A., Malowerschnig, B., Sass, O., & Vacik, H. (2014). Using multi variate data mining techniques for estimating fire susceptibility of Tyrolean forests. *Applied Geography, 53*, 258–270.

Baker, P. (2023). NICC to Offer Program in Indigenous Environmental Health. *Tribal College, 35*(1), 1–2.

Boonchut, P. (2005). Decision support for hazardous material routing. *Masters Degree Thesis in Urban Planning and Land Administration, International Institute for Geo-information Science and Earth Observation.*

Breiman, L. (1996). Bagging predictors. *Machine learning, 24*(2), 123–140.

Bui, D. T., Bui, Q. T., Nguyen, Q. P., Pradhan, B., Nampak, H., & Trinh, P. T. (2017). A hybrid artificial intelligence approach using GIS-based neural-fuzzy inference system and particle swarm optimization for forest fire susceptibility modeling at a tropical area. *Agricultural and Forest Meteorology, 233*, 32–44.

Cai, Y. F., Zhang, H., Feng, Z., & Shen, S. Z. (2021). Intensive wildfire associated with volcanism promoted the vegetation changeover in Southwest China during the Permian–Triassic transition. *Frontiers in Earth Science, 9*, Article 615841.

Cayuela, L., Benayas, J. M. R., & Echeverría, C. (2006). Clearance and fragmentation of tropical montane forests in the Highlands of Chiapas, Mexico (1975–2000). *Forest Ecology and Management, 226*(1–3), 208–218.

Chandra, S., & Arora, M. K. (2006). Forest fire risk zonation mapping using remote sensing technology. In *Disaster forewarning diagnostic methods and management* (Vol. 6412, pp. 61–70). SPIE.

Chauhan, J. S., Gautam, A. S., & Negi, R. S. (2018). Natural and anthropogenic impacts on forest structure: a case study of Uttarakhand state. *The Open Environmental Research Journal, 11*(1).

Chuvieco, E., Aguado, I., Yebra, M., Nieto, H., Salas, J., Martín, M. P., & Zamora, R., et al. (2010). Development of a framework for fire risk assessment using remote sensing and geographic information system technologies. *Ecological Modelling, 221*(1), 46–58.

Cochrane, M. A. (2003). Fire science for rainforests. *Nature, 421*(6926), 913–919.

Conedera, M., Torriani, D., Neff, C., Ricotta, C., Bajocco, S., & Pezzatti, G. B. (2011). Using Monte Carlo simulations to estimate relative fire ignition danger in a low-to-medium fire-prone region. *Forest Ecology and Management, 261*(12), 2179–2187.

Crutzen, P. J., & Lelieveld, J. (2001). Human impacts on atmospheric chemistry. *Annual Review of Earth and Planetary Sciences, 29*(1), 17–45.

De Vasconcelos, M. P., Silva, S., Tome, M., Alvim, M., & Pereira, J. C. (2001). Spatial prediction of fire ignition probabilities: Comparing logistic regression and neural networks. *Photogrammetric Engineering and Remote Sensing, 67*(1), 73–81.

Eugenio, F. C., dos Santos, A. R., Fiedler, N. C., Ribeiro, G. A., da Silva, A. G., dos Santos, Á. B., & Schettino, V. R., et al. (2016). Applying GIS to develop a model for forest fire risk: A case study in Espírito Santo, Brazil. *Journal of Environmental Management, 173*, 65–71.

Fornacca, D., Ren, G., & Xiao, W. (2018). Evaluating the best spectral indices for the detection of burn scars at several post-fire dates in a mountainous region of Northwest Yunnan, China. *Remote Sensing, 10*(8), 1196.

FSI. (2015). https://fsi.nic.in

Gigović, L., Pourghasemi, H. R., Drobnjak, S., & Bai, S. (2019). Testing a new ensemble model based on SVM and random forest in forest fire susceptibility assessment and its mapping in Serbia's Tara National Park. *Forests, 10*(5), 408.

Gómez-González, S., Ojeda, F., & Fernandes, P. M. (2018). Portugal and Chile: Longing for sustainable forestry while rising from the ashes. *Environmental Science & Policy, 81*, 104–107.

Gruber, S. (2025). *Analysis of forest change: a comparison of global and national perspectives* (Doctoral dissertation, University of Graz).

Ham, J., Chen, Y., Crawford, M. M., & Ghosh, J. (2005). Investigation of the random forest framework for classification of hyperspectral data. *IEEE Transactions on Geoscience and Remote Sensing, 43*(3), 492–501.

Hong, H., Jaafari, A., & Zenner, E. K. (2019). Predicting spatial patterns of wildfire susceptibility in the Huichang County, China: An integrated model to analysis of landscape indicators. *Ecological Indicators, 101*, 878–891.

Jaafari, A., Mafi-Gholami, D., Thai Pham, B., & Tien Bui, D. (2019). Wildfire probability mapping: Bivariate vs. multivariate statistics. *Remote Sensing, 11*(6), 618.

Jaafari, A. (2018). LiDAR-supported prediction of slope failures using an integrated ensemble weights-of-evidence and analytical hierarchy process. *Environmental Earth Sciences, 77*(2), 42.

Jaafari, A., Gholami, D. M., & Zenner, E. K. (2017). A Bayesian modeling of wildfire probability in the Zagros mountains, Iran. *Ecological Informatics, 39*, 32–44.

Jaafari, A., Pazhouhan, I., & Bettinger, P. (2021). Machine learning modeling of forest road construction costs. *Forests, 12*(9), 1169.

Kanga, S., Tripathi, G., & Singh, S. K. (2017). Forest fire hazards vulnerability and risk assessment in Bhajji forest range of Himachal Pradesh (India): A geospatial approach. *Journal of Remote Sensing & GIS, 8*(1), 1–16.

Kayet, N., Chakrabarty, A., Pathak, K., Sahoo, S., Dutta, T., & Hatai, B. K. (2020). Comparative analysis of multi-criteria probabilistic FR and AHP models for forest fire risk (FFR) mapping in Melghat Tiger Reserve (MTR) forest. *Journal of Forestry Research, 31*(2), 565–579.

Khatun, S., Saha, A., Gogoi, P., Saha, S., & Sarkar, R. (2024). Assessment of landslide vulnerability using statistical and machine learning methods in Bageshwar District of Uttarakhand, India. In *Geomorphic risk reduction using geospatial methods and tools* (pp. 99–116). Singapore: Springer Nature Singapore. https://doi.org/10.1007/978-981-99-7707-9_6

Maeda, E. E., Formaggio, A. R., Shimabukuro, Y. E., Arcoverde, G. F. B., & Hansen, M. C. (2009). Predicting forest fire in the Brazilian Amazon using MODIS imagery and artificial neural networks. *International Journal of Applied Earth Observation and Geoinformation, 11*(4), 265–272.

Mambile, C., Kaijage, S., & Leo, J. (2024). Application of deep learning in forest fire prediction: a systematic review. *IEEE Access.*

Merghadi, A., Yunus, A. P., Dou, J., Whiteley, J., ThaiPham, B., Bui, D. T., & Abderrahmane, B., et al. (2020). Machine learning methods for landslide susceptibility studies: A comparative overview of algorithm performance. *Earth-Science Reviews, 207*, 103225.

Micheletti, N., Foresti, L., Robert, S., Leuenberger, M., Pedrazzini, A., Jaboyedoff, M., & Kanevski, M. (2014). Machine learning feature selection methods for landslide susceptibility mapping. *Mathematical Geosciences, 46*(1), 33–57.

Mina, U., Dimri, A. P., & Farswan, S. (2023). Forest fires and climate attributes interact in central Himalayas: An overview and assessment. *Fire Ecology, 19*(1), 1–18.

Moresi, F. V., Maesano, M., di Cristofaro, M., ScarasciaMugnozza, G., & Brunori, E. (2025). Evaluating shallow landslide prediction mapping by using two different GIS-based models: 4SLIDE and SHALSTAB. *ISPRS International Journal of Geo-Information, 14*(4), 144.

Morgan, J., Artemieva, N., & Goldin, T. (2013). Revisiting wildfires at the K-Pg boundary. *Journal of Geophysical Research: Biogeosciences, 118*(4), 1508–1520.

Nguyen, D., Belval, E. J., Wei, Y., Short, K. C., & Calkin, D. E. (2024). Dataset of united states incident management situation reports from 2007 to 2021. *Scientific Data, 11*(1), 23.

Nikhil, S., Danumah, J. H., Saha, S., Prasad, M. K., Rajaneesh, A., Mammen, P. C., & Kuriakose, S. L., et al. (2021). Application of GIS and AHP method in forest fire risk zone mapping: A study of the Parambikulam Tiger Reserve, Kerala, India. *Journal of Geovisualization and Spatial Analysis, 5*(1), 14.

Oliveira, S., & Oehler, F. (2012). San-Miguel-Ayanz J, Camia A, Pereira JM. Modeling spatial patterns of fire occurrence in Mediterranean Europe using multiple regression and random forest. *Forest Ecology and Management, 275*, 117–129.

Ormeno, E., Cespedes, B., Sanchez, I. A., Velasco-García, A., Moreno, J. M., Fernandez, C., & Baldy, V. (2009). The relationship between terpenes and flammability of leaf litter. *Forest Ecology and Management, 257*(2), 471–482.

Pal, S., & Mandal, I. (2021). Noise vulnerability of stone mining and crushing in Dwarka river basin of Eastern India. *Environment, Development and Sustainability, 23*(9), 13667–13688.

Pastor, E., Zárate, L., Planas, E., & Arnaldos, J. (2003). Mathematical models and calculation systems for the study of wildland fire behaviour. *Progress in Energy and Combustion Science, 29*(2), 139–153.

Pawluszek, K., & Borkowski, A. (2017). Impact of DEM-derived factors and analytical hierarchy process on landslide susceptibility mapping in the region of Rożnów Lake, Poland. *Natural Hazards, 86*(2), 919–952.

Pham, B. T., Bui, D. T., Prakash, I., & Dholakia, M. B. (2017). Hybrid integration of Multilayer Perceptron Neural Networks and machine learning ensembles for landslide susceptibility assessment at Himalayan area (India) using GIS. *CATENA, 149*, 52–63.

Pham, B. T., Tien Bui, D., Prakash, I., & Dholakia, M. B. (2016). Rotation forest fuzzy rule-based classifier ensemble for spatial prediction of landslides using GIS. *Natural Hazards, 83*(1), 97–127.

Pourghasemi, H. R. (2016). GIS-based forest fire susceptibility mapping in Iran: A comparison between evidential belief function and binary logistic regression models. *Scandinavian Journal of Forest Research, 31*(1), 80–98.

Pourghasemi, H. R., Beheshtirad, M., & Pradhan, B. (2016a). A comparative assessment of prediction capabilities of modified analytical hierarchy process (M-AHP) and Mamdani fuzzy logic models using Netcad-GIS for forest fire susceptibility mapping. *Geomatics, Natural Hazards and Risk, 7*(2), 861–885.

Pourghasemi, H. R., Beheshtirad, M., & Pradhan, B. (2016b). A comparative assessment of prediction capabilities of modified analytical hierarchy process (M-AHP) and Mamdani fuzzy logic models using Netcad-GIS for forest fire susceptibility mapping. *Geomatics, Nat.*

Pourghasemi, H. R., Yousefi, S., Kornejady, A., & Cerdà, A. (2017). Performance assessment of individual and ensemble data-mining techniques for gully erosion modeling. *Science of the Total Environment, 609*, 764–775.

Pourtaghi, Z. S., Pourghasemi, H. R., Aretano, R., & Semeraro, T. (2016). Investigation of general indicators influencing on forest fire and its susceptibility modeling using different data mining techniques. *Ecological Indicators, 64*, 72–84.

Ramanathan, V. C. P. J., Crutzen, P. J., Kiehl, J. T., & Rosenfeld, D. (2001). Aerosols, climate, and the hydrological cycle. *Science, 294*(5549), 2119–2124.

Randerson, J. T., Liu, H., Flanner, M. G., Chambers, S. D., Jin, Y., Hess, P. G., & Zender, C. S., et al. (2006). The impact of boreal forest fire on climate warming. *Science, 314*(5802), 1130–1132.

Román-Cuesta, R. M., Retana, J., & Gracia, M. (2004). Fire trends in tropical Mexico: A case study of Chiapas. *Journal of Forestry, 102*(1), 26–32.

Saha, S., Saha, A., Hembram, T. K., Kundu, B., & Sarkar, R. (2022). Novel ensemble of deep learning neural network and support vector machine for landslide susceptibility mapping in Tehri region, Garhwal Himalaya. *Geocarto International, 37*(27), 17018–17043. https://doi.org/10.1080/10106049.2022.2120638

Saha, S., Kundu, B., Saha, A., Mukherjee, K., & Pradhan, B. (2023). Manifesting deep learning algorithms for developing drought vulnerability index in monsoon climate dominant region of West Bengal, India: Manifesting deep learning algorithms for developing drought vulnerability index in monsoon climate dominant region of West Bengal, India. *Theoretical and applied climatology, 151* (1), 891–913.

Saha, S., Saha, A., Santosh, M., Kundu, B., Sarkar, R., & Hembram, T. K. (2024). Integrating deep learning neural network and M5P with conventional statistical models for landslide susceptibility modelling. *Bulletin of Engineering Geology and the Environment, 83* (1), 12. https://doi.org/10.1007/s10064-023-03498-5

Sakr, G. E., Elhajj, I. H., & Huijer, H. A. S. (2010). Support vector machines to define and detect agitation transition. *IEEE Transactions on Affective Computing, 1*(2), 98–108.

Saloum, J., & Abdou, H. (2016). Statistical modeling for conservation the vegetation of the land in Al Kadmous area from rainfall erosion. *Latakia University Journal-Arts and Humanities Sciences Series, 38*(3).

Satendra, K. A. (2014). *Forest fire disaster management*. National Institute of Disaster Management, Ministry of Home Affairs, New Delhi.

Satir, O., Berberoglu, S., & Donmez, C. (2016). Mapping regional forest fire probability using artificial neural network model in a Mediterranean forest ecosystem. *Geomatics, Natural Hazards and Risk, 7*(5), 1645–1658.

Schimel, D., & Baker, D. (2002). The wildfire factor. *Nature, 420*(6911), 29–30.

Sharma, S. K., & Kumar, K. (2025). Forest fire effects on structure, composition, and functioning of ecosystem. In *Forest fire and climate change: Insights into science* (pp. 79–99). Cham: Springer Nature Switzerland.

Singh, S., Sharma, C. M., Kumar, A., Purohit, S., & Negi, M. S. (2023). GIS-based forest fire risk zonation mapping of Kirtinagar block of Garhwal Himalaya, Uttarakhand, India. *Papers in Applied Geography,* 1–18.

Shen, L., & Li, C. (2010). Water body extraction from Landsat ETM+ imagery using adaboost algorithm. In *2010 18th International Conference on Geoinformatics* (pp. 1–4). IEEE.

Sinha, A., Nikhil, S., Ajin, R. S., Danumah, J. H., Saha, S., Costache, R., & Abioui, M., et al. (2023). Wildfire risk zone mapping in contrasting climatic conditions: an approach employing AHP and F-AHP models. *Fire, 6*(2), 44.

Sturtevant, B. R., Miranda, B. R., Yang, J., He, H. S., Gustafson, E. J., & Scheller, R. M. (2009). Studying fire mitigation strategies in multi-ownership landscapes: Balancing the management of fire-dependent ecosystems and fire risk. *Ecosystems, 12*(3), 445–461.

Strobl, C., Malley, J., & Tutz, G. (2009). An introduction to recursive partitioning: Rationale, application, and characteristics of classification and regression trees, bagging, and random forests. *Psychological Methods, 14*(4), 323.

Tien Bui, D., Le, K. T. T., Nguyen, V. C., Le, H. D., & Revhaug, I. (2016). Tropical forest fire susceptibility mapping at the Cat Ba National Park Area, Hai Phong City, Vietnam, using GIS-based kernel logistic regression. *Remote Sensing, 8*(4), 347.

Tiwari, A., Shoab, M., & Dixit, A. (2021). GIS-based forest fire susceptibility modeling in Pauri Garhwal, India: A comparative assessment of frequency ratio, analytic hierarchy process and fuzzy modeling techniques. *Natural Hazards, 105*(2), 1189–1230.

UN-Spider. https://www.un-spider.org/category/disaster-type/forest-fire

Van Der Werf, G. R., Randerson, J. T., Collatz, G. J., Giglio, L., Kasibhatla, P. S., Arellano Jr., A. F., Kasischke, E. S., et al. (2004). Continental-scale partitioning of fire emissions during the 1997 to 2001 El Nino/La Nina period. *Science, 303*(5654), 73–76.

Veena, H. S., Ajin, R. S., Loghin, A. M., Sipai, R., Adarsh, P., Viswam, A., & Jayaprakash, M., et al. (2017). Wildfire risk zonation in a tropical forest division in Kerala, India: a study using geospatial techniques. *International Journal of Conservation Science, 8*(3).

Verma, R. K., Pandro, V., Mishra, S. N., Raj, D., & Asaiya, A. J. K. (2019). Sal forest: A source of wild edible mushrooms for livelihood support to tribal people of Dindori district, Madhya Pradesh, India. *International Journal of Current Microbiology and Applied Sciences, 8*(1), 563–575.

Vilar, L., Woolford, D. G., Martell, D. L., & Martín, M. P. (2010). A model for predicting human-caused wildfire occurrence in the region of Madrid, Spain. *International Journal of Wildland Fire, 19*(3), 325–337.

Wang, X., Zhang, Y., Luo, M., Xiao, K., Wang, Q., Tian, Y., & Li, H., et al. (2021). Radium and nitrogen isotopes tracing fluxes and sources of submarine groundwater discharge driven nitrate in an urbanized coastal area. *Science of the Total Environment, 763*, 144616.

Wu, H. L., Cheng, W. C., Shen, S. L., Lin, M. Y., & Arulrajah, A. (2020). Variation of hydro-environment during past four decades with underground sponge city planning to control flash floods in Wuhan, China: An overview. *Underground Space, 5*(2), 184–198.

Yadav, I. C., & Devi, N. L. (2019). Biomass burning, regional air quality, and climate change. *Encyclopedia of Environmental Health, 2*, 386–391.

Yariyan, P., Janizadeh, S., Van Phong, T., Nguyen, H. D., Costache, R., Van Le, H., & Tiefenbacher, J. P., et al. (2020). Improvement of best first decision trees using bagging and dagging ensembles for flood probability mapping. *Water Resources Management, 34*(9), 3037–3053.

Youssef, A. M., Pradhan, B., Jebur, M. N., & El-Harbi, H. M. (2015). Landslide susceptibility mapping using ensemble bivariate and multivariate statistical models in Fayfa area, Saudi Arabia. *Environmental Earth Sciences, 73*(7), 3745–3761.

Zhang, K., Ali, A., Antonarakis, A., Moghaddam, M., Saatchi, S., Tabatabaeenejad, A., & Moorcroft, P., et al. (2019). The sensitivity of North American terrestrial carbon fluxes to spatial and temporal variation in soil moisture: An analysis using radar-derived estimates of root-zone soil moisture. *Journal of Geophysical Research: Biogeosciences, 124*(11), 3208–3231.

Chapter 13
Landslide Susceptibility Mapping of Kalimpong Region, Darjeeling Himalayas Using GIS-Based Frequency Ratio Model

Vaishnavi Bansal and Raju Sarkar

Abstract Landslides are among the most catastrophic natural hazards in mountainous areas, posing serious risks to human life, infrastructure, and the environment. The Kalimpong region of the Darjeeling Himalayas is prone to slope failures due to steep terrain, heavy monsoon rainfall, complex geology, and human activities such as road construction and deforestation. The purpose of this study is to identify landslip susceptibility zones (LSZ) in Kalimpong using a Geographic Information System (GIS)-based Frequency Ratio (FR) model. A comprehensive landslip inventory of 126 landslip locations was created using the Geological Survey of India's Bhukosh portal and divided into training (75%) and testing (25%) datasets. Six major landslip causes—slope, aspect, elevation, distance to road, distance to lineament, annual rainfall were chosen based on data relevance and availability. FR values were calculated for each class, and the factors were weighted based on relative frequency and prediction rates. The integrated susceptibility index map divided the territory into five zones: very low, low, moderate, high, and very high susceptibility. The findings show that regions with steep slopes, significant rainfall, and closeness to roadways are more susceptible. The resulting LSM contains useful information for hazard management, land-use planning, and slope stabilisation measures. Future research that incorporates geotechnical parameters could improve model precision and promote sustainable development in vulnerable areas.

Keywords Kalimpong · Landslide susceptibility mapping · Frequency ratio · GIS · Darjeeling Himalayas · Slope stability · Hazard zonation

V. Bansal (✉)
Department of Civil Engineering, JSS Academy of Technical Education, Noida, India
e-mail: vaishnavibansal@jssaten.ac.in

R. Sarkar
Department of Civil Engineering, Delhi Technological University, Delhi, India
e-mail: rajusarkar@dce.ac.in

R. Sarkar et al. (eds.), *Climate Change, Environmental Hazards and Community-Based Resilience*, Disaster Risk Reduction, https://doi.org/10.1007/978-981-95-9485-6_13

Introduction

Property and human life are extensively destroyed by landslides. The landslip event causes several ongoing infrastructure projects to come to a halt. An estimated 18,000 individuals died as a result of landslides that harmed over 4.8 million people worldwide between 1998 and 2007. Studying landslides has proven to be difficult due of their complexity. For an effective forecast and mitigation, it requires a great deal of cooperation from several domains (Bharti et al., 2025a; Shukla et al., 2024). As a result, landslip susceptibility map data can be quite helpful. A landslip is defined as an occurrence of an event where soil, debris, or rock mass falls down the slope. Slide and flow are characteristics of its mechanism. The NEHR's Kalimpong is extremely vulnerable to landslides, especially from July to September when the rainy season is in full swing. The district is extremely vulnerable to landslides due to its sharply sloping hilly terrain and abundant rainfall. The Teesta, Relli, Neora, Geesh, Leesh, Jaldhaka, and Murti are among the rivers and streams that drain the area, and major settlements are located on ridges close to the Teesta River. Kalimpong receives between 119 and 417 cm of rain on average per month during the monsoon season (https://worldclim.org). Roads, settlements, and hydropower projects are examples of infrastructure projects that frequently destabilise slopes by removing vegetation, which makes it possible for loose debris to slide downslope with little water lubrication (Bansal & Shukla, 2022; Bansal et al., 2025; Bharti et al., 2022a, 2022b, 2023; Rawat et al., 2024; Shukla et al., 2023). Kalimpong is a crucial location for landslip study because of these factors. To reduce hazards to individuals and property, it is essential to identify areas that are prone to danger by identifying geotechnical parameters and putting precautions in place (Bharti et al., 2022; Roy et al., 2022). Given the seriousness of slope failures, identifying impacted regions and identifying susceptible slopes is crucial for evaluating highway risks. Key data about landslides and their predictions depend on geological and geotechnical investigations (Bharti et al., 2025b). For both short-term and long-term strategies to mitigate the effects of severe slope collapse, this data is essential. Stabilising these slopes through a disaster management approach that depends on the distribution, availability, and effective use of current landslip field data is one of the study's objectives. Mitigation methods could include raising awareness. Using Geographic Information System (GIS) techniques and capabilities, landslides that have previously happened can be accurately and efficiently mapped (Audisio et al., 2009; Bansal & Sarkar, 2023, 2024a; Mandal & Mondal, 2019; Nagavi et al., 2024; Yalcin, 2008; Yilmaz & Keskin, 2009). They can be of many different kinds, including anthropogenic, physical, morphological, and geological. They either directly or indirectly cause the landslip to occur. An place may occasionally become more vulnerable to landslides due to man-made or artificial elements including dams, volcanoes, and blasting. According to Pourghasemi et al. (2012a, 2012b), a landslide inventory map basically includes information about the locations of landslides that have previously occurred in a region. A thorough examination of current landslide inventory maps is essential before developing susceptibility maps for upcoming landslides (Bansal & Sarkar, 2024b; Guzzetti et al., 2012; Regmi et al., 2014; Paliaga et al., 2018). For this study, 126 landslip locations in Kalimpong, West Bengal, were obtained in point shapefile format from the Geological Survey of India's Bhukosh Portal. The Himalayas are the world's youngest mountain range, and because of their geodynamic activity and delicate, unstable soil slopes, the Himalayan environment is susceptible to frequent landslides (Chauhan et al., 2010). According to Raman and Punia (2012), landslides are quite common in India, where they can occur in around 15% of the country's entire geographical area. In India, landslides killed over 6779 people in 958 according to data till 2019. The lower North-eastern Himalayan region is made up of states like Assam, Arunachal Pradesh, Sikkim, and North Bengal, while the North-Western Himalayan region includes the Indian states of Jammu and Kashmir, Himachal Pradesh, and Uttarakhand.

One of the main causes of landslip tragedies is heavy rainfall, which occurs during the monsoon season Glacial lake outburst floods (GLOF) and earthquakes are additional elements that raise the risk of landslides. The environmental effects of development projects intended to shelter a large population and tourists can contribute to landslides. Information on previous landslides is essential for making landslip susceptibility maps because it helps calculate the likelihood of a future landslip occurrence in the chosen area or places with comparable or identical features (Polykretis et al., 2019; Westen et al., 2008). By combining several probabilistic and machine learning models with GIS mapping software, several methods have been put forth for a thorough assessment of landslip vulnerability (Lee et al., 2017). Bivariate frequency ratio model was also applied to create a landslip hazard zonation map for Uttarakhand's Chamoli district after assessing the consequences of the 1999 Chamoli landslip before to and following the earthquake. A geographic information system was used to generate a landslide hazard zonation map that included the six contributing factors to landslides: Elevation, Slope, Aspect, Distance to road, Distance to lineament and Annual Rainfall. Pathak (2016) delineated the landslip susceptibility zone (LSZ) in the Himalayan Chamoli-Joshimath zone after doing considerable research there. The elements that caused or triggered the most were rated highest, while those that caused or triggered the least were ranked lowest. Professional field observation of the area under consideration was used to determine these rankings. These elements affect the probability of a landslip incident either directly or indirectly. Landslip susceptibility maps (LSMs) were produced by statistical model frequency ratio based on these variables. Bivariate models are the ones used in this investigation.

Study Region Investigation

General

Situated on the border with Nepal, Kalimpong is a small peninsula town in the Indian state of West Bengal. It is well known for its lovely natural surroundings and pleasant temperatures, and it is located 1,250 m above sea level. Kalimpong is renowned for its tea plantations, flowers, and Himalayan views. It has verdant hills all around it. The Relli and Teesta rivers encircle it on the east and west, respectively. This region has temperatures ranging from 27 to 5 °C on average. Every year, the terrible floods caused by the powerful monsoons in this area block off Kalimpong from the rest of the state. Kalimpong's location and natural features make it susceptible to landslides, just as many other hilly areas. Because of the steep slopes and unstable soil, the area experiences a lot of rainfall throughout the summer, which regularly causes landslides. It still poses a serious threat to Kalimpong and the nearby towns in spite of multiple efforts. To prevent and lessen its effects on the community, local authorities and citizens must exercise caution and take the necessary steps (Das et al., 2022). A mix of natural and man-made causes make the Kalimpong hill station in West Bengal particularly vulnerable to landslides.

The region has considerable monsoon rainfall, ranging from 2000 to 2500 mm per year, which leads to slope failures. Slopes become even more unstable as a result of human activities like increasing urbanisation and deforestation, which alter natural drainage patterns, increase surface runoff, and remove vegetation cover. Additionally, because Kalimpong is situated in a seismically active area, landslides are regularly caused by earthquakes that weaken the slope's material and cause ground movement. Effective mitigation strategies, such as improved drainage and slope stabilisation, afforestation, and suitable land use planning, must be implemented in Kalimpong to lower the danger of landslides.

Charting Landslip Vulnerability

One of the most important techniques for locating and ranking important areas at landslide-prone areas is landslide susceptibility mapping (Saraswat et al., 2020; Pourghasemi et al., 2012a, 2012b). First and foremost, comprehensive data collection is crucial. This entails obtaining geological information, including fault lines, rock formations, and soil types, in addition to high-resolution digital elevation models (DEMs), in order to comprehend the region's topography. A susceptibility map is created by superimposing these characteristics using Geographic Information Systems (GIS) and other sophisticated geospatial techniques. Critical locations at the site are then identified using the susceptibility map that is produced. High-risk areas are prioritised for more research and mitigation strategies.

For the current study's landslip susceptibility mapping, 126 landslip locations were made available via the Geological Survey of India's Bhukosh portal. The landslip inventory was collected in point shapefile format. A training dataset and a testing dataset were created by randomly dividing the collected data. This is because without properly segmenting the dataset, it is hard to demonstrate the suitability of the chosen model for landslip susceptibility mapping (Chung & Fabbri, 2003). The "Geostatistical Analyst" function in GIS was used to separate the points. Of the 126 points collected in the landslip inventory, 75% (94 points) were taken from the "training" dataset, and the remaining 25% (32 points) were taken from the "testing" dataset as

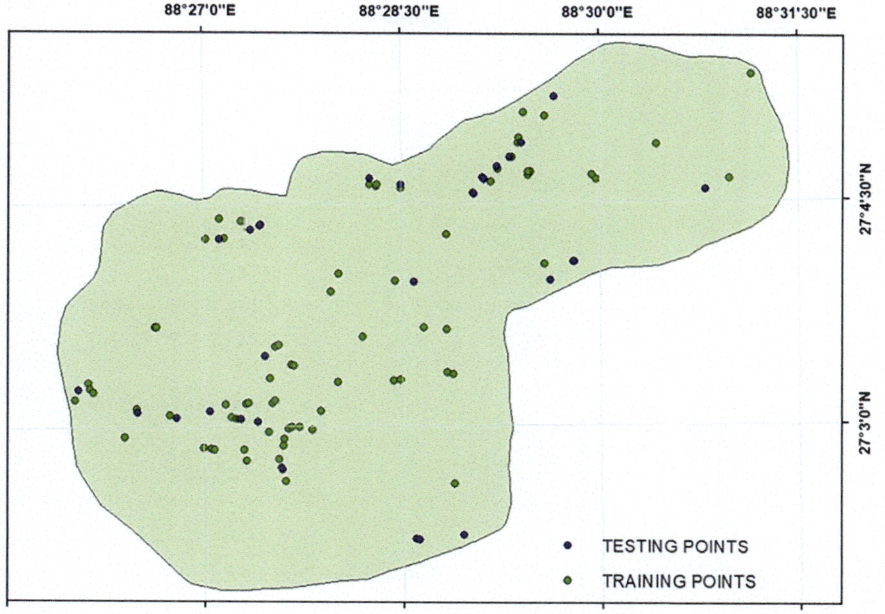

Fig. 13.1 Map showing testing and training points in Kalimpong dataset

shown in Fig. 13.1. One of the most used statistical techniques for landslip susceptibility mapping (LSM) is the frequency ratio (FR) model, which was used to construct the landslide susceptibility map and its flowchart is shown in Fig. 13.2. It helps determine the relationship between landslip episodes and conditioning factors such height, slope, rainfall, geology, and land use. The FR model's interpretability, computational economy, and ease of use make it particularly advantageous.

Frequency Ratio (FR) Method

A popular bivariate statistical technique for assessing landslip susceptibility is the frequency ratio (FR) method, which measures the geographical correlation between landslip occurrences and other conditioning factors. The approach is predicated on the idea that some classes of conditioning variables have a greater impact on the incidence of landslides, and that the spatial distribution of previous landslides identifies regions that are more likely to see failures in the future. It operates by contrasting the overall distribution of landslides in the research area with the spatial frequency of landslides within particular classes of each factor. The percentage of landslip pixels in a class divided by the percentage of all pixels in that class within the research region is known as the frequency ratio (Mohammady et al., 2012; Youssef et al., 2015). It is expressed mathematically as:

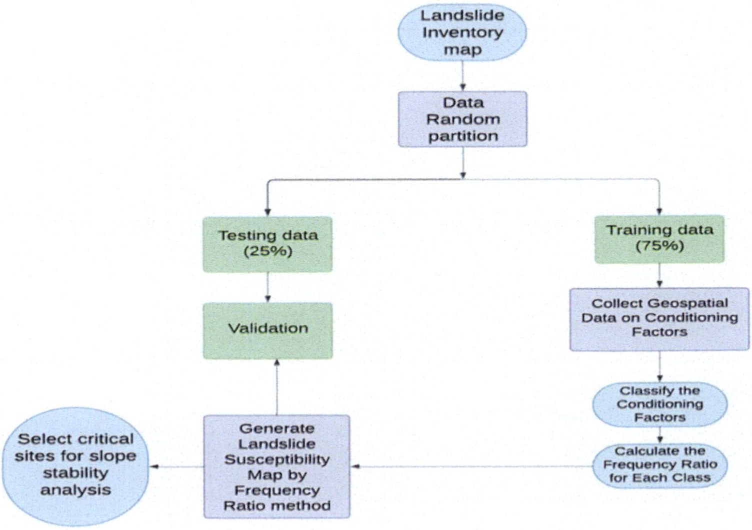

Fig. 13.2 Flowchart for Landslide susceptibility mapping

$$\text{FR} = (\%\ \text{Landslide Pixels} \div \%\,\text{Class Pixels})$$

where, FR is the frequency ratio, % Landslide Pixels represents the proportion of landslide occurrences that fall within a specific class of the factor and % Class Pixels represents the proportion of the total study area that the same class occupies.

An FR value greater than 1 indicates a strong positive correlation between the factor class and landslide occurrence, suggesting higher susceptibility, whereas an FR value less than 1 indicates a weak or negative correlation (Pradhan, 2010; Ramesh & Anbazhagan, 2015). To standardize these values, FRs are converted into relative frequency (RF) values ranging from 0 to 1 using the following equation (Li et al., 2016; Sabatakakis et al., 2013; Ullah & Zhang, 2020):

$$\text{RF} = \text{FR}/\text{Total sum of Frequency Ratios for that causative factor}$$

The RF values are then multiplied by 100 to assign relative weights to each factor class. The prediction rate (PR) is then computed to evaluate the relative contribution of each factor to landslide susceptibility (Althuwaynee et al., 2016; Ghosh, 2011). It is given by:

$$\text{PR for a LCF} = \left[\{\text{RF}_{\text{max}} - \text{RF}_{\text{min}}\} \div \{(\text{RF}_{\text{max}} - \text{RF}_{\text{min}})_{\text{min}}\}\right]$$

where RF_{max} and RF_{min} represent the maximum and minimum RF values within a factor, and $(RF_{max} - RF_{min})_{min}$ denotes the smallest difference among all factors. Finally, the landslide susceptibility index (LSI) is obtained by integrating the PR and RF values of all factors (Baral et al., 2021) as follows:

$$\text{LSM} = \sum (\text{PR for a LCF}) * (\text{Corresponding LCF})$$

In this study, FR values were first computed for each class of every conditioning factor using Microsoft Excel and shown in Table 13.1 and 13.2 with corresponding prediction rate graph in Fig. 13.4. The corresponding RF and PR values were then determined to assign appropriate weights. Using the raster calculator in ArcGIS, the weighted factor layers were integrated to produce the final landslide susceptibility map (LSM), which delineates areas with varying degrees of susceptibility based on the spatial relationship between landslide occurrences and conditioning factors.

Results and Discussions

Landslide Causative Factors (LCFs)

The district's required area and the reasons behind landslides were mapped using the GIS system software ArcMap after the data required to further this research study was gathered from various sources. Six landslip causal elements are included in this study. Elevation, slope, aspect, distance to road, distance to lineament, annual rainfall are some of the factors that influence the likelihood of a landslip event in a given location. The study's information came from remotely sensed data from a number of trustworthy sources, including the Environmental Systems Research Institute (ESRI), Climatic Research Unit (CRU), Geological Survey of India (GSI), Shuttle Radar Topography Mission (SRTM), and Advances in Geographical Research (AIGR).

Elevation

Elevation is a fundamental topographic factor derived directly from the digital elevation model (DEM), representing the height of the terrain above mean sea level. for this study, the elevation map was clipped for the delineated research area, and classified into five classes ranging from 378.31 to 1649.733 m. elevation significantly influences landslide occurrence, both directly and indirectly.

From a direct standpoint, higher elevations are typically associated with steeper slopes, higher relief energy, and increased potential energy, all of which contribute to slope instability and mass movement. Moreover, areas at higher altitudes are more exposed to intense rainfall and weathering processes, which further weaken slope materials in contrast, lower elevations often accumulate eroded material and experience different hydrological conditions, sometimes leading to debris flow or soil creep rather than large-scale landslides.

Indirectly, elevation controls land-use patterns, vegetation cover, and infrastructure development. For instance, mid-elevation zones often attract road networks and

Table 13.1 Calculation table for frequency ratio model for all factors (FR and RF)

Factor	Factor classes	Landslide pixels	% landslide pixels	Class pixels	% class pixels	Frequency ratio	Relative frequency
Slope	0.469799966- 13.66061902	15,300	22.37	8976	23.29	0.96	0.21
	13.88061903–20.34654964	23,400	34.21	13,371	34.70	0.99	0.21
	20.34654965–27.7703959	18,900	27.63	8977	23.29	1.19	0.26
	27.77039591–36.87059455	9000	13.16	4731	12.28	1.07	0.23
	36.87059456–61.53692245	1800	2.63	2482	6.44	0.41	0.09
	Total	68,400	100.00	38,537	100.00	4.61	1.00
Aspect	−1	0	0.00	0	0.00	0.00	0.00
	0–22.5	2700	3.95	1240	3.22	1.23	0.14
	22.5–67.5	2700	3.95	2663	6.91	0.57	0.07
	67.5–112.5	9000	13.16	5720	14.84	0.89	0.10
	112.5–157.5	9900	14.47	5856	15.20	0.95	0.11
	157.5–202.5	7200	10.53	6199	16.09	0.65	0.07
	202.5–247.5	6300	9.21	3259	8.46	1.09	0.12
	247.5–292.5	12,600	18.42	6124	15.89	1.16	0.13
	292.5–337.5	16,200	23.68	5913	15.34	1.54	0.18
	337.5–360	1800	2.63	1563	4.06	0.65	0.07
	Total	68,400	100.00	38,537	100.00	8.73	1.00
Distance to lineament	3000.149902–4401.581623	3600	5.26	3674	9.53	0.55	0.12
	4401.581624–5420.804693	7200	10.53	7056	18.31	0.57	0.12
	5420.804694–6376.326321	16,200	23.68	8083	20.97	1.13	0.24
	6376.326322–7289.380321	27,900	40.79	9675	25.11	1.62	0.35

(continued)

Table 13.1 (continued)

Factor	Factor classes	Landslide pixels	% landslide pixels	Class pixels	% class pixels	Frequency ratio	Relative frequency
	7289.380322–8414.772461	13,500	19.74	10,049	26.08	0.76	0.16
	Total	68,400	100.00	38,537	100.00	4.64	1.00
Distance to road	0–301.8014074	43,200	63.16	15,470	40.14	1.57	0.44
	301.8014075–686.8583755	15,300	22.37	10,521	27.30	0.82	0.23
	686.8583756–1144.763959	6300	9.21	6769	17.56	0.52	0.15
	1144.76396–1737.959829	2700	3.95	3912	10.15	0.39	0.11
	1737.95983–2653.770996	900	1.32	1865	4.84	0.27	0.08
	Total	68,400	100.00	38,537	100.00	3.58	1.00
Elevation	378.3127747–717.358372	6300	9.21	5016	13.02	0.71	0.15
	717.3583721–936.7408173	9900	14.47	8688	22.54	0.64	0.14
	936.7408174–1131.193439	16,200	23.68	11,530	29.92	0.79	0.17
	1131.19344–1345.58992	32,400	47.37	9961	25.85	1.83	0.40
	1345.589921–1649.733765	3600	5.26	3342	8.67	0.61	0.13
	Total	68,400	100.00	38,537	100.00	4.58	1.00
Rainfall	537.7427979–14,352.70518	6300	9.21	5915	15.35	0.60	0.12
	14,352.70519–28,167.66755	18,000	26.32	10,000	25.95	1.01	0.21
	28,167.66756–41,982.62993	10,800	15.79	7836	20.33	0.78	0.16
	41,982.62994–55,797.59231	31,500	46.05	7690	19.95	2.31	0.48
	55,797.59232–69,612.55469	1800	2.63	6999	18.16	0.14	0.03
	Total	68,400	100	38,440	99.74829385	4.843465059	1

Table 13.2 Calculation table for frequency ratio model for all factors (PR)

Feature	PR	Net PR
Aspect	1.05	105
Slope	1	100
Distance from lineament	1.35	135
Elevation	1.59	159
Distance from road	2.12	212
Annual avg rainfall	2.65	265

settlements due to moderate slopes and accessibility, which can disturb slope equilibrium and trigger landslides through excavation, loading, and deforestation in contrast, extremely high or low elevations may remain relatively undisturbed, affecting the frequency of anthropogenic triggers (Dai & Lee, 2003; Koukis et al., 1994; Sifa et al., 2020).

Furthermore, climatic variations with altitude—such as temperature gradients and precipitation patterns—play a crucial role in soil moisture regimes and vegetation stability. Therefore, elevation acts as an integrative factor influencing both natural geomorphological processes and human activities, making it indispensable in landslide susceptibility mapping (LSM) models.

Aspect

Aspect refers to the direction a slope faces, measured in degrees from 0° (north) to 360°, indicating the azimuthal orientation of terrain surfaces. Using the spatial analyst tool in ARCGIS, aspect was extracted from the DEM, and classified into directional categories representing different slope orientations.

It has a strong bearing on microclimatic conditions such as solar radiation, temperature, evaporation rates, and soil moisture. For instance, in the northern hemisphere, south-facing slopes receive greater solar insolation, resulting in drier soils, sparser vegetation, and weaker root reinforcement, thereby enhancing landslide susceptibility. Conversely, north-facing slopes tend to retain more moisture and vegetation, which may stabilize slopes but also increase pore-water pressure during heavy rainfall events.

Additionally, aspect influences weathering intensity and vegetation type, which collectively affect soil cohesion and surface runoff. The variation in moisture and temperature regimes across aspects leads to differential soil strength and infiltration capacity, key components in slope failure mechanisms.

Furthermore, rainfall distribution and wind exposure may vary with slope orientation, influencing erosion rates and the likelihood of saturation during storms. Thus, aspect acts as a secondary geomorphological factor modifying the effects of other primary variables like slope and rainfall in determining landslide potential.

Distance to Roads

Distance to roads is a critical anthropogenic factor, as road construction and maintenance significantly modify natural slope conditions. The data were acquired from the Bhukosh portal in shapefile format, processed in Arcmap, and converted to raster using the Euclidean distance tool, with distances ranging from 0 to 2653.77 m.

Road networks often traverse hilly terrain through cutting, excavation, and filling, disturbing slope equilibrium. Slopes adjacent to roads are highly vulnerable due to toe cutting, drainage obstruction, and increased surface runoff. Moreover, roads often concentrate water flow along their edges, leading to gully formation and progressive erosion areas closer to roads (0–500 m) tend to exhibit higher landslide density due to direct mechanical disturbances and vibrations from vehicular movement. In contrast, distant zones are less affected by such human interventions. However, poor road drainage systems can induce slope failures even at moderate distances through hydrological connectivity.

Therefore, distance to roads serves as a proxy indicator of anthropogenic disturbance intensity and slope modification, both of which are essential contributors to landslide hazard evaluation in mountainous terrains.

Slope (in Degrees)

Slope angle is one of the most dominant controlling factors in landslide initiation. derived from the DEM using ARCGIS Spatial Analyst tool, slope was categorized into five classes: 0–13.88°, 13.88°–20.36°, 20.36°–27.77°, 27.77°–36.87°, and 36.87°–61.53°.

a steeper slope implies a greater component of gravitational force acting downslope, reducing shear resistance and increasing shear stress on slope materials thus, as slope angle increases, slope stability decreases, especially when cohesion and internal friction are insufficient to counteract gravitational pull.

In gentle slopes (0–15°), landslides are relatively rare, dominated by soil creep or minor erosion. Moderate slopes (15–30°) may witness shallow translational slides, especially during intense rainfall. Steep slopes (>30°) often experience frequent rockfalls, debris slides, and rotational failures due to reduced stability margins.

Additionally, slope gradient influences runoff velocity, infiltration capacity, and vegetation anchorage, all of which are critical to slope stability. Steep slopes encourage rapid drainage and erosion, while gentle slopes tend to accumulate surface water, potentially causing pore pressure build-up. Therefore, the slope factor is indispensable in both empirical and statistical LSM approaches, serving as a primary physical determinant of landslide potential.

Distance to Faults/Lineaments

Faults and lineaments represent zones of structural weakness within bedrock, often associated with fractures, joints, and shear zones. Data extracted from the Bhukosh portal were processed using the Euclidean distance tool to generate a raster-based map at 30 m resolution, with values ranging from 0 to 8414.77 m proximity to lineaments significantly elevates landslide risk due to rock mass discontinuities, reduced shear strength, and enhanced permeability. Fault zones are also prone to seismic activity, where even minor tremors can destabilize already weakened slopes.

Moreover, fractures facilitate groundwater infiltration, increasing pore-water pressure and promoting weathering along discontinuity planes. Areas closer to faults (<1 km) are therefore more susceptible to slope deformation, mass wasting, and rock detachment.

Lineaments also act as preferential pathways for erosion and hydrological flow, leading to localized instability. Thus, distance to faults/lineaments serves as an important geological factor representing the tectonic influence on slope instability and landslide occurrence.

Annual Rainfall

Rainfall is the most common triggering factor for landslides in tropical and subtropical regions. The hydrological effect of rainfall manifests through infiltration, pore pressure build-up, and reduction of matric suction, ultimately leading to slope failure (Rahardjo et al., 1995).

in this study, rainfall data (2011–2020) were obtained from the climatic research unit (CRU) in NetCdf format and spatially interpolated using inverse distance weighted (IDW) method in Arcmap 10.8. Annual rainfall in the study area ranges from 537.74 to 69,612.55 mm, indicating strong spatial variability.

High rainfall intensities lead to soil saturation, loss of cohesion, and decrease in effective stress, making slopes prone to shallow landslides and debris flows. Long-duration rainfall events contribute to deep-seated failures by allowing water to penetrate subsurface layers, while short, intense storms cause rapid surface runoff and erosion.

Additionally, rainfall interacts synergistically with other factors like slope, land cover, and geology, amplifying instability. Hence, annual rainfall is not merely a climatic variable but a dynamic triggering mechanism, crucial for temporal prediction and spatial susceptibility assessment in LSM frameworks. Thematic maps for all factors is shown in Fig. 13.3 (Fig. 13.4).

To develop the landslide susceptibility model (LSM) for Kalimpong, six primary landslide causative factors—elevation, slope, aspect, distance to roads, distance to faults/lineaments, and annual rainfall—were compiled from trusted remote-sensing and geospatial sources (SRTM DEM, Bhukosh shapefiles, CRU, GSI, ESRI datasets) and processed in ArcMap. Topographic derivatives (elevation, slope, aspect) were

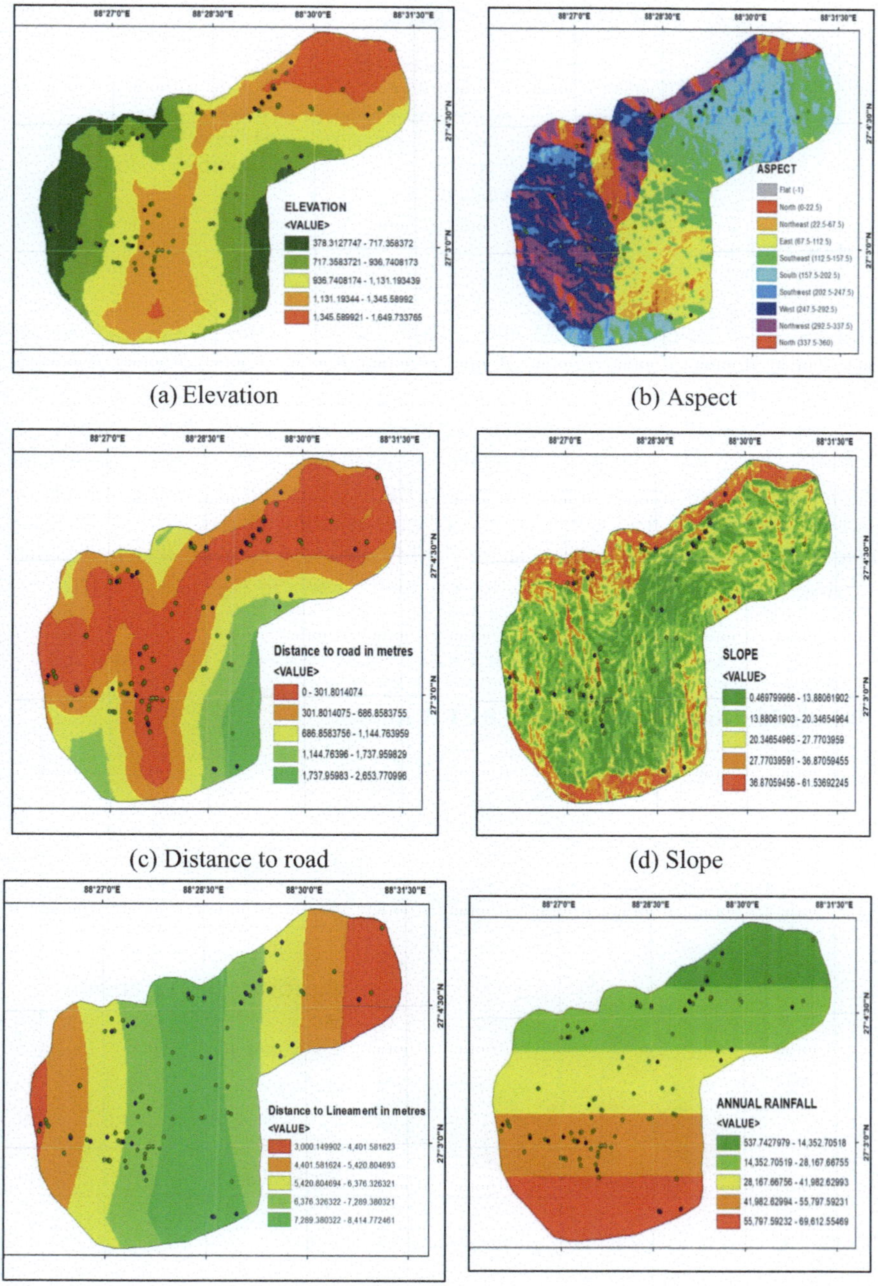

(a) Elevation

(b) Aspect

(c) Distance to road

(d) Slope

(e) Distance to lineament

(f) Annual Rainfall

Fig. 13.3 Thematic map for various landslide causative factors

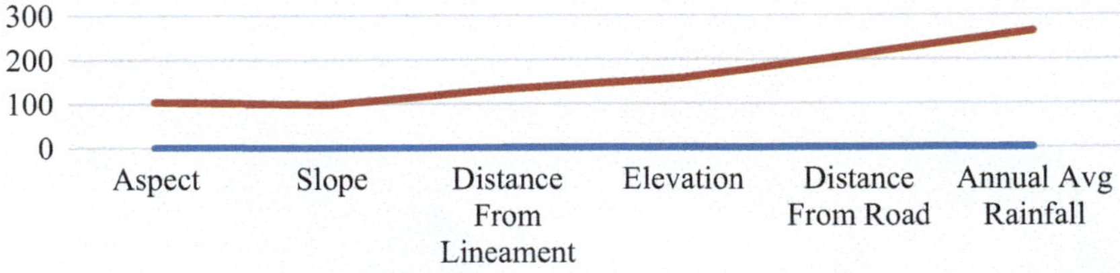

Fig. 13.4 Prediction rate graph of various factors

extracted from the clipped DEM and classified into meaningful intervals, hydrological and structural proximities (distance-to-roads, distance-to-lineaments) were generated with the Euclidean distance tool, and annual rainfall (2011–2020 CRU) was interpolated using IDW to produce a continuous raster. Each thematic layer was reclassified into classes informed by field knowledge and literature, and class-specific weights were quantified using the Frequency Ratio (FR) method to represent relative landslide propensity. The weighted rasters were combined pixel-wise in the raster calculator to produce a continuous susceptibility surface, which was then reclassified into discrete hazard zones (very low to very high) shown in Fig. 13.5.

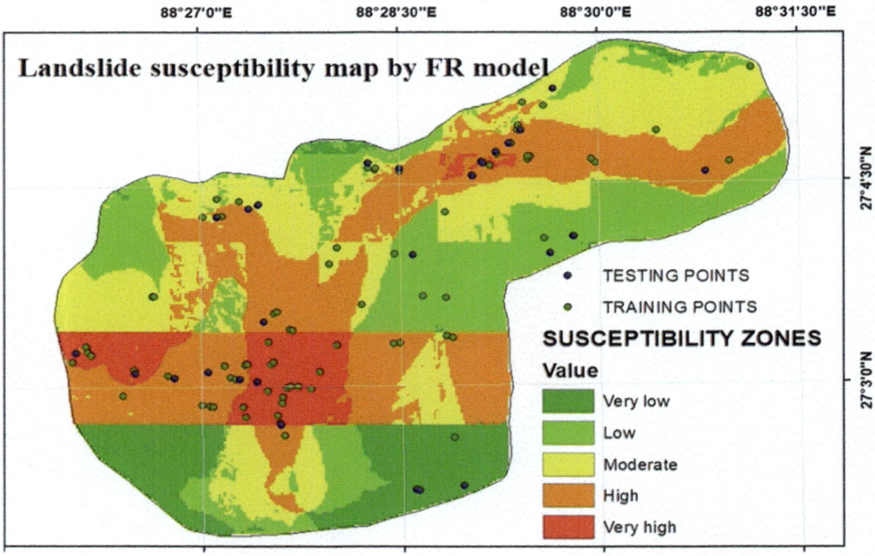

Fig. 13.5 Landslide susceptibility map by frequency ratio method

Conclusions

With its steep slopes, rough terrain, and severely dissected topography, the Kalimpong region of the Darjeeling Himalayas in India is home to the current study site. The area's elevation spans from roughly 378–1650 m, and because of the region's complicated geology, heavy monsoonal precipitation, and human activities like road cutting and slope alteration, landslides caused by rainfall are common there. This study's primary goal is to create a landslip susceptibility map (LSM) for efficient hazard management and planning by assessing the spatial distribution of landslip susceptibility using the Frequency Ratio (FR) model. Using field validation, satellite imagery interpretation, and historical landslide data, a thorough landslide inventory was created, revealing a total of 126 landslides. To assess the performance of the model, 70% of these were used as training data and 30% as testing data. Based on the relevance and availability of data, eight primary landslide-causing elements were chosen: rainfall, slope, aspect, elevation, lithology, land use/cover, distance to drainage, and distance to roadways. The frequency ratio (FR) values, which indicate the probability of landslip occurrence within each class, were calculated by classifying each of these parameters and evaluating their spatial association with the training landslip inventory. The frequency ratio for each class was calculated using the ratio between the landslide density within a specific class and the total landslide density in the study area. The derived FR values for all factor classes were combined in a GIS environment using the raster calculator in ArcGIS to generate the Landslide Susceptibility Index (LSI). The final LSI values were classified into five susceptibility zones: very low, low, moderate, high, and very high, as shown in the generated LSM.

The susceptibility map generated in this study provides valuable insights for regional land-use planning, slope management, infrastructure development, and disaster risk reduction. It can assist local authorities and planners in prioritizing high-risk zones for mitigation measures such as slope stabilization, afforestation, surface drainage improvement, and restriction of construction activities in vulnerable areas. However, it is important to note that this study primarily relies on GIS-based spatial analysis and does not incorporate detailed geotechnical investigations of soil and rock properties. Future studies should integrate geotechnical and geophysical parameters to enhance the accuracy of susceptibility assessments and to design effective landslide prevention and mitigation strategies tailored to the geomechanical conditions of the Kalimpong region.

References

Althuwaynee, O. F., Pradhan, B., & Lee, S. (2016). A novel integrated model for assessing landslide susceptibility mapping using CHAID and AHP pairwise comparison. *International Journal of Remote Sensing, 37*(5), 1190–1209.

Audisio, C., Nigrelli, G., & Lollino, G. (2009). A GIS tool for historical instability processes data entry: An approach to hazard management in two Italian Alpine River Basins. *Computers & Geosciences, 35*, 1735–1747.

Bansal, V., & Sarkar, R. (2024a). Comparative analysis of slope stability for Kalimpong region under dynamic loading using limit equilibrium method and machine benchmark learning classifiers. *Iranian Journal of Science Technology, Transactions of Civil Engineering, 48*, 2785–2807.

Bansal, V., & Sarkar, R. (2024b). Prophetical modeling using limit equilibrium method and novel machine learning ensemble for slope stability gauging in kalimpong. *Iranian Journal of Science Technology, Transactions of Civil Engineering, 48*, 411–430.

Bansal, V. & Sarkar, R. (2023). Numerical analysis and implications of fluctuating water levels on slope stability. In *International Conference on Sustainable Infrastructure: Innovation, Opportunities and Challenges* (pp. 407–419). Springer.

Bansal, V., & Shukla, B. K. (2022, May). Application of urea formaldehyde raisin in pavement subgrade stabilization at Hindon Barrage. *Journal of Physics: Conference Series* (Vol. 2267, No. 1, p. 012150). IOP Publishing.

Bansal, V., Bharti, G., Hurukadli, P., Singh, H., & Shukla, B. K. (2025). Investigation of geotechnical properties of soil using shredded Polyethylene Terephthalate (PET) wastes. *Lecture Notes in Civil Engineering, 355*, 119–129. https://doi.org/10.1007/978-981-97-8393-9_13

Baral, N., Karna, A. K., & Gautam, S. (2021). Landslide susceptibility assessment using modified frequency ratio model in Kaski District, Nepal. *International Journal of Engineering Management and Research, 11*(1), 167–177.

Bharti, G., Hurukadli, P., Shukla, B. K., Tripathi, A., Sharma, P. K., Singh, A., Singh, G., Patel, H. & Nigam, K. (2025a). Advanced geotechnical analysis and mitigation of slope stability in Dharchula, Pithoragarh, Uttarakhand, using GeoStudio software. In *Intelligent infrastructure and smart materials: Sustainable technologies for a greener future* (pp. 233–254). Cham: Springer Nature Switzerland. https://doi.org/10.1007/978-3-031-92421-7_14

Bharti, G., Hurukadli, P., Shukla, B. K., Tripathi, A., Sharma, P. K., Singh, A., Singh, G., Patel, H. & Nigam, K. (2025b). Advanced geotechnical analysis and mitigation of slope stability in Dharchula, Pithoragarh, Uttarakhand, using GeoStudio software. In *Intelligent infrastructure and smart materials: sustainable technologies for a greener future* (pp. 233–254). Cham: Springer Nature Switzerland.

Bharti, G., Shukla, B. K., Srinivasan, V., & Bansal, V. (2022a). Experimental study on eccentrically loaded rectangular footing on sand for embedded condition to enhance serviceability of structures. *Materials Today: Proceedings, 61*, 517–522.

Bharti, G., Shukla, B. K., Srinivasan, V., & Bansal, V. (2022b). Experimental study on eccentrically loaded rectangular footing on sand for embedded condition to enhance serviceability of structures. *Materials Today: Proceedings, 61*, 517–522. https://doi.org/10.1016/j.matpr.2021.12.532

Bharti, G., Singh, A., Shukla, B. K., Varadharajan, S., Kirthanashri, S. V., Bishetti, P., & Tripathi, A. (2023). Effect of self-healing concrete on modular structure. *Lecture Notes in Civil Engineering, 281*, 323–331. https://doi.org/10.1007/978-981-19-4731-5_31

Chauhan, S., Sharma, M., & Arora, M. K. (2010). Landslide susceptibility zonation of the Chamoli region, Garhwal Himalayas, using logistic regression model. *Landslides, 7*(4), 411–423.

Chung, C. F., & Fabbri, A. G. (2003). Validation of spatial prediction models for landslide hazard mapping. *Natural Hazards, 30*(3), 451–472.

Dai, F. C., & Lee, C. F. (2003). A spatiotemporal probabilistic modelling of storm-induced shallow landslides using aerial photographs and logistic regression. *Earth Surf Process Landforms, 28*(5), 527–545.

Das, S., Sarkar, S., & Kanungo, D. P. (2022). GIS-based landslide susceptibility zonation mapping using the analytic hierarchy process (AHP) method in parts of Kalimpong Region of Darjeeling Himalaya. *Environmental Monitoring Assessment, 194*(4), 234. https://doi.org/10.1007/s10661-022-09851-7

Ghosh, S. (2011). *Knowledge guided empirical prediction of landslide hazard*. PhD thesis, University of Twente, Netherlands.

Guzzetti, F., Mondini, A. C., Cardinally, M., Fiorucci, F., Santangelo, M., & Chang, K. T. (2012). Landslide inventory maps: New tools for an old problem. *Earth-Science Reviews, 112*(1–2), 42–66.

Koukis, G., Tsiambaos, G., & Sabatakakis, N. (1994). Slope movements in the Greek territory: a statistical approach. In *International Congress of the International Association of Engineering. Geology* (pp. 4621–4628).

Lee, J. H., Sameen, M. I., Pradhan, B., & Park, H. J. (2017). Modeling landslide susceptibility in data-scarce environments using optimized data mining and statistical methods. *Geomorphology, 303*, 284–298.

Li, L., Lan, H., Guo, C., Zhang, Y., Li, Q., & Wu, Y. (2016). A modified frequency ratio method for landslide susceptibility assessment. *Landslide, 14*(2), 727–774.

Mandal, S., & Mondal, S. (2019). Probabilistic approaches and landslide susceptibility. Geoinformatics and modelling of landslide susceptibility and risk. In *Environmental science and engineering, Springer book series (ESE)* (pp. 145–163).

Mohammady, M., Pourghasemi, H. R., & Pradhan, B. (2012). Landslide susceptibility mapping at Golestan Province, Iran: A comparison between frequency ratio, Dempster-Shafer, and weights-of-evidence models. *Journal of Asian Earth Sciences, 60*, 221–236.

Nagavi, J. C., Shukla, B. K., Bhati, A., Rai, A., & Verma, S. (2024). Harnessing Geospatial technology for sustainable development: A multifaceted analysis of current practices and future prospects. In *Sustainable development and geospatial technology: Volume 1: Foundations and innovations* (pp. 147–170). Cham: Springer Nature Switzerland. https://doi.org/10.1007/978-3-031-65683-5_8

Paliaga, G., Luino, F., Turconi, L., & Faccini, F. (2018). Inventory of geohydrological phenomena in Genova municipality (NW Italy). *Journal of Maps, 15*(4), 1–11.

Pathak, D. (2016). Knowledge based landslide susceptibility mapping in the Himalayas. *Geoenviron Disasters*.

Polykretis, C., Chalkias, C., & Ferentinou, M. (2019). Adaptive neuro-fuzzy inference system (ANFIS) modeling for landslide susceptibility assessment in a Mediterranean hilly area. *Bulletin of Engineering Geology and the Environment, 78*, 1173–1187.

Pourghasemi, H. R., Mohammady, M., & Pradhan, B. (2012a). Landslide susceptibility mapping using index of entropy and conditional probability models in GIS: Safarood Basin. *Iran Catena, 97*, 71–84.

Pourghasemi, H. R., Pradhan, B., Candan Gokceoglu, K., & Moezzi, D. (2012b). Landslide susceptibility mapping using a spatial multi criteria evaluation model at Haraz Watershed, Iran. In P. Biswajeet & B. Manfred (Eds.), *Terrigenous mass movements* (pp. 23–49). Springer.

Pradhan, B. (2010). Landslide susceptibility mapping of a catchment area using frequency ratio, fuzzy logic and multivariate logistic regression approaches. *Journal of the Indian Society of Remote Sensing, 38*, 301–320.

Rahardjo, H., Lim, T. T., Chang, M. F., & Fredlund, D. G. (1995). Shear-strength characteristics of a residual soil. *Canadian Geotechnical Journal , 32*(1), 60–77.

Raman, R., & Punia, M. (2012). The application of GIS-based bivariate statistical methods for landslide hazards assessment in the upper Tons-river valley, Western Himalaya. *India Georisk, 6*(3), 145–161.

Ramesh, V., & Anbazhagan, S. (2015). Landslide susceptibility mapping along Kolli hills Ghat road section (India) using frequency ratio, relative effect and fuzzy logic models. *Environment and Earth Science, 73*, 8009–8021.

Rawat, S., Shukla, B. K., Sihag, P., Srivastav, Y., & Singh, H. K. (2024, February). Environmental effects and sustainable hydropower development: A review of deployment research, environmental valuation and IoT sensors integration. In *AIP Conference Proceedings* (Vol. 3050, No. 1, p. 040008). AIP Publishing LLC.

Regmi, A. D., Yoshida, K., Pourghasemi, H. R., Dhital, M. R., & Pradhan, B. (2014). Landslide susceptibility mapping along Bhalubang-Shiwapur area of mid-western Nepal using frequency ratio and conditional probability models. *Journal of Mountain Science, 11*(5), 1266–1285.

Roy, P., Ghosal, K., & Paul, P. K. (2022). Landslide susceptibility mapping of Kalimpong in Eastern Himalayan Region using a Rprop ANN approach. *Journal of Earth System Science, 131*(2), 130. https://doi.org/10.1007/s12040-022-01877-2

Sabatakakis, N., Koukis, G., Vassiliades, E., & Lainas, S. (2013). Landslide susceptibility zonation in Greece. *Natural Hazards, 65*, 523–543.

Saraswat, R., Parkash, S., & Pal, S. (2020). Landslide susceptibility assessment due to construction of buildings in Garhwal area using GIS. In S. Ahmed, S. Abbas & H. Zia (Eds.), *Smart cities—Opportunities and challenges. Lecture notes in civil engineering* (Vol. 58). Springer, Singapore. https://doi.org/10.1007/978-981-15-2545-2_70

Shukla, B. K., Bansal, V., Varadharajan, S., Bharti, G., Shenoy, A., Singh, U. K., & Kirthanashri, S. V. (2023). Identification and analysis of factors of delay in road construction projects. *Lecture Notes in Civil Engineering, 280*, 505–511. https://doi.org/10.1007/978-981-19-4739-1_47

Shukla, B. K., Saraswat, R., Bharti, G., Sharma, P. K., Singh, D., Srivastava, P. K., Yadav, P., & Mishra, S. (2024). Soil stabilization through microbial bioenzymes: A comprehensive review of recent developments in eco-friendly approaches. *Lecture Notes in Civil Engineering, 508*, 75–95. https://doi.org/10.1007/978-981-97-3823-6_7

Sifa, S. F., Mahmud, T., Tarin, M. A., & Haque, D. M. E. (2020). Event-based landslide susceptibility mapping using weights of evidence (WoE) and modified frequency ratio (MFR) model: A case study of Rangamati district in Bangladesh. *Geol Ecol Landscapes, 4*(3), 222–235.

Ullah, K., & Zhang, J. (2020). GIS-based flood hazard mapping using relative frequency ratio method: A case study of Panjkora River Basin, eastern Hindu Kush, Pakistan. *Plos ONE., 15*(3), 1–18.

Westen, C. J. V., Castellanos, E., & Kuriakose, S. L. (2008). Spatial data for landslide susceptibility, hazard, and vulnerability assessment: An overview. *Engineering Geology, 102*(3–4), 112–131.

Yalcin, A. (2008). GIS-based landslide susceptibility mapping using analytical hierarchy process and bivariate statistic in Ardesen (Turkey): Comparison of results and confirmations. *CATENA, 72*, 1–12.

Yilmaz, I., & Keskin, I. (2009). GIS based statistical and physical approaches to landslide susceptibility mapping (Sebinkarahisar, Turkey). *Bulletin of Engineering Geology and the Environment, 68*, 459–471.

Youssef A. M., Pourghasemi, H. R., El-Haddad, B. A., & Dhahry, B. K. (2015). Landslide susceptibility maps using different probabilistic and bivariate statistical models and comparison of their performance at Wadi Itwad basin, Asir region, Saudi Arabia. *Bulletin of Engineering Geology and the Environment.*

Part III
Community-Based Resilience

Chapter 14
Strengthening Community Resilience Through Agroforestry Systems in Southern Assam

Aditi Nath, Animekh Hazarika, and Arun Jyoti Nath

Abstract Agroforestry systems in southern Assam serve as a vital source of livelihood for indigenous and smallholder communities. This research investigates the contribution of community-driven resilience in sustaining pineapple-based agroforestry productivity and maintaining ecological balance within these systems. Employing field surveys, participatory rural appraisal (PRA), and stakeholder interviews across selected villages, the research underscores the value of local adaptive strategies, traditional ecological knowledge, and collective action in fostering resilience. Findings reveal that community-driven land restoration efforts, integrated cropping systems, soil conservation measures, and improved market access significantly contribute to livelihood security and environmental sustainability. Strong social networks, institutional partnerships, and community-led innovations also enhance the capacity to withstand climate-related and economic shocks. The study advocates for the recognition and empowerment of grassroots institutions, promoting agroecological practices, and formulating policies responsive to local contexts. By fostering participatory approaches, pineapple-based agroforestry systems can be transformed into resilient models that support ecological conservation and ensure long-term socioeconomic stability. The insights offered a replicable framework for similar ecologically sensitive and socio-economically vulnerable regions across Northeast India.

Keywords Adaptive capacity · Slash and burn · Soil health · Socioeconomic stability

---

A. Nath (✉)
Department of Social Work, Assam University, Silchar, India
e-mail: aditinath1@gmail.com

A. Hazarika · A. J. Nath
Department of Ecology and Environmental Science, Assam University, Silchar, India

R. Sarkar et al. (eds.), *Climate Change, Environmental Hazards and Community-Based Resilience*, Disaster Risk Reduction, https://doi.org/10.1007/978-981-95-9485-6_14

Introduction

Agroforestry, the purposive inclusion of crops and/or livestock with shrubs and trees on the same unit of land, has been an intrinsic part of traditional land use in Assam (Reang et al., 2022a). With a landscape richly endowed with biodiversity, cultural diversity, and varied agro-climatic conditions, Assam provides a unique setting where multiple agroforestry systems have evolved over generations (Nath et al., 2021). Traditional agroforestry in Assam is explicitly rooted in the lifestyle of various ethnic communities. The practice of *Jhum* or shifting cultivation, although considered unsustainable in its modern form due to shortened fallow cycles, originally evolved as a form of rotational agroforestry where forests could regenerate naturally. In recent years, many tribal communities have gradually moved towards more stable agroforestry systems such as homegardens, silvipastoral, and horticulture-based models. One of Assam's most prevalent and ecologically sound traditional agroforestry systems is the homegarden system, particularly among rural households in both plains and hill districts. Species like banana (*Musa* spp.), areca nut (*Areca catechu*), bamboo (*Bambusa* spp.), betel vine (*Piper betle*), jackfruit (*Artocarpus heterophyllus*), mango (*Mangifera indica*), guava (*Psidium guajava*) and citrus are commonly found (Reang et al., 2022)a. In the southern plains and hill districts of Assam, such as Cachar, Karbi-Anglong, and Dima Hasao, and ethnic communities like the *Hmar*, *Karbi*, and *Dimasa* have adopted pineapple-based (*Ananas comosus*) agroforestry systems as an alternative to *Jhum* cultivation. Pineapple, interplanted with nitrogen-fixing trees, banana, and forest species, provides a perennial cover, stabilizes soil, and offers substantial income. These models have gained attention as climate-resilient alternatives, capable of restoring degraded *Jhum* lands and reducing dependence on forest biomass (Reang et al., 2021). The integration of fruit trees, forest species, and contour bunding practices in these hill agroforestry systems also aids in water retention and biodiversity conservation (Reang et al., 2022)a.

Across Northeast India and other parts of South and Southeast Asia, there has been a gradual transition toward agroforestry systems and high-value cropping models. These systems are widely recognized for their capacity to provide long-term ecological stability, diversified livelihoods, and economic resilience. Among them, pineapple agroforestry systems (PAFS) have emerged as a particularly viable model in the region (Hazarika et al., 2024a). Therefore, we ask the following questions to understand the community resilience through agroforestry systems in Southern Assam. How do community-driven resilience strategies contribute to the productivity and ecological balance of pineapple-based agroforestry systems in southern Assam? In addition, what role do traditional ecological knowledges and local adaptive strategies play in sustaining these systems? The answer to the above research questions, the chapter aims to fulfil the following objectives: (i) to document and assess the role of traditional ecological knowledge, adaptive practices, and collective action in fostering system resilience, and (ii) to propose strategies for building resilience and livelihoods among the Hmar community.

Data Collection

A comprehensive review of studies on PAFS in Southern Assam was undertaken to synthesize information on management practices, traditional ecological knowledge, and the role of local adaptive strategies and collective action in enhancing system resilience. We conducted extensive field investigations across selected villages to complement this desk-based review. These included structured field surveys, participatory rural appraisals (PRA), and in-depth stakeholder interviews with farmers, community leaders, and local institutions. This mixed-methods approach provided nuanced insights into the socio-ecological dimensions of pineapple cultivation, particularly the challenges farmers encounter in sustaining productivity, adapting to climatic variability, addressing soil fertility decline, and accessing markets. The combination of literature review and primary field engagement thus allowed for a holistic understanding of the traditional wisdom embedded in local practices and the contemporary issues shaping the resilience of pineapple-based agroforestry systems in the region.

About the Pineapple Agroforestry Systems

Pineapple agroforestry systems are a dominant land use in the Indian Eastern Himalayas and are prevalent in countries like the Philippines, Vietnam, and Thailand. These systems typically involve the cultivation of pineapple in rows with a spacing of 0.7–0.8 m, in association with a variety of multipurpose tree species such as *Albizia procera*, *Artocarpus heterophyllus*, *Gmelina arborea*, and fruit trees like mango, guava, and litchi (Reang et al., 2022b). Increases of sucker and slips growth resulted in 5200–5400 individual pineapple plants over 0.1 ha at 15 years of stand age. The pineapple agroforestry system of the *Hmar* community in southern Assam is shown in Fig. 14.1. The presence of nitrogen-fixing species like *Albizia lebbeck* and *Leucaena leucocephala* enhances soil fertility. At the same time, the multi-tier canopy optimises sunlight utilisation and reduces erosion. Planting *Areca catechu* and *Musa* species along the boundaries of the farmland acts as natural fencing, mitigates soil erodibility, and serves as shelterbelts and windbreaks. Furthermore, the inclusion of cash crops like *Hevea brasiliensis*, *Areca catechu*, *Aquilaria malaccensis,* and *Parkia timoriana* into pineapple agroforestry systems substantially enhances year-round economic returns. The tree-pineapple combinations enhance ecological functionality by improving soil structure, nutrient cycling, and carbon sequestration while also offering shade, microclimate regulation, and habitat for beneficial fauna. Additionally, the synergistic management of crops and trees showcases the value of farmers' ecological understanding in prioritising tree species, leading to economic and ecological benefits.

Fig. 14.1 Pineapple agroforestry system of *Hmar* community in southern Assam (Photo credit: Animekh Hazarika)

About the Ethnic Hmar *Community*

The *Hmar* are an indigenous tribal community of Northeast India whose population is estimated to be fewer than one hundred thousand individuals (Lalzawmlien, 2018), placing them among the country's most vulnerable tribal groups. They are part of the larger Mizo–Kuki–Chin ethno-linguistic family and speak a language belonging to the Tibeto–Burman group. Ethno-historical accounts suggest that the *Hmar* trace their ancestral origins to Central China, from where they are believed to have migrated southwards over several centuries, eventually settling across parts of the Indo-Myanmar borderlands (https://en.wikipedia.org/wiki/Hmar_people).At

present, their distribution spans over multiple states of Northeastern Indian (NEI), viz., Assam, Manipur, Meghalaya, Mizoram, and Tripura, as well as the Chittagong Hill Tracts of Bangladesh and regions of upper Myanmar.

Despite their long historical presence in the region, the *Hmar* remain relatively unknown to the broader Indian populace. Academic and popular literature on their identity, culture, and livelihoods is sparse, and until recent decades, there has been little systematic ethnographic or anthropological documentation of the community (Lalzawmlien, 2018). The absence of a written historical tradition has further contributed to their limited visibility in mainstream narratives. Instead, the *Hmar* have preserved their collective memory through oral traditions, myths, legends, folktales, and ceremonial songs, which serve as repositories of cultural history (Khawbung, 2019). Consequently, even after more than seventy years of India's independence, the name Hmar is still unknown to a large section of the country's population (Lalzawmlien, 2018).

Culturally, the *Hmar* possess a distinctive and vibrant heritage, marked by extensive traditional knowledge systems encompassing agriculture, forest resource management, and ethno-medicine. These knowledge systems, refined through generations of experiential learning, have played a critical role in sustaining their livelihoods in diverse and often challenging ecological landscapes. The *Hmar* are a patrilineal society, with inheritance and kinship systems structured around male lineage (Khawbung, 2019). However, their relatively small population size, combined with their dispersed and often remote settlements, particularly in the hilly tracts of Cachar district, Assam, has contributed to a history of socio-political marginalisation and neglect within the wider regional framework (Lalzawmlien, 2018).

Currently, *Hmar* households continue to maintain and expand indigenous pineapple-based agroforestry systems for dual purposes: ensuring food security through home consumption and improving economic well-being by marketing surplus produce. These systems also demonstrate strong climate resilience, making them well-suited to the region's fragile hill ecosystems, which are vulnerable to erratic rainfall, land degradation, and biodiversity loss (Hazarika et al., 2024a).

The Context: From Jhum Cultivation to Sustainable Agroforestry

Jhum cultivation has historically been the dominant agricultural practice across the hilly regions of Northeast India, with over 0.44 million indigenous communities and serves as a primary means of livelihood. However, in recent decades, it has become increasingly unsustainable due to a drastic reduction in the fallow cycle (from 20 years to 3–5 years) driven by rising population pressures and shrinking land availability (Debnath et al., 2024). Shortened fallow periods lead to rapid depletion of soil fertility, increased soil erosion, reduced biomass regeneration, and declining agricultural productivity (Nath et al., 2016; Schröder et al., 2023). The influence of socioeconomic settings, growth of human population, increasing market demands, and government initiatives resulted in enhancing the transformation of short-fallowed cultivation to economically viable land uses like plantations of cash crops. The shift toward sedentary agriculture through intentionally integrating trees with crops and/or livestock at the farm or landscape level is emerging as a viable strategy for climate change mitigation. This approach enhances farmers' resilience to climate risks, while delivering biophysical and socioeconomic benefits and preserving cultural identities (Nath et al., 2020; Quandt et al., 2017; Rosenstock et al., 2019). The Intergovernmental Panel on Climate Change (IPCC, 2022) identifies agroforestry as a key agroecological strategy for climate change adaptation, noting its potential to deliver a range of co-benefits. These include long-term carbon storage, improved food and nutritional security, livelihood diversification, sustained ecosystem services, and biodiversity conservation (Duguma et al., 2023; Tschora & Cherubini, 2020). By combining trees with crops and, in some cases, livestock, agroforestry systems create multifunctional landscapes that enhance resilience to climatic variability while supporting productivity and ecological stability. An illustrative case is found among the Pnar communities in the Indian sub-Himalayan region. Here, the twin challenges of declining productivity and reduced land availability under shifting cultivation encouraged the search for more sustainable and economically viable alternatives. One such response has been the adoption of Piper-based agroforestry, in which black pepper (*Piper nigrum*) is cultivated by training vines on live support trees (Hazarika et al., 2024b). This system not only provides an additional source of income but also reflects the adaptive capacity of local communities in balancing ecological and economic goals.

Pineapple-based agroforestry represents a locally adapted innovation that simultaneously strengthens livelihoods and ecological resilience. By combining a high-value cash crop with increased tree cover, it addresses the interconnected challenges of climate change, land degradation, and resource scarcity. Among the Hmar communities of the Indian sub-Himalayas, this transition has been particularly significant. Traditionally dependent on shifting cultivation, the Hmar have, over the past four decades, increasingly converted portions of their land into pineapple-based agroforestry systems. Expanding market demand, the crop's reliable economic returns, and its deep cultural roots within Hmar tradition have driven this shift. When practised with multipurpose tree species, such systems not only generate timber, fuelwood, and fodder but also provide soil conservation, biodiversity maintenance, and carbon sequestration benefits. Thus, Hmar pineapple agroforestry offers a compelling model of how indigenous knowledge and adaptive strategies can transform land-use practices into climate-smart, economically viable, and culturally embedded systems (Lal et al., 2021).

Ecological and Environmental Benefits

In southern Assam, the pineapple agroforestry system is notable for integrating a wide range of tree species that provide diverse ecological and economic benefits. According to Reang et al., (2022a), this system hosts 20 tree species from 15 flowering plant families. Among them, Fabaceae is the most represented family (15%), followed by Bignoniaceae (10%), Euphorbiaceae (10%), and Lamiaceae (10%). Most tree species (65%) supply fuelwood, while 55% provide shade and timber, contributing to improved soil health and fertility. Additionally, 45% generate substantial income for farmers through the sale of their produce. Conservation assessments reveal that the majority of species (75%) fall under the Least Concern category, with 5% listed as Data Deficient in the IUCN Red List of Threatened Species. Including Parkia timoriana, a valuable tree bean, and Aquilaria malaccensis, a critically endangered species no longer found in the wild, underscores the system's role in conserving threatened species within managed agroecosystems. Moreover, this agroforestry system functions as a foraging habitat for rare and threatened wildlife, including *Hystrix indica*, *Macaca mulatta*, and *Sus scrofa cristatus*.

The retention of mature and large trees during vegetation clearing provides shade for understory crops and helps prevent soil erosion. In pineapple cultivation, agricultural phases often include intercropping with annual crops such as cereals, spices, rhizomatous plants, and legumes, alongside fallow trees at the onset of the first monsoon (May–June). After harvest, crop residues are left on the field as mulch, which enriches soil organic carbon stocks (Muchane et al., 2020), enhances nutrient content, and conserves soil moisture for pineapple growth. Inadequate shade has been associated with 15–20% fruit damage in pineapple (Das et al., 2011).

Farmers frequently plant Musa species and Areca catechu along field edges to mitigate soil erosion and use bamboo, timber logs, and burned tree remains as protective barriers (Craswell, Sajjapongse, Howlett, & Dowling, 1998). These plantings also function as natural fences, windbreaks, and shelterbelts. Nitrogen-fixing *Albizia* species are preferred for shade during the early years of system establishment, as they also suppress weeds, similar to practices observed among the Garo tribes of Bangladesh (Khaleque & Gold, 1993).

Incorporating tree bean (*Parkia timoriana*) further improves soil fertility by fostering beneficial bacteria such as *Agrobacterium fabrum*, *Bacillus subtilis*, and *Pseudomonas fluorescens*. These bacteria can be developed into native bio-inoculants for rehabilitating degraded soils (Singh, 2019). Trees such as Mangifera indica and cash crops like Aquilaria malaccensis and Hevea brasiliensis contribute to nutrient cycling and biomass decomposition, increasing soil organic carbon, nitrogen, and phosphorus levels (Deng et al., 2016). Combining tree components enhances organic matter accumulation in the topsoil, safeguards soil structure, and improves aggregate stability (Tarafdar & Ray, 2005).

Minimum tillage, mulching, green manuring, and other soil conservation measures help regulate soil temperature, conserve soil moisture, reduce evaporation, enhance infiltration, and improve soil structure (Jabran, 2019). Reang et al.

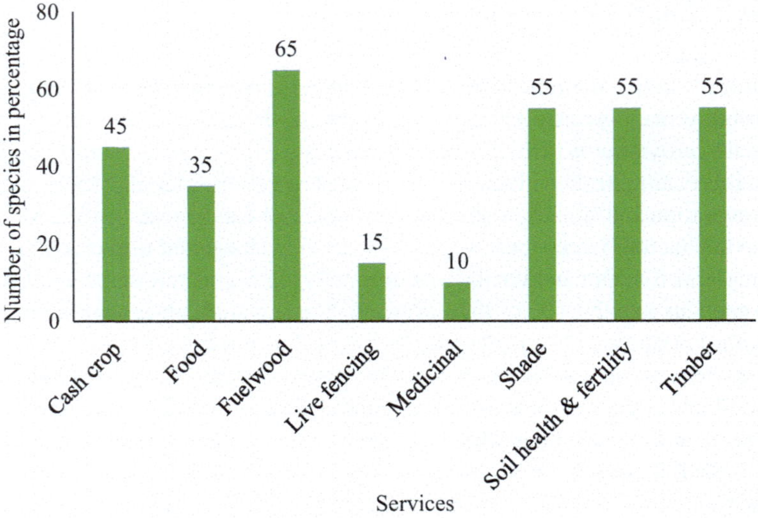

Fig. 14.2 Number of tree species (in percent) under different services from pineapple agroforestry adoption

(2021) recorded a stand density of 190 trees per hectare along with 55,400 pineapple plants per hectare, resulting in an ecosystem carbon storage of 247.46 Mg ha^{-1}. A schematic representation of tractional changes in SOC stocks under two agricultural scenarios is shown in Fig. 14.2. Hazarika et al., (2024a) reported that clay ratios and modified clay ratios decrease with the increasing age of pineapple agroforestry stands, contributing to greater organic matter aggregation in the topsoil, thereby promoting soil structure and stability (Tarafdar & Ray, 2005). They also outlined soil organic carbon (SOC) stocks ranging from 169.7 to 177 Mg C ha^{-1} in pineapple agroforestry systems of different ages, with stratification ratios (SR > 2) for SOC and macronutrients indicating improved soil quality comparable to that of nearby forested lands (Fig. 14.3)

Building Resilience and Livelihoods

The average landholding under PAFS is about 2.63 ha (Reang et al., 2021). Farming in these systems is predominantly rain-fed, with both men and women of the household actively participating in cultivation and crop management, including activities such as land preparation and harvesting. On average, farmers earn around US$138.00 per month from PAFS, which serves as a supplementary income alongside their primary occupations, whether in farming, government service, or private employment. Thus, PAFS plays an important role in enhancing household income and livelihood security (Reang et al., 2021). PAFS integrates diverse crops and multipurpose

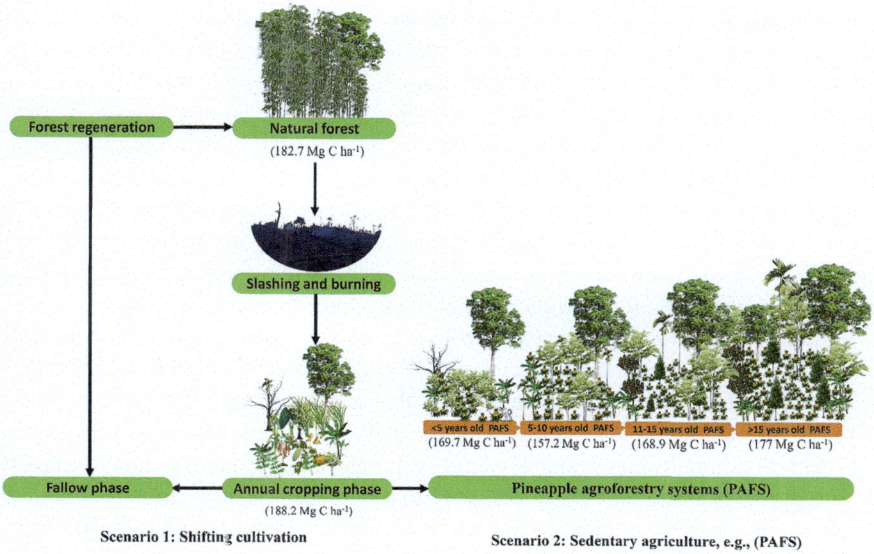

Fig. 14.3 Schematic representation of transitional changes in land use and estimated SOC stocks under two agricultural scenarios in the study region. Scenario 1 illustrates the different stages of shifting cultivation, while Scenario 2 shows the transition to sedentary pineapple agroforestry systems (PAFS) of varying ages (Values of SOC adopted from Hazarika et al., 2024b)

trees (MPTs) such as *Aquilaria malaccensis*, *Parkia timoriana*, *Areca catechu*, and *Hevea brasiliensis*, which provide year-round cash income. Studies across Northeast India highlight their economic potential: pineapple cultivation yielded about US$ 2681 ha^{-1} annually (Swett & Bera, 2018); *A. malaccensis* generated US$ 600–18,000 per family per year (Saikia & Khan, 2014), with returns reaching up to US$ 100,000 kg^{-1} depending on quality (Naef, 2011); *P. timoriana* fetched US$ 100–200 per tree annually (Rocky et al., 2004); *A. catechu* earned US$ 261 per acre (Aditya et al., 2014); and rubber remains an important commercial crop. Additional income also comes from occasional sales of fruits, livestock and their products in local markets. These findings underscore the substantial economic potential of PAFS for the Hmar community. Therefore, intercropping pineapple with arecanut, banana, citrus, bamboo, timber, and shade trees offers a vital pathway to environmental management and diversified incomes for pineapple growers.

Yet these farmers face intertwined risks: erratic rainfall and dry spells, soil erosion on undulating terrain, periodic pest and disease surges and limited post-harvest infrastructure. Hence, resilience strategy must therefore integrate ecological design, climate-smart practices, value addition, finance, and inclusive institutions to enhance resilience and advance food security for the farmers.

Global forest cover continues to decline, making it imperative to strengthen tree cover beyond conventional forests. Trees outside forests—particularly those integrated into farm forestry by smallholders—offer a promising alternative. In this context, shifting cultivation lands could be systematically converted into PAFS under a mission-oriented approach. Traditional resource-poor farmers should be supported through incentives that recognize ecosystem services, along with mechanisms for fair pricing, credit, insurance, and market access. Equally important is the identification and promotion of underutilized "Cinderella" species (Nair et al., 2017) that are traditionally domesticated within PAFS and suited to diverse agro-ecological zones. Policies that secure farmers' ownership and tree-use rights must form the backbone of agroforestry development, alongside enabling frameworks for implementation. Furthermore, enhancing PAFS with a broader mix of multipurpose tree species will improve system structure, diversity, and resilience—strengthening climate change mitigation, sustainable food production, and long-term forest conservation.

Field diversification plays a crucial role in enhancing ecological stability within pineapple agroforestry systems. Nitrogen-fixing hedgerows like *Flemingia* and *Gliricidia* boost soil fertility and provide biomass that can serve as fodder, green manure, or fuelwood. The combination of pineapple with fruit or timber trees maximizes sunlight utilization, fosters a beneficial microclimate, and mitigates production risks. During dry spells, soil and water conservation methods—such as trenches, ponds, and lined pits—can be supplemented with cost-effective irrigation solutions like drip, pitcher, or gravity-fed systems, thereby enhancing water availability for farmers.

In addition to ecological measures, improving livelihood resilience is essential for the sustainability of pineapple agroforestry. A significant issue is the high level of post-harvest losses prevalent in pineapple value chains. Educating farmers on optimal harvest timing, grading, and careful handling techniques can lead to a marked increase in the proportion of marketable produce. Joint investments in shared facilities—such as pack houses, cold storage, and pre-cooling units—can lower losses and transaction expenses while enhancing farmers' negotiating power by reducing dependence on intermediaries. Moreover, local processing initiatives create benefits and broaden income sources by transforming fresh pineapples into products like juices, jams, dried slices, or candies, which create opportunities for small businesses, women's groups, and youth-led initiatives.

Improving smallholder farmers' market access requires more than physical connectivity; food safety, quality assurance, and consumer trust are equally critical. Measures such as barcoded labelling, digital traceability, and transparent supply chains may build confidence while enabling farmers to stand out in competitive markets. Certifications like Good Agricultural Practices and Participatory Guarantee Systems further open access to premium markets, institutional procurement, and online retail. By reducing dependence on uncertain farm-gate sales, these systems integrate farmers into structured, profitable value chains. Aligning high-quality production with assured markets encourages sustainable land management, stabilizes incomes, strengthens rural economies, and advances food security goals.

Conclusions

The traditional pineapple-based agroforestry system provides significant livelihood benefits, generating reliable income and improving household food security. Beyond its economic role, it also reflects the farmers' deep ecological knowledge, particularly their ability to prioritize and manage diverse tree species. This knowledge contributes directly to the conservation of tree diversity within agricultural landscapes, thereby strengthening ecological resilience. In the study area, PAFS practiced by the Hmar community illustrate how indigenous land-use strategies can restore degraded soils, safeguard biodiversity, and enhance productivity while simultaneously delivering wider social and environmental gains. The system not only integrates cash crop cultivation with tree cover but also supports sustainable management of natural resources in fragile hill ecosystems. Given these multiple benefits, there is strong justification for promoting PAFS more widely across suitable habitats of Northeast India, where it can serve as a replicable model for resilient livelihoods, sustainable intensification, and climate-smart land management.

References

Aditya, K. S., Kammardi, T. P., Kumar, R. K., Patil, R., & Pavithra, H. K. (2014). Stability of returns ensuring livelihood and economic security of areca farmers in Karnataka: An empirical analysis. *Research Journal of Agriculture & Forestry Sciences, 5*(5), 1043–1046.

Craswell, E. T., Sajjapongse, A., Howlett, D. J. B., & Dowling, A. J. (1998). Agroforestry in the management of sloping lands in Asia and the Pacific. *Agroforestry Systems, 38*, 121–137. https://doi.org/10.1023/A:1005960612386

Das, S. C., Prakash, J., Suresh, C. P., Das, A., & Bhattacharjee, T. (2011). Pineapple cultivation in hilly Tripura with year-round production: Improving livelihood opportunities in rural areas of Tripura. *Acta Horticulturae, 902*, 291–298.

Debnath, N., Laskar, S. Y., Nath, A. J., Pandey, R., & Das, A. K. (2024). Systematic review on shifting cultivation in North-East India: Emerging research directions. In S. K. Tripathi & F. Q. Brearley (Eds.), *Shifting cultivation systems: From indigenous practices to soil fertility in North-East India*. Springer. https://doi.org/10.1007/978-3-031-70388-1_15

Deng, J., Sun, P., Zhao, F., Han, X., Yang, G., Feng, Y., & Ren, G. (2016). Soil C, N, P and its stratification ratio affected by artificial vegetation in subsoil loess plateau China. *PLoS One, 11*(3), e0151446.

Duguma, D. W., Law, E., Shumi, G., Rodrigues, P., Senbeta, F., Schultner, J., & Fischer, J. (2023). Spatial predictions for the distribution of woody plant species under different land-use scenarios in southwestern Ethiopia. *Landscape Ecology, 38*(5), 1249–1263.

Hazarika, A., Kurmi, B., Francaviglia, R., Sileshi, G. W., Paramesh, V., Das, A. K., & Nath, A. J. (2024a). The transition from shifting cultivation to indigenous agroforestry as nature-based solution for land restoration in the Indian Eastern Himalayas. *Ecological Indicators, 162*, 112031.

Hazarika, A., Nath, A. J., Pandey, R., Pebam, R., Devi, N. B., & Das, A. K. (2024b). Climate change vulnerability of tribe managing Piper agroforestry systems in the Indian sub-Himalayan region. *Agricultural Systems, 216*, 103914. https://doi.org/10.1016/j.agsy.2024.103914

IPCC. (2022). *Impacts, adaptation and vulnerability. Contribution of working group II to the sixth assessment report of the intergovernmental panel on climate change* (pp. 3–33). Cambridge

University Press. https://www.ipcc.ch/report/ar6/wg2/downloads/report/IPCC_AR6_WGII_SummaryForPolicymakers.pdf

Jabran, K. (2019). *Role of mulching in pest management and agricultural sustainability*. Springer.

Khaleque, K., & Gold, M. A. (1993). Pineapple agroforestry: An indigenous system among the Garo community of Bangladesh. *Society & Natural Resources, 6*(1), 71–78. https://doi.org/10.1080/08941929309380808

Khawbung, M. L. (2019). Introduction to origin of Hmar tribe. *International Journal of Innovative Research & Advanced Studies, 6*(9), 59–63.

Lal, R., Bouma, J., Brevik, E., Dawson, L., Field, D. J., Glaser, B., Hatano, R., Hartemink, A. E., Kosaki, T., Lascelles, B., & Monger, C. (2021). Soils and sustainable development goals of the United Nations: An international union of soil sciences perspective. *Geoderma Regional, 25*, e00398.

Lalzamlien. (2018). Tribal poverty: Occupation, income and savings with special reference to the Hmar tribe of Cachar district. *Assam. Journal of International Academic Research for Multidisciplinary, 6*(11), 1–13.

Muchane, M. N., Sileshi, G. W., Gripenberg, S., Jonsson, M., Pumarino, L., & Barrios, E. (2020). Agroforestry boosts soil health in the humid and sub-humid tropics: A meta-analysis. *Agriculture, Ecosystems & Environment, 295*, 106899. https://doi.org/10.1016/j.agee.2020.106899

Naef, R. (2011). The volatile and semi-volatile constituents of agarwood, the infected heartwood of Aquilaria species: A review. *Flavour & Fragrance Journal, 26*(2), 73–87. https://doi.org/10.1002/ffj.2034

Nair, P. K. R., Viswanath, S., & Lubina, P. A. (2017). Cinderella agroforestry systems. *Agroforestry Systems, 91*(5), 901–917. https://doi.org/10.1007/s10457-016-9966-3

Nath, A. J., Brahma, B., Lal, R., & Das, A. K. (2016). Soil and jhum cultivation. *The Encyclopedia of Soil Science*, 1–9. https://doi.org/10.1081/e-ess3

Nath, A. J., Kumar, R., Devi, N. B., Rocky, P., Giri, K., Sahoo, U. K., & Pandey, R. (2021). Agroforestry land suitability analysis in the eastern Indian Himalayan region. *Environmental Challenges, 4*, 100199. https://doi.org/10.1016/j.envc.2021.100199

Nath, A. J., Sahoo, U. K., Giri, K., Sileshi, G. W., & Das, A. K. (2020). Incentivizing hill farmers for promoting agroforestry as an alternative to shifting cultivation in Northeast India. *Agroforestry for Degraded Landscapes, 1*, 425–444.

Quandt, A., Neufeldt, H., & McCabe, J. T. (2017). The role of agroforestry in building livelihood resilience to floods and drought in semiarid Kenya. *Ecology & Society, 22*(3).

Reang, D., Hazarika, A., Sileshi, G. W., Pandey, R., Das, A. K., & Nath, A. J. (2021). Assessing tree diversity and carbon storage during land use transitioning from shifting cultivation to indigenous agroforestry systems: Implications for REDD+ initiatives. *Journal of Environmental Management, 298*, 113470. https://doi.org/10.1016/j.jenvman.2021.113470

Reang, D., Nath, A. J., Sileshi, G. W., Hazarika, A., & Das, A. K. (2022). Post-fire restoration of land under shifting cultivation: A case study of pineapple agroforestry in the sub-Himalayan region. *Journal of Environmental Management, 305*, 114372. https://doi.org/10.1016/j.jenvman.2021.114372.

Reang, D., Sahoo, U. K., Giri, K., Hazarika, A., & Nath, A. J. (2022). Ethnic homestead forests of North-East India revealed as diverse land-use systems. *Agroforestry systems, 96*(3), 465–478. https://doi.org/10.1007/s10457-021-00700-z

Rocky, P., Sahoo, U. K., & Thapa, H. S. (2004). Livelihood generation through tree bean (Parkia roxburghii g. Don) in Imphal. *Journal of Non-Timber Forest Products, 11*(2), 135–139.

Rosenstock, T. S., Dawson, I. K., Aynekulu, E., Chomba, S., Degrande, A., Fornace, K., et al. (2019). A planetary health perspective on agroforestry in sub-Saharan Africa. *One Earth, 1*(3), 330–344.

Saikia, P., & Khan, M. L. (2014). Homegardens of upper Assam, Northeast India: A typical example of on-farm conservation of Agarwood (*Aquilaria malaccensis* lam.). *The International Journal of Biodiversity Science, Ecosystem Services & Management, 10*(4), 262–269. https://doi.org/10.1080/21513732.2014.977953

Schröder L. S, Rasche L, Jantke K, Mishra G, Lange S, Eschenbach A, Schneider UA (2023). Combined effects of climate change and agricultural intensification on soil erosion in uphill shifting cultivation in Northeast India. *Land Degradation & Development*, 1–17. https://doi.org/10.1002/ldr.4944

Singh, M. K. (2019). Potential of underutilized legume tree Parkia timoriana (DC.) Merr. In eco-restoration of jhum fallows of Manipur. *Journal of Pharmacognosy & Phytochemistry, 8*, 1685–1687.

Swett, B. S., & Bera, B. (2018). An economic viability analysis of pineapple cultivation in Ri Bhoi district of Meghalaya. *The Journal of Crop & Weed, 14*(2), 33–39.

Tarafdar, P. K., & Ray, R. (2005). Effect of trees on improvement of physical environment and fertility in soils of West Bengal. *Bulletin of the National Institute of Ecology, 16*, 129–136.

Tschora, H., & Cherubini, F. (2020). Co-benefits and trade-offs of agroforestry for climate change mitigation and other sustainability goals in West Africa. *Global Ecology & Conservation, 22*, e00919. https://doi.org/10.1016/j.gecco.2020.e00919

Chapter 15
Drought and Climate Change in Kenya and Ethiopia: Economic Implications, Livelihood Vulnerability and Policy Responses

Morami Dutta and Raghunath Prasad Saket

Abstract Drought remains one of the most critical climate-related challenges in Kenya and Ethiopia. Rising temperatures and decreasing rainfall patterns are causing drought more frequent, severe, and widespread. The complex connection between drought dynamics and climate variability in both countries is examined in this chapter, along with their significant financial consequences and livelihood vulnerabilities. Through a narrative review of peer-reviewed literature and policy documents, the study reveals that drought frequency in Kenya's Arid and Semi-Arid Lands has significantly increased from once every ten years (1960s–1970s) to once every two to three years (1990s). In 2010 and 2011, the drought alone caused Kenya's GDP to decline by 2.8% per year, resulting in losses of USD 12.1 billion, while millions of Ethiopians needed emergency food assistance. Vulnerability patterns demonstrate significant gender, age, and wealth disparities, with women and children disproportionately bearing drought burdens. Due to a lack of institutional capacity, insufficient funding, and poor coordination, implementation gaps continue despite the existence of policy frameworks, such as early warning systems and national adaptation plans. In order to increase community resilience against growing drought risks in East Africa, the chapter highlights the critical need for improved drought monitoring that incorporates atmospheric evaporative demand, climate-resilient infrastructure, livelihood diversification strategies, and strengthened institutional coordination.

Keywords Drought · Climate change · Livelihood vulnerability · Water scarcity · East Africa

M. Dutta (✉)
Department of African Studies, University of Delhi, Delhi, India
e-mail: mdutta@as.du.ac.in

R. P. Saket
Department of Economics, Satyawati College, University of Delhi, Delhi, India
e-mail: raghunathjnu@gmail.com

R. Sarkar et al. (eds.), *Climate Change, Environmental Hazards and Community-Based Resilience*, Disaster Risk Reduction, https://doi.org/10.1007/978-981-95-9485-6_15

Introduction

Drought is widely regarded as one of the most challenging climatic phenomena affecting societies across the world (Sivakumar 2014; Van Loon & Van Lanen, 2015). It is a gradual natural hazard that unfolds slowly over months or years, often impacts large geographical areas and deeply influences water resources, livelihoods and ecosystems. In most cases, drought typically happens when the annual rainfall decreases below sustained or normal averages (Wilhite, 2005). Depending on climate, geography and exposure different locations are affected by drought in different ways. Therefore, it is challenging to create a single, universally accepted definition of drought as its characteristics are highly multidimensional and context specific (Van Loon, 2015). Academic literature distinguishes several drought typologies such as meteorological forms linked to rainfall shortfalls, agricultural variants characterized by insufficient soil moisture, ecological types involving vegetation water stress, and hydrological types marked by depleted surface and groundwater reserves. Broadly, drought represents an extended phase of water scarcity that occurs worldwide, but manifests differently across space and time (Wilhite & Glantz, 1985). From a hydrological perspective, there is a fundamental link of drought to water imbalance; it's a situation in which water availability is below normal levels. (Van Loon, 2015).

Drought has affected human civilisations throughout history, with early references appearing in Biblical stories and ancient literature such as the Epic of Gilgamesh (Heathcote, 2016). Sediment-core evidence reveals mega droughts in tropical Africa between approximately 135,000 and 75,000 years ago, likely to contribute to early human migrations out of Africa (Schaebitz et al., 2021b). More recently, the Indian subcontinent experienced over 18 meteorological and 16 hydrological drought events between 1870 and 2016, including the devastating 1876–1878 famine (Mishra et al., 2019). Throughout history, droughts have fundamentally shaped settlement patterns, agricultural practices, and human migration (IPCC, 2024).

This study focuses on Kenya and Ethiopia because both the countries exhibit high vulnerability to drought and climate variability compared to other East African nations. Second, their economies depend heavily on climate-sensitive livelihoods such as pastoralism, agro-pastoralism, and smallholder agriculture which have been severely impacted by recurrent droughts. Third, their shared experiences and regional proximity provide a robust comparative framework for analysing drought impacts and adaptation strategies applicable across East Africa (Fig. 15.1).

Through a narrative assessment of relevant research, this chapter summarizes the current understanding of drought, climate variability, and livelihood implications in Kenya and Ethiopia. The analysis draws on peer-reviewed journal articles, national policy documents, and assessments from key international organizations including the IPCC, FAO, World Bank, USAID, and National Drought Management Authority of Kenya (NDMA). Sources were identified through targeted searches using relevant terms such as drought, changes in climate, vulnerability, and adaptation in the Eastern part of Africa. Although not systematic, the review prioritized

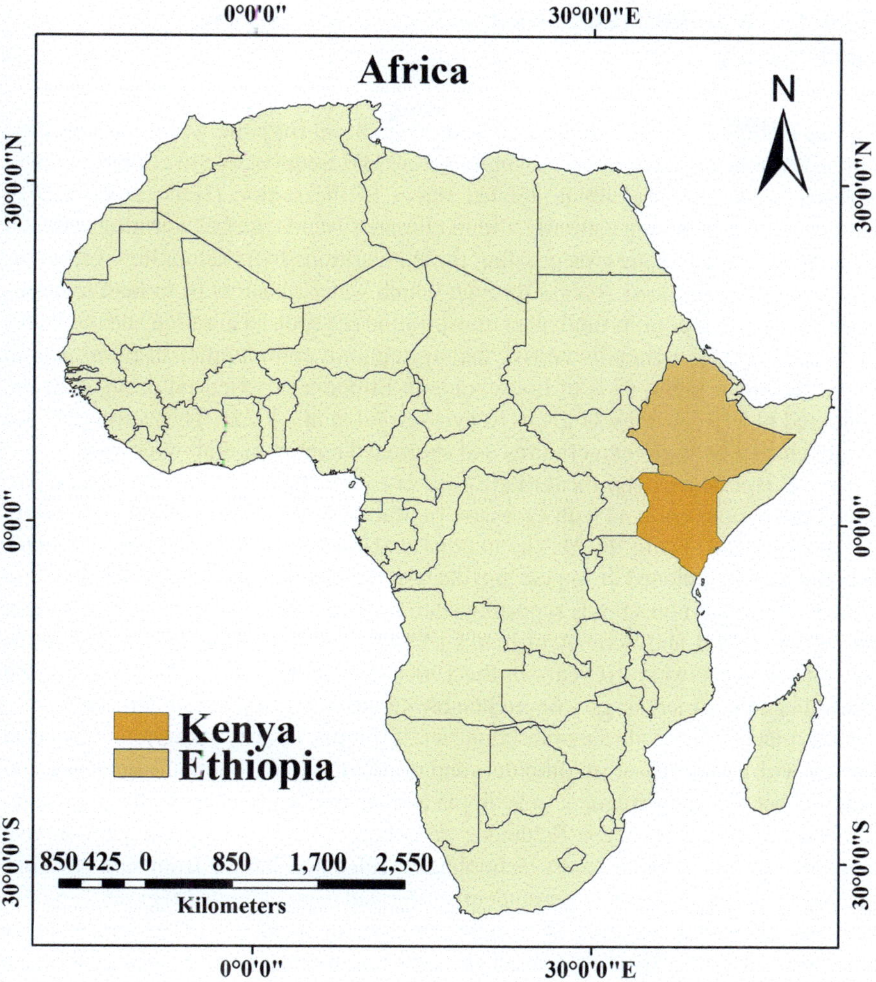

Fig. 15.1 Map of Africa, highlighting Ethiopia and Kenya

studies published from 2010 onwards to capture recent drought events, updated climate projections, and current policy frameworks. Earlier literature was included where it provided essential historical or conceptual context. The chapter follows an integrative structure: climate–drought relationships are examined first, followed by economic impacts, livelihood vulnerabilities, and adaptation responses. It is a comparison between Kenya and Ethiopia, highlighting both shared regional challenges and country-specific dynamics. The aim is to synthesise the most credible and policy-relevant evidence to inform ongoing discussions on drought resilience in East Africa.

Climate Variability and Drought Patterns in Kenya and Ethiopia

Climate change is closely associated with drought all over the world. Across East Africa, drought has become increasingly severe and frequent due to climate change, representing a primary climate-related threat to the region (Haile et al. 2020). Although drought occurs naturally within climate systems, global warming-induced temperature increases are exacerbating these conditions by accelerating evapotranspiration—the combined process through which water transfers from land to atmosphere via both evaporation and plant transpiration (Evapotranspiration and the Water Cycle, 2019). The main climate risk that agricultural communities in Ethiopia and Kenya face is drought. 41% of households in Ethiopia experienced drought stress, compared to 71% of households in Kenya. (Aryal et al., 2021). These droughts are mainly caused by high temperatures and shortened annual rainfall windows.

Kenya experiences significant hailstorm events (affecting 56% of households), while both nations contend with excessive rainfall patterns that contribute to flooding (Aryal et al., 2021) (Fig. 15.2). The interaction between drought and climate manifests through heightened crop pest and disease incidence, with 73% of Kenyan and 25% of Ethiopian households reporting such impacts (Aryal et al., 2021). Drought frequency in Arid and Semi-Arid Lands (ASALs) of Kenya has increased significantly from once every 10 years in the 1960s–70s to once every 2–3 years in the 1990s, becoming increasingly unpredictable since 2000 (Aryal et al., 2021).

Like many other Sub-Saharan countries, Ethiopia is also an agriculture-based country with a majority of smallholders and rain-fed farms. Climate-change-induced droughts become very frequent in Ethiopia and adversely affect agricultural production and household incomes (Schlenker & Lobell, 2010). As a result, people have adopted different climate-smart technologies, relocated labour from farm to non-farm work and temporarily or permanently migrated to less affected regions. Ethiopia possesses an extended history of droughts and a rising incidence of extreme weather conditions (Mera, 2018). Prolonged drought has mainly brought two outcomes; firstly, it enhances the probability of non-agricultural self-employment and diminishes prospects for agricultural wage labor. Secondly, it shifts time allocation to off-farm self-employment (OFSE)and decreases hours devoted to both agricultural wage work and on-farm independent activities. Small farmholders can't afford new technologies due to lack of credit and proper information. Drought reduces the wages and compels rural labourers to shift to non-agricultural sectors (Jessoe et al., 2018; Jayachandran 2006). Drought also worsens income inequality in Ethiopia, making lower-income groups more vulnerable to drought. When temperature effects are taken into account using the Standardized Precipitation-Evapotranspiration Index (SPEI) (Shelton & Dixon, 2023), the results show a very different reality that drought conditions are getting worse throughout East Africa, even though precipitation-based assessments using the Standardized Precipitation Index (SPI) indicate declining drought trends (Haile et al., 2019). It establishes the fact that temperature has a

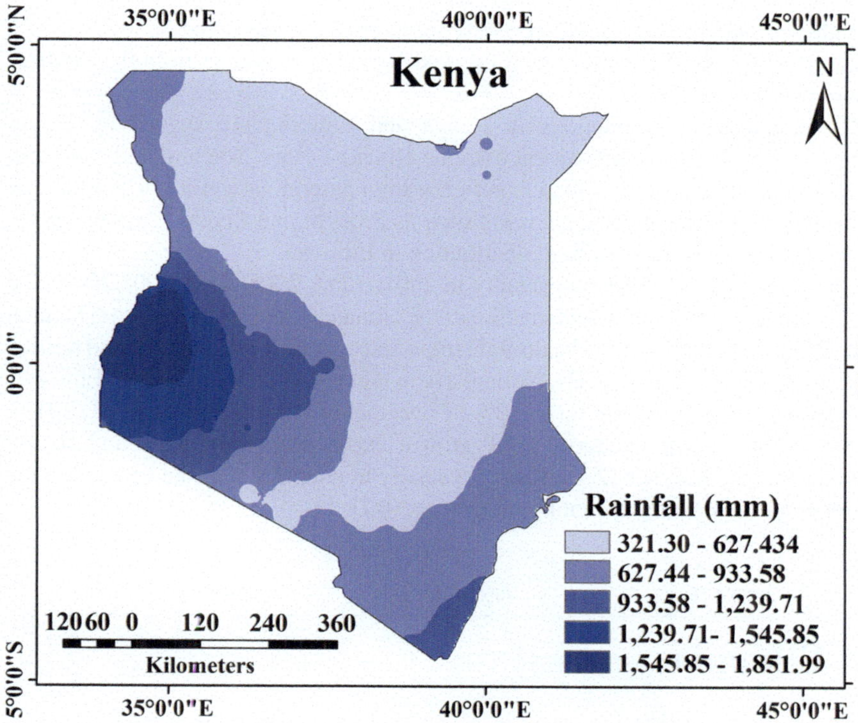

Fig. 15.2 Rainfall distribution in Kenya

dominant influence on drought development because it increases potential evapotranspiration (PET), which in turn magnifies moisture deficits across the region. In the context of global warming, drought is no longer a purely "natural hazard" since human activities have fundamentally changed drought characteristics through greenhouse gas emissions, which are the major cause of regional warming and drying conditions (Haile et al., 2019). Under climate change scenarios, droughts become more frequent, intense, and prolonged. For example, projections indicate that the drought-affected area in East Africa could expand by about 16%, 36%, and 54% under different Representative Concentration Pathways (RCP) by the end of the twenty-first century (Haile et al., 2019). Spatially, the Horn of Africa such as Somalia, Eritrea, eastern Ethiopia, and someportions of Tanzania appears to experience a majorexpansion in drying conditions in terms of duration and frequency. Although pockets of relative wetting appear in regions such as Uganda, Kenya, and parts of the Ethiopian highlands, the general trend shows widespread and persistent drought conditions. Thus climate change imposes substantial pressure on ecosystems, agriculture, water resources, and livelihoods in East Africa.

In Ethiopia, the same drought left 4.6 million people requiring emergency food assistance, with pastoral areas in the south and southeast suffering the worst impacts. Existing evidence indicates that when there is change in climate, the likelihood of drought increases in terms of "severity, frequency, duration, and spatial extent", fundamentally altering long-term water and livelihood systems in both countries (Duguma et al., 2017).

Climate–drought interactions in Kenya and Ethiopia are shaped by highly changing rainfall and very frequent extreme climate events. The arid and semi-arid lands (ASALs) of Eastern Africa experience high natural variability of rainfall and recurrences of extreme climate events such as drought and floods (Ongoma et al., 2018). Figure 15.3 shows rainfall distribution in Ethiopia.

Major droughts occurred repeatedly in 2001, 2003, 2005/2006, 2008/2009, and 2011, expressing the intensifying nature of the climate system in the region (Duguma et al., 2017). The 2010/2011 The drought impacted over 13 million people and pushed food insecurity to famine levels, particularly in the pastoral borderlands of Ethiopia, Kenya, and Somalia, where up to 80% of livestock was lost (Headey et al., 2013). In Kenya, this drought reduced GDP growth by an average of 2.8% and caused losses estimated at USD 12.1 billion, especially in counties with the lowest human development indicators (Republic of Kenya, 2012).

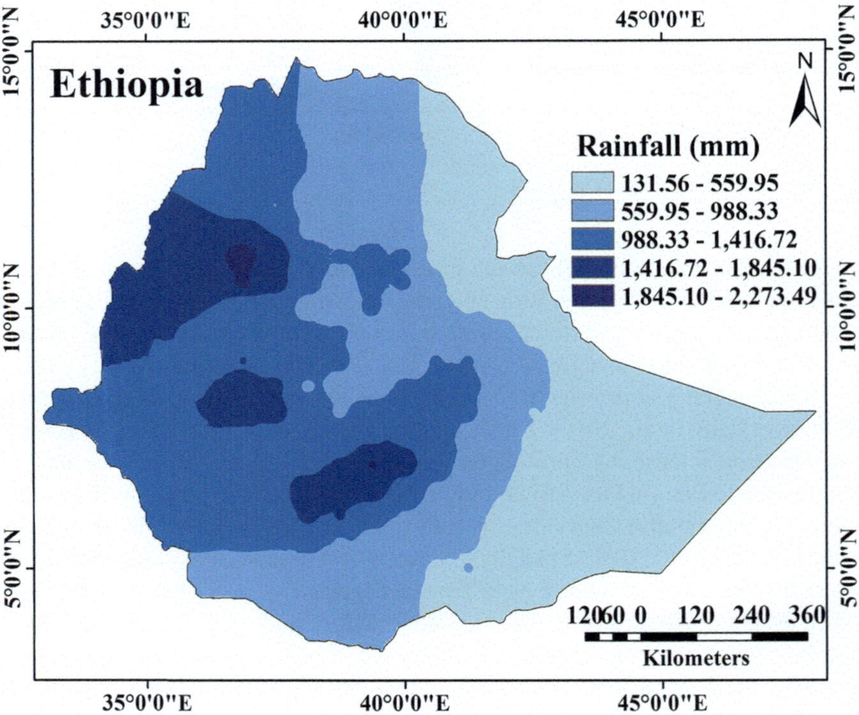

Fig. 15.3 Rainfall distribution in Ethiopia

The Economic Implications of Increasing Drought Risks

Drought in Kenya and Ethiopia has serious and significant economic effects that affect household assets, production systems, and the trajectory of national development. The draught between 2008 and 2011 in Kenya affected the nation, reducing GDP by an average of 2.8% annually and causing an estimated USD 12.1 billion in damage and losseswith extremely low levels of human development (Republic of Kenya, 2012). During that time, Ethiopia experienced major cereal market disruptions with price hikes above 2008 levels that intensified food insecurity among the poorest populations (World Bank, 2011).

Drought shocks in Kenya significantly weakened livelihood conditions, especially among the people who live in villagesand are highlyrelying on rain-dependent farming and pastoralism. Due to recurrent and prolonged droughts, there is significantly low agricultural yield and high cattle mortality and thus lower the primary sources of household food and income (FAO, 2023; NDMA, 2022). As a result, food insecurity intensified and affected populations with large-scale malnutrition and rising poverty levels (WFP, 2022). Livelihood vulnerability also multiplied due to declining agricultural productivity and disrupted local markets. Extreme water scarcity has made it more difficult to obtain irrigation supplies, livestock water sources, and safe drinking water, further straining already precarious livelihoods (NDMA, 2023). Kenya experienced significant declines in the production of major staple crops such as maize. This is attributed to increased evapotranspiration from higher temperatures and unfavorable rainfall patterns. Most farmers in Kenya experienced a decline in average rainfall over the past two decades, and about 51% of them observed that drought periods between rainy seasons have become significantly prolonged (Deresse, 2023). Table 15.1 shows the effects of change in climate on water resource in Kenya and Ethiopia.

Communities in both Kenya and Ethiopia face profound vulnerabilities due to drought-induced water scarcity, shaped by intertwined climatic, socio-economic, and infrastructural factors. In Ethiopia, rising temperatures, declining and erratic rainfall, and increasingly severe drought cycles have significantly reduced water availability across many regions, intensifying household vulnerability (Emiru et al., 2021; Gelete et al., 2020). Women and children are disproportionately affected due to long daily journeys to collect water, often up to six hours. As a result, psychological distress and limiting educational opportunities (Gurara et al., 2021) among them are very prominent. National sectors are also at risk from water scarcity; for

Table 15.1 Effects of change in climate on water resource in Kenya and Ethiopia (Deresse, 2023)

Name of the Nation	Effects of change in climate
Kenya	Flooding in the time of rain and ongoing drought
Ethiopia	Periodic earthquakes, volcanic eruptions, floods, and droughts carried out by the geological activity of the great Rift Valley

example, extreme events like floods harm water infrastructure and raise sediment loads in dams, while decreased river flows and unpredictable hydrology destabilise hydropower generation (De Trincheria et al., 2016). Similarly, Kenya has experienced worsening water scarcity, with per capita water availability projected to decline sharply as droughts intensify and rainfall variability increases (Mogaka, 2006). Rapid population growth, deforestation, inadequate storage, and uneven water distribution further increased community vulnerability, especially in arid and semi-arid regions. Women and children in rural Kenya also face the heavy burden of water collection, with many households spending over an hour daily to secure water for domestic use (Deresse, 2023). Climate-induced declines in water availability threaten agriculture, tourism, and energy sectors and adversely affect national development goals (Republic of Kenya, 2012). In both the countries, climate change multiplies pre-existing vulnerabilities such as poverty, weak infrastructure, and limited adaptive capacity. As a result, the communities are highly exposed to water scarcity, food insecurity, and reduced resilience (UNHCR, 2020).

Livestock losses constitute one of the largest economic shocks. As already mentioned, along the Kenya–Ethiopia–Somalia pastoral corridor, up to 80% of livestock died during the 2010/2011 drought, destroying household capital and eroding long-term livelihood resilience (Headey et al., 2013).

The economic and human costs of drought are significantly multiplied when humanitarian response is delayed. Despite early warning systems issued alerts as early as August 2010 regarding deteriorating conditions inEast Africa, the responses of theinternational community remained inadequate until the formal declaration of famine in July 2011 (Hillbruner & Moloney, 2012). For example, the prior alert of famine in Somalia in 2011 was prompt, precise, and actionable, still the large-scale emergency response in southern Somalia was significantly delayed (Hillbruner & Moloney, 2012). This delayed action resulted in tragic consequences: an estimated 258,000 lives were lost, with nearly half of the victims being children under five years of age (Maxwell & Fitzpatrick, 2012). Overall, drought in both countries is not an episodic crisis but a structural economic challenge that reduces productive capacity, destabilises markets, and weakens assets essential for recovery and long-term development (Duguma et al., 2017).

In response to drought conditions, farmers in Kenya have adopted various strategies. In the arid zone of Garissa, the main adaptation was moving animals to areas with more rainfall for grazing. In order to deal with shorter rainy seasons, farmers in Participatory Rural Appraisals (PRAs) often reported cultivating more early-growing and drought- tolerant crops, such as cassava, sweet potatoes, and pigeon peas (Bryan et al. 2013). But there is always a need for greater investment in drought-tolerant crop varieties and integrating management of drought and readiness programswith-agricultural growth initiatives to build community resilience against future climate crises.

The economic outcomes of the escalating frequency of the dry season in Kenya and Ethiopia are severe and comprehensive. Massive crop failures across countries, large-scale unemployment and the declining family incomes of the pastoral communities made these communities exceptionally vulnerable. Farmers in both countries have relatively larger landholdings (East Africa) compared to South Asia but maintain lower adaptive capacity due to inadequate governance, inefficient institutions, and limited funding availability and markets, restricting their investment capacityto adaptation measures (Aryal et al., 2021).

Current policy responses in Kenya and Ethiopia remain insufficient to address the escalating drought risks driven by climate change. Although both countries have introduced national adaptation plans, early warning systems, and water management initiatives, the document shows that rising temperatures, declining rainfall, and increasing changes of climate continue to outpace the advantage of existing plans. In Ethiopia, recurrent droughts, intensifying water scarcity and heavy burden on female folk and children to fetch water covering a long way indicate that current interventions have not adequately improved access or reduced vulnerability. The dominance of hydropower and the high sensitivity of river systems further reveal gaps in climate-resilient infrastructure (De Trincheria et al., 2016). Kenya faces similar shortcomings: worsening water scarcity, poor water distribution, rapid population growth, and inadequate storage facilities continue to strain communities despite existing policies). Furthermore, increased drought frequency, reduced per capita water availability, and impacts on agriculture and tourism show that adaptation measures have not sufficiently strengthened resilience. Overall, while policy frameworks exist, the existing studies suggest a wide implementation gap, limited resources, and structural weaknesses, making current policy responses inadequate for the scale of future drought risks.

Patterns of Livelihood Vulnerability

Livelihood vulnerability in Kenya and Ethiopia is closely tied to environmental stress, limited assets, and institutional weaknesses. Recent evidence from (IATI Datastore Search, n.d.) shows that recurrent drought continues to "threaten food security and livelihoods" in the ASAL regions of both countries. In Ethiopia, the 2016 drought left over ten million people requiring emergency food assistance, with pastoral and agro-pastoral regions experiencing high livestock mortality and weakened coping systems. These vulnerabilities arise from environmental degradation, low land productivity, and weak household asset bases, which limit resilience to repeated shocks.

In Kenya, vulnerability is most severe in ASAL counties, where limited access to water, dependence on rain-fed livelihoods, and chronic poverty intensify drought impacts. It is already mentioned that delayed humanitarian responses worsen vulnerability and undermine household resilience, increase mortality, and deepen long-term exposure. In fact, vulnerability emerges from the interaction of climatic shocks with structural deficits in assets, governance, infrastructure, and access to timely support.

Vulnerability patterns reveal significant gender, age, and wealth disparities in both Kenya and Ethiopia. Male-headed households demonstrate greater capacity to adopt diverse adaptation strategies, including changing farming practices (39% in Kenya, 17% in Ethiopia). They are ableto seek additional employment, and implement sustainable land management systems. On the other hand, female-headed households predominantly rely on consumption reduction and savings/borrowing. According to the existing studies, adaptation strategies employed include farming practice modifications, sustainable land management (12% in Kenya), livelihood diversification, and livestock mobility. Institutional support remains critically inadequate in East Africa, with minimal government or NGO assistance (0–4% coverage). Key determinants of adaptation capacity include education level, possession of farming fields and cattle, availability of loans and training and membership in village organisations. Policy implications emphasise the necessity for enhanced institutional support mechanisms, improved market access, gender-inclusive interventions, social protection programmes for elderly farmers, and investments in climate information services and alternative employment opportunities (Aryal et al., 2021).

Another important point is that shrinking natural resources has intensified and occasionally triggered localised conflicts and displacement, particularly in dry and semi-dryareas (UN OCHA, 2022). Women and children disproportionately carry the burden of shock of drought because of their responsibilities in household water and food provision (UNICEF, 2021). Repeated drought cycles also erode household assets and diminish resilience, creating long-term development setbacks (FAO, 2023). Although government and humanitarian interventions such as food assistance, emergency water trucking, and destocking programmes have provided relief, these efforts are often insufficient to address the structural drivers of vulnerability (NDMA, 2022). Given predictions that Kenya's drought frequency and intensity may rise due to variation in the climate and climate change, strengthening adaptive capacity through sustainable water management, livelihood diversification, and robust early warning systems is essential to protecting communities from future shocks (IPCC, 2022).

Current Policy Responses and Adaptation Strategies

Though implementation is still limited, policy frameworks in Ethiopia and Kenya have changed dramatically toward more proactive and systematic drought-risk management. Both nations have created comprehensive strategies that prioritize long-term resilience building, coordination systems, early warning, and emergency preparation. Kenya has created intergovernmental platforms and policy tools to improve drought governance in dry and semi-dryareas, while Ethiopia has developed multi-sectoral institutions and included disaster-risk management into its national development strategy. These initiatives mark a significant departure from crisis-driven reactions and toward proactive strategies.

However, the efficiency of the systems is still limited, due to limited institutional capacity, coordination, and resource distribution. Consistent investment, prompt decision-making, and improved links between national policy and local practice ensure constant adaptation. Without these, policy aspirations run the risk of surpassing real results and leaves communities at risk in spite of legal frameworks (Duguma et al., 2017).

Turkana pastoralists in Kenya employ both interimmanagingstrategies and prolongedplans of adjustment. Some short-term coping mechanisms are herd splitting and migration of livestockin search of water and food to diversify risk, selling animals before complete asset depletion, and relying on indigenous knowledge of vegetation patterns (particularly Acacia tortilis protein-rich pods called sakaram). Long-term adaptations taken into account are livelihood diversification into fishing, charcoal production, and petty trade; composition of cattle diversification to get advantage of drought and disease tolerance across species (camels, cattle, sheep, and goats); investing in children's formal education as future income insurance; and utilizing social networks for information sharing. However, these traditional mechanisms are increasingly constrained by land tenure changes, resource-based conflicts, and environmental degradation (Opiyo et al., 2015). According to recent data, high-resolution datasets like CHIRPS and MSWEP in Kenya and Ethiopia, together with drought monitoring techniques that are sensitive to atmospheric evaporative demand (AED), like SPEI, are also necessary for effective adaptation (Gebrechorkos et al., 2020). The rapid expansion of drought-affected areas since 2018 highlights the urgency for stronger socio-economic and environmental adaptation measures across both countries (Gebrechorkos et al., 2020).

Conclusion

The drought in Kenya and Ethiopia is getting worse because of rising temperatures, decreasing rainfall, and increased atmospheric evaporative demand with severe consequences on livelihoods, water security, and national development. The study indicates that droughts caused by climate change disrupt agricultural and pastoral systems, destroy assets, and increase structural and gendered vulnerabilities. Strengthening AED-sensitive drought monitoring, funding climate-resilient water and livelihood systems, extending early-response measures, and enhancing institutional coordination are some of the policy proposals. To better enhance resilience planning in both nations, future research should focus on high-resolution climate modelling, mapping socioeconomic vulnerability, and assessing long-term adaptation plans.

References

Aryal, J. P., Sapkota, T. B., Rahut, D. B., Marenya, P., & Stirling, C. M. (2021). Climate risks and adaptation strategies of farmers in East Africa and South Asia. *Scientific Reports, 11*, 10489.

Branco, D., & Féres, J. (2021). Weather shocks and labor allocation: Evidence from rural Brazil. *American Journal of Agricultural Economics, 103*(4), 1359–1377.

Bryan, E., Ringler, C., Okoba, B., Roncoli, C., Silvestri, S., & Herrero, M. (2013). Adapting agriculture to climate change in Kenya: Household strategies and determinants. *Journal of environmental management, 114*, 26–35.

Deresse, T. (2023). The vulnerability of communities to climate Change-induced water scarcity in Ethiopia and Kenya: A systematic review. *Research Square (Research Square).* https://doi.org/10.21203/rs.3.rs-3812532/v1

De Trincheria, J., Wibbing, J., Leal Filho, W., & Otterpohl, R. (2016, November). Practical recommendations to prevent, restore and rehabilitate silted-up sand storage dams in arid and semi-arid areas. In *RWSN Forum. Abidjan, Cote d'Ivoire.*

Duguma, M. K., Brüntrup, M., & Tsegai, D. (2017). Policy options for improving drought resilience and its implication for food security: The cases of Ethiopia and Kenya. *Econstor (Econstor), 98.* http://hdl.handle.net/10419/199223

Emiru, N. C., Recha, J. W., Thompson, J. R., Belay, A., Aynekulu, E., Manyevere, A.,... & Solomon, D. (2021). Impact of climate change on the hydrology of the Upper Awash River Basin, Ethiopia. *Hydrology, 9*(1), 3.

Evapotranspiration and the Water Cycle. (2019, September 8). USGS. https://www.usgs.gov/water-science-school/science/evapotranspiration-and-water-cycle?utm_source=chatgpt.com.

Food and Agriculture Organization of the United Nations. (2023). *Drought impact and response in the horn of Africa.* FAO.

Gebrechorkos, S. H., Hülsmann, S., & Bernhofer, C. (2020). Analysis of climate variability and droughts in East Africa using high-resolution climate data products. *Global and Planetary Change, 186*, 103130.

Gelete, G., Gokcekus, H., & Gichamo, T. (2020). Impact of climate change on the hydrology of Blue Nile basin, Ethiopia: a review. *Journal of Water and Climate Change, 11*(4), 1539-1550.

Gurara, M. A., Jilo, N. B., & Tolche, A. D. (2021). Impact of climate change on potential evapotranspiration and crop water requirement in Upper Wabe Bridge watershed, Wabe Shebele River Basin, Ethiopia. *Journal of African Earth Sciences, 180*, 104223.

Habte, M., Eshetu, M., Maryo, M., Andualem, D., & Legesse, A. (2022). Effects of climate variability on livestock productivity and pastoralists perception: The case of drought resilience in southeastern Ethiopia. *Veterinary and Animal Science, 16*, 100240.

Haile, G. G., Tang, Q., Sun, S., Huang, Z., Zhang, X., & Liu, X. (2019). Droughts in East Africa: Causes, impacts and resilience. *Earth-Science Reviews, 193*, 146–161. https://doi.org/10.1016/j.earscirev.2019.04.015

Haile, G. G., Tang, Q., Hosseini-Moghari, S. M., Liu, X., Gebremicael, T. G., Leng, G., & Yun, X. (2020). Projected impacts of climate change on drought patterns over East Africa. *Earth's Future, 8*(7), e2020EF001502. https://doi.org/10.1029/2020ef001502

Headey, D., Taffesse, A. S., & You, L. (2013). Diversification and development in pastoralist Ethiopia. *World Development, 56*, 200–213. https://doi.org/10.1016/j.worlddev.2013.10.015

Heathcote, R. L. (2016). *Drought and the human story: braving the bull of heaven.* Routledge.

Hillbruner, C., & Moloney, G. (2012). When early warning is not enough—Lessons learned from the 2011 Somalia famine. *Global Food Security, 1*(1), 20–28. https://doi.org/10.1016/j.gfs.2012.08.001

Hodell, D. A., Brenner, M., & Curtis, J. H. (2005). Terminal classic drought in the northern Maya lowlands inferred from multiple sediment cores in Lake Chichancanab (Mexico). *Quaternary Science Reviews, 24*(12–13), 1413–1427. https://doi.org/10.1016/j.quascirev.2004.10.013. https://www.researchgate.net/publication/318040874

IATI Datastore Search. (n.d.).https://datastore.iatistandard.org/activity/CA-3-D002977002

Intergovernmental Panel on Climate Change. (2022). *Climate change 2022: Impacts, adaptation and vulnerability.* IPCC Sixth Assessment Report.

Intergovernmental Panel on Climate Change (IPCC). (2014). Climate Change 2014: Impacts, adaptation, and vulnerability. Part a: Global and Sectoral aspects. In *Contribution of working group II to the fifth assessment report of the intergovernmental panel on climate change.* Cambridge University Press.

Jayachandran, S. (2006). The jeffords effect. *The Journal of Law and Economics, 49*(2), 397–425.

Jessoe, K., Manning, D. T., & Taylor, J. E. (2018). Climate change and labour allocation in rural Mexico: Evidence from annual fluctuations in weather. *The Economic Journal, 128*(608), 230–261.

Maxwell, D., & Fitzpatrick, M. (2012). The 2011 Somalia famine: Context, causes, and complications. *Global Food Security, 1*(1), 5–12. https://doi.org/10.1016/j.gfs.2012.07.003

Mera, G. A. (2018). Drought and its impacts in Ethiopia. *Weather and Climate Extremes, 22*, 24–35. https://doi.org/10.1016/j.wace.2018.10.002

Mishra, V., Tiwari, A. D., Aadhar, S., Shah, R., Xiao, M., Pai, D. S., & Lettenmaier, D. (2019). Drought and famine in India, 1870–2016. *Geophysical Research Letters, 46*(4), 2075–2083. https://doi.org/10.1029/2018gl081477

Mogaka, H. (2006). *Climate variability and water resources degradation in Kenya: improving water resources development and management* (Vol. 69). World Bank Publications.

National Drought Management Authority. (2022). *Kenya drought early warning bulletin*. Government of Kenya.

National Drought Management Authority. (2023). *Drought situation & response update*. Government of Kenya.

Opiyo, F., Wasonga, O., Nyangito, M., Schilling, J., & Munang, R. (2015). Drought adaptation and coping strategies among the Turkana pastoralists of northern Kenya. *International Journal of Disaster Risk Science, 6*(3), 295–309.

Ongoma, V., Chen, H., & Omony, G. W. (2018). Variability of extreme weather events over the equatorial East Africa, a case study of rainfall in Kenya and Uganda. *Theoretical and applied climatology, 131*(1–2), 295-308. https://doi.org/10.1007/s00704-016-1973-9

Republic of Kenya. (2012). Kenya post-disaster needs assessment (PDNA): 2008–2011 drought. Ministry of Finance. https://www.gfdrr.org/sites/default/files/publication/pda-2011-kenya.pdf

Scholz, C. A., Johnson, T. C., Cohen, A. S., King, J. W., Peck, J. A., Overpeck, J. T., Talbot, M. R., Brown, E. T., Kalindekafe, L., Amoako, P. Y. O., Lyons, R. P., Shanahan, T. M., Castañeda, I. S., Heil, C. W., Forman, S. L., McHargue, L. R., Beuning, K. R., Gomez, J., & Pierson, J. (2007). East African megadroughts between 135 and 75 thousand years ago and bearing on early-modern human origins. *Proceedings of the National Academy of Sciences, 104*(42), 16416–16421. https://doi.org/10.1073/pnas.0703874104

Schlenker, W., & Lobell, D. B. (2010). Robust negative impacts of climate change on African agriculture. *Environmental research letters, 5*(1), 014010. https://doi.org/10.1088/1748-9326/5/1/014010

Schaebitz, F., Asrat, A., Lamb, H. F., Cohen, A. S., Foerster, V., Duesing, W., & Trauth, M. H. (2021). Hydroclimate changes in eastern Africa over the past 200,000 years may have influenced early human dispersal. *Communications Earth & Environment, 2*(1), 123.

Shelton, S., & Dixon, R. D. (2023). Moisture transport associated with southwest monsoon rainfall over Sri Lanka in relatively wet and dry rainfall years. *Meteorology and Atmospheric Physics, 135*(5), 46.

United Nations Children's Fund. (2021). Impact of drought on women and children in Kenya. *UNICEF*.

United Nations Office for the Coordination of Humanitarian Affairs. (2022). *Horn of Africa: Drought response situation report*. UN OCHA.

UNHCR - The UN Refugee Agency. (n.d.). Global Report 2020 | UNHCR. UNHCR. https://www.unhcr.org/media/global-report-2020

Van Loon, A. F. (2015). Hydrological drought explained. *Wiley Interdisciplinary Reviews: Water, 2*(4), 359–392. https://doi.org/10.1002/wat2.1085

Wilhite, D. A. (2005). *Drought and water crises*. https://doi.org/10.1201/9781420028386

Wilhite, D. A., & Glantz, M. H. (1985). Understanding: The drought phenomenon: The role of definitions. *Water International, 10*(3), 111–120. https://doi.org/10.1080/02508068508686328

World Food Programme. (2022). *Kenya food security outlook*. WFP.

World Bank. (2011). Food price watch, November 2011. worldbank.org

Chapter 16
Analysis of Developmental Activities with Respect to Landslides in Garhwal Region for Two Assessment Years Using GIS

Ruchi Saraswat, Suvandan Saraswat, Gaurav Bharti, and Vaishnavi Bansal

Abstract Rapid urbanization and infrastructural development in the Garhwal Himalaya have significantly altered the fragile geomorphic conditions, exacerbating the risk of landslides and related hazards. This study presents a comparative assessment of developmental activities between the years 2008 and 2018 in the Uttarkashi district, Garhwal Himalaya, using Geographic Information System (GIS) and remote sensing techniques. High-resolution satellite imagery and topographic maps were analyzed to delineate changes in land use, particularly focusing on residential, commercial, industrial, and road infrastructure. The spatial overlay of these developmental features with Landslide Hazard Zonation (LHZ) maps revealed a pronounced increase in built-up areas within high and very high susceptibility zones. Between 2008 and 2018, residential buildings increased by 954 units, commercial structures by 541, and major roads by 20 units, indicating rapid landscape modification. The findings highlight that nearly two-thirds of the total built-up area lies within high-risk zones, underscoring the need for stricter land-use planning and slope stabilization measures. This study provides crucial spatial insights for sustainable development and disaster mitigation planning in the seismically active and geomorphologically sensitive Garhwal Himalaya.

Keywords Landslide susceptibility · Developmental activities · Change detection · Garhwal Himalaya · GIS and remote sensing

R. Saraswat · G. Bharti · V. Bansal (✉)
Department of Civil Engineering, JSS Academy of Technical Education, Noida, India
e-mail: vaishnavibansal@jssaten.ac.in

R. Saraswat
e-mail: rsaraswat@jssaten.ac.in

G. Bharti
e-mail: gauravbharti000@gmail.com

S. Saraswat
Department of Mechanical Engineering, JSS University, Noida, India
e-mail: suvandan.saraswat@jssaten.ac.in

R. Sarkar et al. (eds.), *Climate Change, Environmental Hazards and Community-Based Resilience*, Disaster Risk Reduction, https://doi.org/10.1007/978-981-95-9485-6_16

Introduction

Due to varying nature of the hilly terrain, various geographic features get hampered due to increase in the developmental activities day by day. So, it is necessary to understand the vulnerability of that area with respect to the hazards which helps in planning future construction work which ultimately helps in finding alternative solution to mitigate the disaster. Therefore, to show the effect of the developmental activities basically construction of buildings and roads in terms of landslides, two assessment years have been taken and according to that change detection of developmental activities can be found out which helps in predicting that how the severity level of the landslides is increasing due to road construction, building construction and hydro power projects.

Comparison of the two assessment years 2008 and 2018 in terms of the developmental activities with respect to hazards has been done using GIS Software and Erdas imagine. In this paper, the main emphasis is on construction of buildings and road construction in hilly terrain. As far as the hydel power is concerned, there is only one hydropower which is existing in this area and in low susceptibility zone. So, its susceptibility is considered as zero. The area of Interest is in the Uttarkashi district, Garhwal Himalaya from Dunda with Latitude and longitude as 30°42“20’ N and 78°21” 00′ E to Bhatwari village latitude30°48“´30’N and longitude 78°37” 15′ E as shown in Fig. 16.1.

Fig. 16.1 Study area location

Literature Review

Landslide hazard evaluation and risk assessment have been extensively studied across various terrains and geological conditions, revealing crucial insights into the mechanisms and impacts of landslides. Anbalagan (1992) initiated a foundational understanding of landslide hazard zonation in mountainous regions through engineering geology methodologies. Subsequent research by Awasthi & Parkash (2001) focused on the morphometry of basins in the Garhwal Himalaya, highlighting the relationship between basin characteristics and landslide potential. Studies such as those by Asthana and Sah (2007) further explored the specific conditions leading to landslides and cloudbursts in the Mandakini Basin, underscoring the vulnerability of Himalayan regions. This body of work has been expanded by remote sensing and GIS applications, such as the efforts by Akgun et al., (2011), which provided advanced techniques for assessing landslide risk in urban areas like Izmir City. The framework for understanding and managing these risks has also evolved to include broader concepts of vulnerability and disaster management. Aven (2007, 2009, 2010) contributed significantly to the development of unified frameworks for risk and vulnerability analysis, addressing both safety and security aspects. This approach has been applied in various contexts, including disaster management strategies in Egypt by Adham Hany Abulnour (2013). The use of advanced technologies like LiDAR for multi-hazard scenarios, as demonstrated by Abdulwahid & Pradhan (2016), and automated geo-spatial systems for socio-economic vulnerability assessments by Bhattacharya et al., (2013), highlight the integration of innovative tools in risk assessment. Bell and Glade (2004) introduced a quantitative risk analysis framework for landslides, emphasizing systematic approaches to hazard assessment. Choudhury et al., (2007) further examined slope behavior under static and seismic conditions using the limit equilibrium method, which is critical for understanding landslide dynamics in seismically active regions. Chong Xu et al., (2012) compared multiple models for susceptibility mapping of earthquake-triggered landslides following the Wenchuan earthquake, providing valuable insights into the efficacy of different analytical techniques. Das et al., (2011) utilized stochastic modeling for assessing landslide vulnerability over time and space in the northern Himalayas, underscoring the importance of temporal factors in hazard evaluation. Geospatial technologies and innovative modeling approaches have enhanced landslide risk assessment. Deshpande et al., (2009) and Kumar et al., (2012) applied remote sensing and GIS techniques for landslide hazard zonation in Uttarakhand, demonstrating the utility of these tools in identifying vulnerable areas. Various studies have been conducted a comparative study of various susceptibility zonation methods, including artificial neural networks and fuzzy logic, in the Darjeeling Himalayas (Bansal & Sarkar, 2023, Bansal & Sarkar, 2024a, Bansal & Sarkar, 2024b, Kanungo et al., 2006). Fuchs et al., (2012) focused on vulnerability assessment in natural hazard and risk analysis, addressing current approaches and future challenges. Recent research by Michael & Samanta (2016) applied weighted linear combination models for landslide vulnerability mapping, further refining methodologies for comprehensive hazard analysis.

Parkash (2012a) provided an overview of national guidelines for landslide management with a focus on infrastructure projects, highlighting the critical need for comprehensive policy frameworks. His subsequent works (2013) on education, training, and capacity development emphasize the importance of equipping communities and professionals with the knowledge and skills to mainstream landslide risk management effectively. Complementing these perspectives, Patra & Devi (2015) assessed landslide hazards in the lesser Himalayas of Himachal Pradesh, focusing on prevention and mitigation measures, which are vital for safeguarding the region's socio-economic stability. Advanced predictive modeling and geospatial analysis have been pivotal in landslide susceptibility assessments. Pradhan & Lee (2009) utilized artificial neural networks for landslide risk analysis, focusing on different training sites to enhance model accuracy. Similarly, Pradhan & Youssef (2010) demonstrated the application of remote sensing data and GIS in spatial-based statistical models for hazard analysis. Pham et al., (2015) employed a GIS-based statistical approach using the frequency ratio method to assess landslide susceptibility in the Uttarakhand Himalayas, highlighting the significance of integrating statistical techniques with geospatial tools. The work of Pistocchi et al., (2002) on predictive modeling for optimal spatial data exploitation in landslide hazard mapping further exemplifies the advancements in this field.

Methodology

The methodological framework was designed to analyze the spatial correlation between developmental activities and landslide hazards across two temporal datasets: 2008 and 2018. The workflow comprised the following steps:

(a) **Data Acquisition and Preparation**

Satellite images for the years 2008 and 2018 were procured from suitable remote sensing sources. Supplementary data including topographic sheets, land-use maps, and field observations were also integrated. The study area extended from Dunda (30°42′20″ N, 78°21′00″ E) to Bhatwari (30°48′30″ N, 78°37′15″ E) in Uttarkashi district, Garhwal Himalaya.

(b) **Preprocessing and Classification**

The satellite images were processed using *Erdas Imagine* software to perform supervised classification. Land-use categories such as residential, industrial, educational, commercial, government, religious buildings, hydro power plants, and major roads were delineated. The classification accuracy was verified through field cross-checking and Google Earth interpretation.

(c) **Change Detection Analysis**

Post-classification comparison was carried out to identify spatial and quantitative changes between the two years. The difference in land-use classes was calculated to determine the extent of developmental expansion (Table 16.2).

Results

Development Activities

The Fig. 16.2 illustrates the spatial distribution of landslide hazard zones (LHZ) in relation to existing buildings within the study area. Using GIS-based analysis, the map categorizes the region into six hazard classes—Very Low, Low, Low-Moderate, Moderate, High, and Very High—each represented by distinct colors. The innermost and darkest zones correspond to Very High Hazard areas, encompassing around 1949 buildings, indicating extreme vulnerability due to steep slopes, unstable lithology, and proximity to drainage or fault zones. Surrounding these are High Hazard zones, containing 1778 buildings, where instability remains considerable. The Moderate and Low-Moderate hazard zones, covering 1221 and 722 buildings respectively, mark transition areas between stable and unstable terrains. In contrast, Low and Very Low Hazard zones, with 229 and 102 buildings, represent comparatively stable areas with minimal risk. Overall, the map demonstrates that nearly two-thirds of the buildings fall within high to very high susceptibility zones, emphasizing the urgent need for targeted slope stabilization, land-use regulation, and infrastructure planning to reduce vulnerability and enhance safety in the landslide-prone upper Garhwal Himalayan region.

Effect of Landslides on Buildings

Manmade structures are also affected by the landslides whether they are near to landslides or not. As the landslides destroy foundations, transportation network, transmission lines, damages of the residential buildings occur Commercial buildings may also face the same after effects as residential buildings but effects will be more in case of commercial structure due to interference in business due to damage of access routes. The Fig. 16.3 presents the distribution of buildings across different landslide hazard zones (series 1) and their corresponding vulnerability percentages (series 2) within the study area. It clearly indicates that the degree of building exposure increases with the hazard level. In the Very Low Hazard Zone, only 102 buildings (1.7%) are located, implying minimal risk and stable terrain conditions. The Low Hazard Zone accommodates 229 buildings (3.82%), reflecting limited vulnerability. In contrast, the Low Moderate Hazard Zone comprises 722 buildings (12.03%), showing a noticeable rise in exposure due to partial instability in certain slopes. The Moderate Hazard Zone, with 1221 buildings (20.35%), represents a significant share of the total built-up area that faces moderate landslide risk. The concentration further escalates in the High Hazard Zone, where 1778 buildings (29.63%) are situated, and peaks in the Very High Hazard Zone, containing 1949 buildings (32.48%). This indicates that nearly two-thirds of the buildings lie within high to very high hazard areas, emphasizing the urgent need for slope stabilization, structural reinforcement, and proper land-use regulations to minimize future damage and enhance community resilience in these vulnerable regions.

Effect of Landslides on Roads

One of the greatest effect of landslides is on the roads which affects maximum number of people in any locality. Connecting roads whether it may be national highway or state highway are also very vulnerable to the effect of landslides.

Vulnerability of Roads with Respect to LHZ

The spatial relationship between road networks and landslide hazard zones (LHZ) within the study area has been shown in Fig. 16.4. Using GIS-based spatial overlay analysis, the map classifies the terrain into six hazard categories—Very Low, Low, Low-Moderate, Moderate, High, and Very High Hazard Zones—each represented by distinct color gradients. The Very High Hazard Zone (shown in dark purple) covers approximately 6.55 km^2 (36.18%) of the area and predominantly aligns along

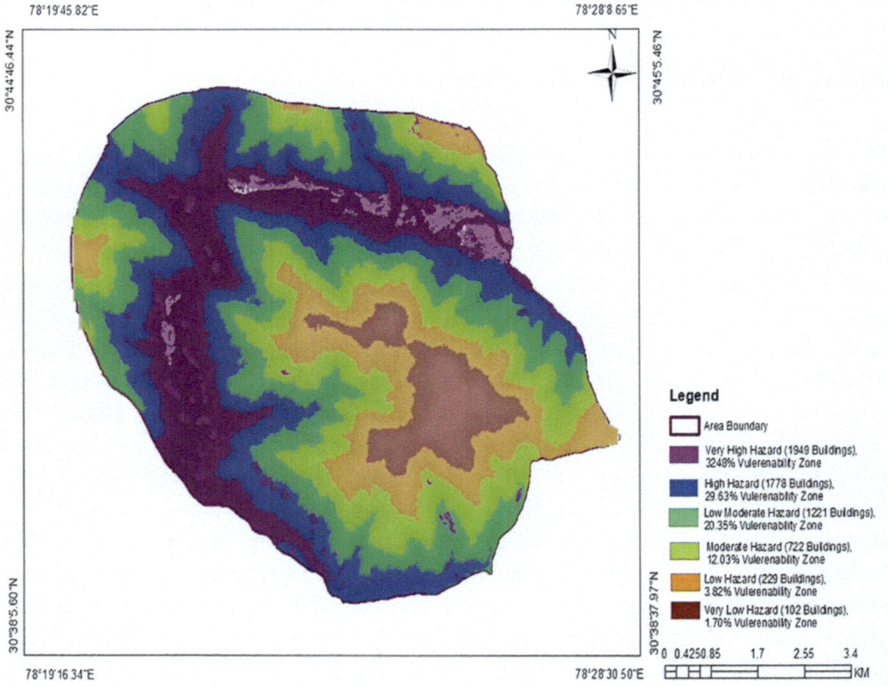

Fig. 16.2 Vulnerability of buildings with LHZ

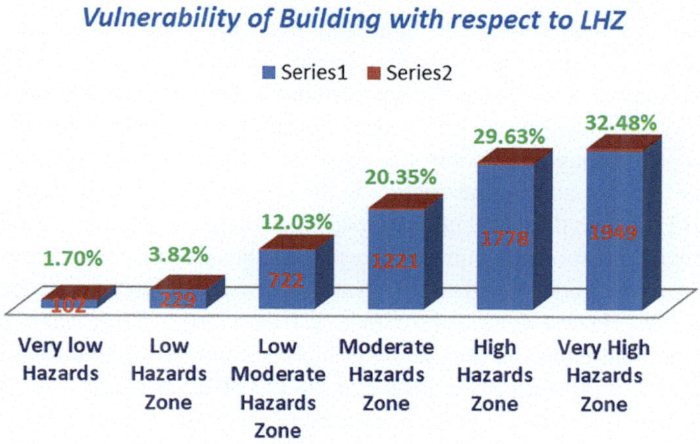

Fig. 16.3 Variation in the number of buildings in different vulnerability zones

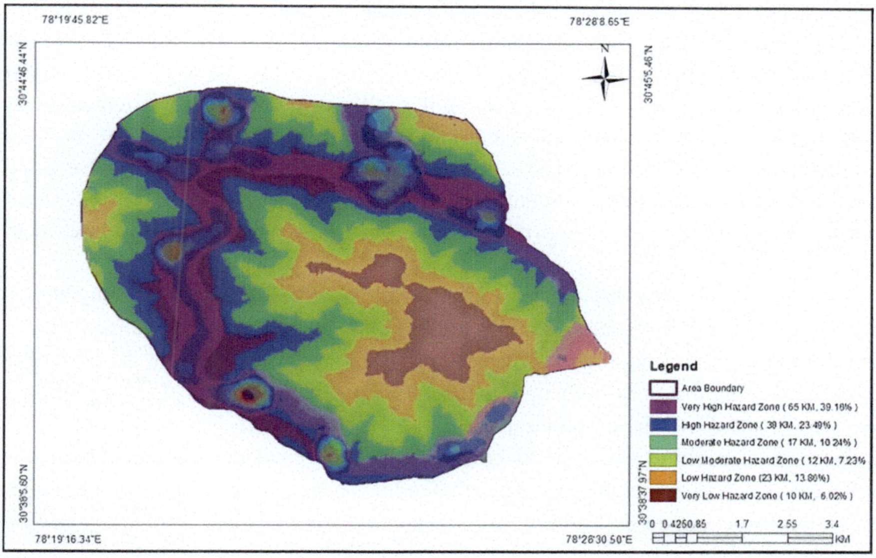

Fig. 16.4 Vulnerability of roads with LHZ

major road corridors and hill-cut slopes, indicating regions of intense vulnerability due to road excavation, slope undercutting, and disturbed drainage patterns. The High Hazard Zone spans 2.09 km^2 (11.58%), representing slopes adjacent to road alignments where moderate instability persists. The Moderate and Low-Moderate Hazard Zones, occupying 2.76 km^2 (15.30%) and 1.36 km^2 (7.52%) respectively, correspond to transitional areas that experience moderate slope stress and partial vegetation loss. The Low Hazard Zone (2.03 km^2; 11.30%) and Very Low Hazard Zone (1.04 km^2; 5.88%) are primarily found in interior or vegetated slopes with stable geomorphic conditions. Overall, the figure highlights that nearly half of the total road length traverses through high to very high hazard zones, emphasizing the significant influence of road construction on slope instability. It underlines the necessity for implementing engineering countermeasures such as retaining structures, bioengineering, proper drainage, and controlled excavation to minimize landslide risks and ensure safer transportation infrastructure in the upper Garhwal Himalayan terrain.

Table 16.1 presents the distribution of road length across different landslide hazard zones and their corresponding vulnerability percentages within the study area. It highlights how road networks are affected by varying degrees of landslide susceptibility. In the Very Low Hazard Zone, approximately 10 km (6.02%) of roads are located, indicating minimal exposure and stable ground conditions suitable for safe transportation. The Low Hazard Zone covers 23 km (13.86%) of roads, reflecting areas with slight instability but generally manageable risk. The Low-Moderate Hazard Zone comprises 12 km (7.23%), marking sections that may experience localized slope movement or erosion. The Moderate Hazard Zone includes 17 km (10.24%), where roads traverse moderately unstable slopes requiring periodic maintenance and drainage control. A significant portion of the network, 39 km (23.49%), falls within the High Hazard Zone, indicating substantial vulnerability due to steep slopes and potential slope failures. The highest exposure is observed in the Very High Hazard Zone, where 65 km (39.16%) of roads lie in highly unstable terrain, often adjacent to steep hill cuts and faulted geological units.Overall, nearly two-thirds of the road length lies within high to very high hazard zones, highlighting a critical need for slope stabilization, drainage improvement, and engineering interventions to mitigate road damage and ensure safe transportation in the landslide-prone Garhwal Himalayan region.

Table 16.2 provides the data of Change Detection with Development Activity Between 2008–2018 presents a comparative analysis of land use changes over a decade, highlighting the extent of urban expansion and infrastructure development across different classes of built-up structures. Between 2008 and 2018, a significant rise was observed in the residential building category, which increased from 2711 to 3665 units, marking a net growth of 954 units. This substantial increase indicates rapid urbanization and population growth, leading to an upsurge in housing demand. Similarly, the commercial building class expanded from 2711 to 3252 units, with a net addition of 541 units, suggesting increased economic activities and commercial

Table 16.1 Categorisation of hazard zone with respect to roads

Sl. No	1	2	3	4	5	6
Type of Hazards	Very low Hazards	Low Hazards Zone	Low Moderate Hazards Zone	Moderate Hazards Zone	High Hazards Zone	Very High Hazards Zone
Road in KM	10	23	12	17	39	65
Vulnerability (in%)	6.02	13.86	7.23	10.24	23.49	39.16

infrastructure development. The industrial building and educational building categories also exhibited positive growth, with increases of 2 and 7 units, respectively. This reflects gradual industrial expansion and improved educational facilities in the study area. Government buildings and religious buildings showed moderate increases of 7 and 17 units, pointing to the parallel development of administrative and socio-cultural infrastructure. The hydro power plant class remained constant at one unit, implying no additional installations during the decade. However, the major roads category nearly doubled, increasing from 21 to 41, with a net change of 20 units. This signifies major infrastructural investments in transportation and connectivity to support urban growth and accessibility. The data clearly depicts a decade characterized by notable residential and commercial expansion, moderate growth in institutional and government sectors, and a significant improvement in road networks, collectively reflecting an accelerated pace of urban and infrastructural development between 2008 and 2018.

Conclusions

The comparative GIS-based analysis of developmental activities between 2008 and 2018 in the Garhwal Himalaya indicates a clear upward trend in urban expansion within landslide-prone zones. The escalation of residential and commercial infrastructure, alongside road widening projects, has significantly increased the vulnerability of the region. The findings emphasize that future developmental planning should strictly adhere to hazard zonation maps and geotechnical assessments to prevent construction in high-susceptibility zones. Implementation of bioengineering techniques, proper drainage systems, slope stabilization, and retaining structures should be prioritized for existing infrastructures in critical zones. For long-term resilience, integration of hazard data into district-level master plans and continuous monitoring through remote sensing and field validation are essential. The methodology and findings of this study can serve as a valuable reference for regional disaster risk reduction strategies and sustainable land-use planning in the Himalayan terrain.

Table 16.2 Change detection with development activity between 2008–2018

	Sl. No	1	2	3	4	5	6	7	8
	Class Name	Residential Building	Industrial Building	Educational Building	Commercial Building	Government Building	Religious Building	Hydro Power Plant	Major Roads
Change Detection with Development Activity Between 2008-2018	Land use 2018	3665	3	29	3252	18	33	1	41
	Land use 2008	2711	1	22	2711	11	16	1	21
	Change Detection	954	2	7	541	7	17	0	20

References

Abdulwahid, M. W., & Pradhan, B. (2016). Landslide vulnerability and risk assessment for multi-hazard scenarios using airborne laser scanning data (LiDAR). *Natural hazards,* Vol. *82*(1), (pp. 1–20). Springer.

Abulnour, A. H. (2013). Towards efficient disaster management in Egypt. *HBRC Journal, 10*(2), 117–126.

Akgun, A., Kincal, C., & Pradhan, B. (2011). Application of remote sensing data and GIS for landslide risk assessment as an environmental threat to Izmir City. *Environmental Monitoring and Assessment, 184*(10), 5453–5470.

Anbalagan, R. (1992). Landslide hazard evaluation and zonation mapping in mountainous terrain. *Engineering Geology, 32*(4), 269–277. https://doi.org/10.1016/0013-7952(92)90053-1992

Asthana, A. K. L., & Sah, M. P. (2007). Landslides and cloudbursts in the Mandakini Basin of Garhwal Himalaya. *Himalayan Geology, 28*(2), 59–67.

Aven, T. (2007). A unified framework for risk and vulnerability analysis covering both safety and security. *Reliability Engineering & System Safety, 92*(6), 745–754.

Aven, T. (2009). Identification of safety and security critical systems and activities. *Reliability Engineering & System Safety, 94*(2), 404–411.

Aven, T. (2010). On how to define, understand and describe risk. *Reliability Engineering & System Safety, 95*(6), 623–631.

Awasthi, A. K., & Parkash, S. (2001). Basin morphometry and landslide potential along Bhagirathi River valley in a part of Uttarkashi District, Garhwal Himalaya, Uttaranchal (India). *Journal of the Indian Association of Sedimentologists, 20*(1), 39–50.

Bansal, V., & Sarkar, R. (2023). Numerical analysis and implications of fluctuating water levels on slope stability. In *International conference on sustainable infrastructure: Innovation, opportunities and challenges* (pp. 407–419). Springer.

Bansal, V., & Sarkar, R. (2024a). Comparative analysis of slope stability for Kalimpong region under dynamic loading using limit equilibrium method and machine benchmark learning classifiers. *Iranian Journal of Science Technology, Transactions of Civil Engineering, 48*, 2785–2807.

Bansal, V., & Sarkar, R. (2024b). Prophetical modeling using limit equilibrium method and novel machine learning ensemble for slope stability gauging in Kalimpong. *Iranian Journal of Science Technology, Transactions of Civil Engineering, 48*, 411–430.

Bell, R., & Glade, T. (2004). Quantitative risk analysis for landslides. *Natural Hazards and Earth System Sciences, 4*(1), 117–131.

Bhattacharya, D., Ghosh, J. K., & Boccardo, P. (2013). Automated geo-spatial system for generalized assessment of socio-economic vulnerability due to landslides in a region. *European Journal of Remote Sensing, 46*(1), 379–399.

Choudhury, D., Basu, S., & Bray, J. D. (2007). Behaviour of slopes under static and seismic conditions by limit equilibrium method. *Proceedings of the American Society of Civil Engineers,* 1–10.

Das, I., Kumar, G., Stein, A., Bagchi, A., & Dadhwal, K. V. (2011). Stochastic landslide vulnerability modeling in space and time in a part of the northern Himalayas. *India. Environmental Monitoring and Assessment, 178*(1–4), 25–35. https://doi.org/10.1007/s10661-010-1668-0

Deshpande, P. K., Patil, J. R., Nainwal, D. C., & Kulkarni, M. B. (2009, December). *Landslide hazard zonation mapping in Gopeshwar, Pipalkoti, and Nandprayag areas of Uttarakhand. Proceedings of the Indian geotechnical conference*. Guntur.

Fuchs, S., Birkmann, J., & Glade, T. (2012). Vulnerability assessment in natural hazard and risk analysis: Current approaches and future challenges. *Natural Hazards, 64*(3), 1969–1975. https://doi.org/10.1007/s11069-012-0352-9

Kanungo, D. P., Arora, M. K., Sarkar, S., & Gupta, R. P. (2006). A comparative study of conventional, ANN black box, fuzzy and combined neural and fuzzy weighting procedures for landslide susceptibility zonation in Darjeeling Himalayas. *Engineering Geology, 85*(3–4), 347–366.

Kumar, K., Devrani, R., Kathait, A., & Aggarwal, N. (2012). Micro-hazard evaluation and validation of landslides in a part of northwestern Garhwal lesser Himalaya, India. *International Journal of Geomatics and Geosciences, 2*(3), 738–748.

Michael, E. A., & Samanta, S. (2016). Landslide vulnerability mapping using weighted linear combination model through remote sensing and GIS techniques. *Modeling Earth Systems and Environment, 2*(88), 1–12.

Parkash, S. (2012a). An overview of the national guidelines for management of landslides with particular reference to infrastructure projects. *Journal of Water & Energy International, 69*(3), 29–37.

Patra, P., & Devi, R. (2015). Assessment, prevention and mitigation of landslide hazard in the lesser Himalaya of Himachal Pradesh. *Environmental and Socio-Economic Studies, 3*(1), 1–11.

Pham, T. B., Bui, T. D., Indra, P., & Dholakia, M. B. (2015). Landslide susceptibility assessment at a part of Uttarakhand Himalaya, India using GIS-based statistical approach of frequency ratio method. *International Journal of Engineering Research & Technology, 4*(11), 1–8.

Pistocchi, A., Luzi, L., & Napolitano, P. (2002). The use of predictive modelling techniques for optimal exploitation of spatial databases: A case study in landslide hazard mapping with expert system-like methods. *Environmental Geology, 58*(2), 251–270.

Pradhan, B., & Lee, S. (2009). Landslide risk analysis using artificial neural network model focusing on different training sites. *International Journal of Physical Sciences, 4*(1), 1–15.

Pradhan, B., & Youssef, A. M. (2010). Manifestation of remote sensing data and GIS on landslide hazard analysis using spatial-based statistical models. *Arabian Journal of Geosciences, 3*(3), 319–326.

Xu, C., Xu, X., Dai, F., & Saraf, A. K. (2012). Comparison of different models for susceptibility mapping of earthquake-triggered landslides related with the 2008 Wenchuan earthquake in China. *Computers & Geosciences, 46*, 317–329.

The manufacturer's authorised representative in the EU is Springer Nature Customer Service Centre GmbH, Europaplatz 3, 69115 Heidelberg, Germany. If you have any concerns regarding our products, please contact ProductSafety@springernature.com

Printed and bound by CPI Group (UK) Ltd, Croydon, CR0 4YY
07/10/2026
02229350-0006